IUTAM SYMPOSIUM ON DIFFRACTION AND SCATTERING
IN FLUID MECHANICS AND ELASTICITY

FLUID MECHANICS AND ITS APPLICATIONS
Volume 68

Series Editor: R. MOREAU
MADYLAM
Ecole Nationale Supérieure d'Hydraulique de Grenoble
Boîte Postale 95
38402 Saint Martin d'Hères Cedex, France

Aims and Scope of the Series

The purpose of this series is to focus on subjects in which fluid mechanics plays a fundamental role.

As well as the more traditional applications of aeronautics, hydraulics, heat and mass transfer etc., books will be published dealing with topics which are currently in a state of rapid development, such as turbulence, suspensions and multiphase fluids, super and hypersonic flows and numerical modelling techniques.

It is a widely held view that it is the interdisciplinary subjects that will receive intense scientific attention, bringing them to the forefront of technological advancement. Fluids have the ability to transport matter and its properties as well as transmit force, therefore fluid mechanics is a subject that is particulary open to cross fertilisation with other sciences and disciplines of engineering. The subject of fluid mechanics will be highly relevant in domains such as chemical, metallurgical, biological and ecological engineering. This series is particularly open to such new multidisciplinary domains.

The median level of presentation is the first year graduate student. Some texts are monographs defining the current state of a field; others are accessible to final year undergraduates; but essentially the emphasis is on readability and clarity.

IUTAM Symposium on Diffraction and Scattering in Fluid Mechanics and Elasticity

Proceedings of the IUTAM Symposium
held in Manchester, United Kingdom,
16–20 July 2000

Edited by

I. DAVID ABRAHAMS
University of Manchester,
Manchester, United Kingdom

PAUL A. MARTIN
Colorado School of Mines,
Golden, Colorado, U.S.A.

and

MICHAEL J. SIMON
University of Manchester,
Manchester, United Kingdom

KLUWER ACADEMIC PUBLISHERS

DORDRECHT / BOSTON / LONDON

A C.I.P. Catalogue record for this book is available from the Library of Congress.

ISBN 1-4020-0590-3

Published by Kluwer Academic Publishers,
P.O. Box 17, 3300 AA Dordrecht, The Netherlands.

Sold and distributed in North, Central and South America
by Kluwer Academic Publishers,
101 Philip Drive, Norwell, MA 02061, U.S.A.

In all other countries, sold and distributed
by Kluwer Academic Publishers,
P.O. Box 322, 3300 AH Dordrecht, The Netherlands.

Printed on acid-free paper

Contents

Part VII   Elastic Wave Propagation

Preface

These Conference Proceedings are intended to summarise the latest developments in diffraction and scattering theory as reported at the IUTAM Symposium on *Diffraction and Scattering in Fluid Mechanics and Elasticity* held in Manchester, England on 16–20 July 2000. This informal meeting was organised to discuss mathematical advances, both from the theoretical and more applied points of view. However, its primary goal was to bring together groups of researchers working in disparate application areas, but who nevertheless share common models, phenomenological features arising in such problems, and common mathematical tools. To this end, we were delighted to have four Plenary Speakers, Professors Allan Pierce, Ed Kerschen, Roger Grimshaw and John Willis FRS, who are undisputed leaders in the four thematic areas of our meeting (these are respectively acoustics, aeroacoustics, water or other free surface waves, elasticity). These Proceedings should offer an excellent vehicle for continuing the dialogue between these groups of researchers.

The participants were invited because of their expertise and recent contributions to this field. Collectively, there were around 90 contributors to the Symposium from some 13 countries located all around the world. These included 45 speakers, 35 co-authors and about 10 other delegates. Individuals came from many of the major international centres of excellence in the field of scattering theory.

I would like to thank my colleagues on the local Organising Committee, and members of the Scientific Committee, for their enormous help and advice on all aspects of this meeting. We are also most grateful to the sponsors of this event, namely IUTAM, the Leverhulme Trust, the London Mathematical Society, and Manchester and Keele Universities. Their financial assistance allowed us to support a good number of students, research assistants and staff from former Soviet Union countries. Finally I offer sincere thanks to Mr Andrew Looms, Learning and Technology Officer at Keele University, for the creation and maintenance of our excellent Symposium Web site.

Organising Committee

Chair: I D Abrahams, University of Manchester, Manchester, UK

Treasurer: M J Simon, University of Manchester, Manchester, UK

C J Chapman, Keele University, Keele, Staffordshire, UK

P A Martin, Colorado School of Mines, Golden CO, USA

G Wilks, Keele University, Keele, Staffordshire, UK

A J Willmott, Keele University, Keele, Staffordshire, UK

International Scientific Committee

M A Hayes, University College, Dublin, Ireland

V M Babich, Steklov Mathematical Institute, St Petersburg, Russia

A Boström, Chalmers University of Technology, Goteborg, Sweden

D V Evans, University of Bristol, Bristol, UK

R H J Grimshaw, Loughborough University, Loughborough, UK

E J Kerschen, University of Arizona, Tucson AZ, USA

A N Norris, Rutgers University, Piscataway NJ, USA

Sponsorship

The organisers extend their thanks to the following for sponsorship of this IUTAM Symposium:

The International Union of Theoretical and Applied Mechanics
The Leverhulme Trust   The London Mathematical Society
The University of Manchester   Keele University

DAVID ABRAHAMS

Introduction

Diffraction and scattering phenomena occur in a multitude of areas of physics and engineering, and consequently the understanding and estimation of such wave behaviour is key in many industrial settings. Typical applications, where specialist knowledge is required, include the ultrasonic nondestructive evaluation of components, medical scanning, wave forces on offshore ocean structures (oil-rigs etc.), detection of underwater vehicles or fish shoals. This area has been the focus for a great deal of attention by the theoretical mechanics community for the best part of a century; the fact that it continues to this day to demand such study is due to the combination of a continuance of new and pressing applications, and the development of improved analytical and numerical tools.

It is clear on examining previous IUTAM symposia lists that it has been some time since there has been one concerned with an examination of scattering and diffraction effects in both solids and fluids. In contrast, there has been a very healthy number of small meetings or workshops encompassing specific areas within the proposed theme, for example special sessions at recent Acoustical Society of America and Society for Industrial and Applied Mathematics meetings; the annual International Day on Diffraction meetings (St. Petersburg, Russia); and the annual International Water Wave Workshops. The profusion and variety of smaller workshops concerned with scattering and diffraction indicates that it is timely to bring together workers from different subject disciplines and industrial focuses, to examine the ways that the various subjects have developed and to stimulate cross-fertilisation of ideas and methods.

To keep within reasonable bounds of size and interests the IUTAM Symposium specialised on the following areas concerned with diffraction and scattering, each of which has a large and identifiable research community:

- diffraction and propagation of free surface and other geophysical waves;

- elastic waves and fluid structural interactions;

- aeroacoustics;

- acoustic phenomena in stationary fluids.

Mathematical techniques, including analytical (exact and asymptotic), numerical and hybrid, and their developments were of primary concern, as was the understanding of phenomenological aspects of new models. Emphasis was placed on theoretical analysis of the mechanics, and so straightforward evaluation of specific problems or the presentation of routine numerical results without global conclusions to the subject was discouraged.

For clarity and coherence of presentation, these Proceedings present papers grouped into seven themes (rather than the four areas indicated above). This better reflects the (overlapping) range of subjects discussed at the meeting and provides a certain continuity of methodology in the grouped papers.

I

DIFFRACTION AND PROPAGATION OF FREE SURFACE AND OTHER GEOPHYSICAL WAVES

THE SCATTERING OF ROSSBY WAVES BY OCEAN RIDGES

I. D. Abrahams
Department of Mathematics
University of Manchester, Manchester M13 9PL, UK
i.d.abrahams@ma.man.ac.uk

G. W. Owen*, A. J. Willmott
Department of Mathematics
Keele University, Keele, Staffordshire ST5 5BG, UK
g.w.owen@maths.keele.ac.uk
a.j.willmott@maths.keele.ac.uk

1.   INTRODUCTION

Overview and equation of motion. Rossby waves, or planetary waves, play a crucial rôle in global oceanic circulation. These waves propagate in regions of non-uniform ambient potential vorticity by conserving the potential vorticity of the flow. Bottom topography and the variation of the Coriolis parameter (a quantity proportional to the normal component of the Earth's angular velocity at the surface; see, for example, Cushman-Roisin [1]) with latitude both give rise to a nonuniform ambient potential vorticity field. Baroclinic Rossby waves typically propagate at low speeds (relative to the inertial Poincaré waves or coastally trapped Kelvin waves) and have periods of the order of six months or longer. Further details may be found in Pedlosky [2]. We are interested in the interactions of these waves with the topography of the ocean floor, in particular ridges.

Scattering of Rossby waves has been considered previously by a number of authors. The diffraction of a Rossby wave by a semi-infinite, surface piercing barrier in a homogeneous ocean has been considered in [3]. Topographic scattering of Rossby waves by ridges, has been in-

---

*The principal author of this work was supported by institutional Leverhulme Trust research grant #F/130/U.

I.D. Abrahams et al. (eds.),
IUTAM Symposium on Diffraction and Scattering in Fluid Mechanics and Elasticity, 3–12.
© 2002 *Kluwer Academic Publishers. Printed in the Netherlands.*

vestigated using layered models, first by Huthnance [4] and later by Wang & Koblinsky [5] amongst others. Similar work has been done in a continously stratified context by Schmidt & Johnson [6].

As our initial model we consider the ocean to be a continuously stratified Boussinesq fluid occupying $-H \leq z \leq 0$, on a mid-latitude β-plane, with background density linear with depth and a linearised dependence of Coriolis parameter on the meridional coordinate, Y, i.e. $f_0 + \beta Y$, where f_0 and β are constants. We shall model the ocean ridge by a thin, bottom-standing barrier occupying the plane $y = 0$ and a fraction of the total depth of the ocean, $-H \leq z < -\mu H$, whose orientation makes an angle θ with the X-axis, as shown in Fig. 1. A westward propagating

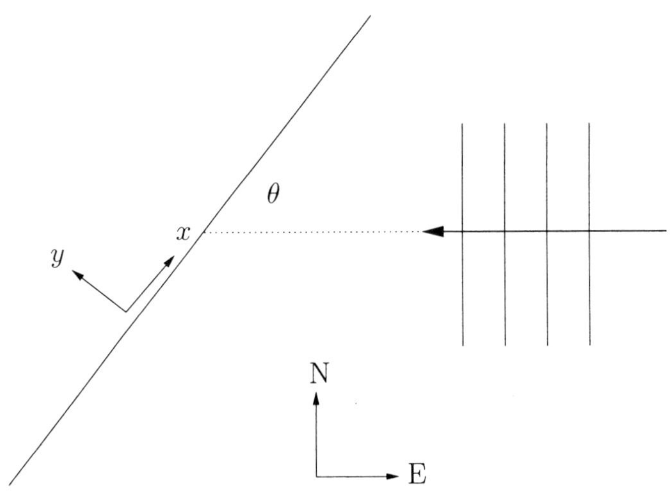

Figure 1   Geometry of barrier and incident wave

long baroclinic Rossby wave of mode 1 is incident on the barrier. We wish to find the scattered field. We assume the flow is quasi-geostrophic, i.e. the horizontal components of the velocity field are given by

$$u = -\frac{1}{\rho_0 f_0}\frac{\partial P}{\partial Y} \quad \text{and} \quad v = \frac{1}{\rho_0 f_0}\frac{\partial P}{\partial X}, \tag{1.1}$$

where $P(X, Y, z, t)$ is the pressure in the fluid. For this, strictly speaking, we require the topography to be of infinitesimal amplitude i.e. $1 - \mu \ll 1$. The pressure field then satisfies the linearised Rossby wave equation:

$$\frac{\partial}{\partial t}\left\{\nabla_H^2 P + \frac{f_0^2}{N_0^2}\frac{\partial^2}{\partial z^2}P\right\} + \beta\frac{\partial P}{\partial X} = 0, \tag{1.2}$$

where N_0 is the constant Brunt-Väisälä frequency and the X and Y axes are aligned with east and north, respectively. A full derivation of the linearised Rossby wave equation may again be found in [2, chap. 3]. We shall seek time harmonic solutions with periods of between 4 months to 2 years, and therefore write $P(X, Y, z, t) = \Re\{p(x, y, z)e^{i\omega t}\}$, where x and y are aligned in the direction parallel to, and normal to, the barrier, respectively and $\omega > 0$ is the constant angular frequency.

Modal decomposition. We shall seek modal solutions of the form

$$p(x, y, z) = e^{ikx}e^{ily}\Gamma(z), \tag{1.3}$$

where k and l are constant wavenumbers. Separating the horizontal and vertical dependence in this way we obtain the Sturm–Liouville eigenvalue problem for Γ, namely

$$(f_0/N_0)^2\,\Gamma''(z) + \lambda^2\Gamma(z) = 0, \qquad \Gamma'(0) = \Gamma'(-H) = 0. \tag{1.4}$$

Due to the simplicity of the assumed stratification (i.e. constant N_0), this eigenvalue problem may be solved analytically and the orthonormal vertical modes, $\Gamma_n(z)$ are given by

$$\Gamma_n(z) = \sqrt{\frac{\epsilon_n}{H}}\cos\frac{n\pi z}{H} \tag{1.5}$$

with $\epsilon_0 = 1$, $\epsilon_n = 2$, $n = 1, 2, 3, \dots$. The associated eigenvalues, λ_n, are

$$\lambda_n = \frac{n\pi f_0}{HN_0} = \frac{\pi}{r_n} = \frac{f_0}{\sqrt{gh_n}}, \tag{1.6}$$

where the r_n are the internal Rossby radii of deformation and the h_n are the effective (equivalent) depths. Substituting these separation constants and the ansatz (1.3) into the equation of motion (1.2) we find that the wavenumbers k_n and l_n for each mode satisfy the dispersion relation

$$\frac{\omega}{\beta} = \frac{l_n\sin\theta - k_n\cos\theta}{k_n^2 + l_n^2 + f_0^2/gh_n}. \tag{1.7}$$

From the symmetry of the geometry of the problem it is clear that the spatially harmonic variation of the pressure in the x direction (i.e. along the ridge) is fixed by the spatial variation of the incident Rossby wave in that direction. Thus the k_n are identical and the dispersion relation may be considered to be a quadratic equation for the wavenumbers l_n. It is well known (see, for example, Pedlosky [2]) that two distinct real roots of the dispersion relation represent long and short Rossby waves which have westward phase velocity but westward and eastward group velocities respectively. We shall denote the wavenumber which decays or has outward-going group velocity in the region $y > 0$ by l_n and that which decays or has outward group velocity in the region $y < 0$ by s_n. Thus, we may write the total pressure field as the sum of the scattered field and the incident Rossby wave by

$$p(x,y,z) = \begin{cases} \displaystyle\sum_{n=0}^{\infty} a_n e^{ik_1 x} e^{is_n y} \Gamma_n(z) + e^{ik_1 x} e^{il_1 y} \Gamma_1(z), & y < 0, \\ \displaystyle\sum_{n=0}^{\infty} b_n \Gamma_n(z) e^{ik_1 x} e^{il_n y}, & y > 0, \end{cases} \tag{1.8}$$

where $\{a_n\}$ and $\{b_n\}$ are the reflection and transmission coefficients, respectively, to be determined.

Boundary conditions.   Above the barrier, in the region $-\infty < x < \infty$, $y = 0$, $-\mu H < z < 0$ we shall impose continuity of both x and y components of the velocity field. From (1.1) these matching conditions are equivalent to continuity of pressure and the velocity component parallel to the barrier. We shall also assume that the barrier, which occupies $-\infty < x < \infty$, $y = 0$, $-H < z < -\mu H$, is impermeable and thus that there is no normal flow through it. We may now, using the quasi-geostrophic approximation given in (1.1), express these boundary conditions in terms of the coefficients a_n and b_n given in (1.8). From the continuity conditions we have, for $-\mu H < z < 0$,

$$\sum_{n=0}^{\infty} a_n \Gamma_n(z) + \Gamma_1(z) \;=\; \sum_{n=0}^{\infty} b_n \Gamma_n(z), \tag{1.9}$$

$$\sum_{n=0}^{\infty} a_n s_n \Gamma_n(z) + l_1 \Gamma_1(z) \;=\; \sum_{n=0}^{\infty} b_n l_n \Gamma_n(z). \tag{1.10}$$

Similarly the impermeability condition gives us, for $-H < z < -\mu H$,

$$\sum_{n=0}^{\infty} a_n \Gamma_n(z) + \Gamma_1(z) \;=\; 0, \tag{1.11}$$

$$\sum_{n=0}^{\infty} b_n \Gamma_n(z) \;=\; 0, \tag{1.12}$$

From (1.9), (1.11) and (1.12) it may be seen that the equation

$$\sum_{n=0}^{\infty} a_n \Gamma_n(z) + \Gamma_1(z) = \sum_{n=0}^{\infty} b_n \Gamma_n(z) \tag{1.13}$$

holds throughout the entire depth of the ocean, $-H < z < 0$.

2. ALGEBRAIC METHOD OF SOLUTION

Algebraic form of boundary conditions. In order to find the transmission and reflection coefficients of the scattered modes it is now necessary to find the coefficients a_n and b_n. Multiplying (1.13) by $\Gamma_m(z)$, integrating over $-H < z < 0$ and using the orthonormality of the eigenfunctions, $\Gamma_j(z)$, we obtain

$$a_n + \delta_{n1} = b_n, \quad n = 0, 1, 2, \ldots, \tag{1.14}$$

where δ_{ij} is the Kronecker delta. Upon multiplying (1.11) by $\Gamma_m(z)$ and integrating over the range of validity we obtain

$$(\mathbf{I} - \mathbf{C})\mathbf{a} + (\mathbf{I} - \mathbf{C})\mathbf{e}_2 = \mathbf{0}, \tag{1.15}$$

where $\mathbf{a}$ denotes the column vector $\{a_n\}$, $\mathbf{e}_n$ denotes the column vector with the zero elements except for a 1 in the nth row, $\mathbf{I}$ is the identity matrix and the infinite symmetric matrix $\mathbf{C}$ is defined by

$$c_{ij} = \int_{-\mu H}^{0} \Gamma_i(z)\,\Gamma_j(z)\,dz = c_{ji}. \tag{1.16}$$

Similarly integrating (1.10) and eliminating the b_n using (1.14) we obtain

$$\mathbf{C}(\mathbf{L} - \mathbf{S})\mathbf{a} = \mathbf{0}, \tag{1.17}$$

where the infinite diagonal matrices $\mathbf{L}$ and $\mathbf{S}$ have the the wavenumbers l_n and s_n as their diagonal entries. It may be seen then, from (1.15) and (1.17), that if either of the operators $\mathbf{C}$ and $\mathbf{I} - \mathbf{C}$ is invertible then the problem does not have a solution. It is, however, easy to show that this is not the case. Denoting the restriction of $\Gamma_n(z)$ to $(-\mu H, 0)$ by $\gamma_n(z)$, i.e.

$$\gamma_n(z) = \begin{cases} \Gamma_n(z), & z \in [-\mu H, 0], \\ 0, & z \in [-H, -\mu H), \end{cases} \tag{1.18}$$

and again using the orthonormality of the vertical eigenfunctions we may express $\gamma_i(z)$ as a generalised Fourier series

$$\gamma_i(z) = \sum_{j=0}^{\infty} \left[\int_{-H}^{0} \gamma_i(s)\Gamma_j(s)ds \right] \Gamma_j(z) = \sum_{j=0}^{\infty} c_{ij}\Gamma_j(z). \tag{1.19}$$

Thus we have

$$\begin{aligned}
c_{ij} &= \int_{-\mu H}^{0} \gamma_i(s)\Gamma_j(s)\,ds = \int_{-\mu H}^{0} \left[\sum_{k=0}^{\infty} c_{ik}\Gamma_k(s) \right] \Gamma_j(s)\,ds \\
&= \sum_{k=0}^{\infty} c_{ik} \left[\int_{-\mu H}^{0} \Gamma_k(s)\Gamma_j(s)\,ds \right] = \sum_{k=0}^{\infty} c_{ik}c_{kj}
\end{aligned}$$

and hence $\mathbf{C}^2 = \mathbf{C}$. Therefore, except in the trivial cases where $\mu = 0$ and $\mu = 1$ ($\mathbf{C} = \mathbf{0}$ and $\mathbf{C} = \mathbf{I}$ respectively), we find that 1 and 0 are the only eigenvalues of $\mathbf{C}$. Equations (1.15) and (1.17) may be rewritten as

$$\mathbf{a} + \mathbf{e}_2 \in \ker(\mathbf{I} - \mathbf{C}) \quad \text{and} \quad (\mathbf{L} - \mathbf{S})\mathbf{a} \in \ker \mathbf{C}, \qquad (1.20)$$

where $\ker \mathbf{C}$ denotes the kernel of $\mathbf{C}$. We shall solve these equations by approximately constructing the kernels by consideration of the eigenvalues and eigenvectors of the truncated system.

Solving the truncated system.   Equations (1.15) and (1.17) may be considered to form two infinite systems of algebraic equations. Suppose we truncate these equations to form two of systems of, say, M equations. We shall denote the truncation of $\mathbf{C}$ to a real symmetric square matrix of size $M \times M$ by $\widetilde{\mathbf{C}}$, and the similarly truncated vectors $\mathbf{a}$ and $\mathbf{e}_2$ by $\widetilde{\mathbf{a}}$ and $\widetilde{\mathbf{e}}_2$. It is now possible to calculate the M eigenvalues, ν_n and eigenvectors $\mathbf{v}_n$, $n = 1, 2, \ldots, M$, of $\widetilde{\mathbf{C}}$. The eigenvalues of such a matrix, with $M = 50$, are shown in Fig. 2. This is a quite typical distribution of

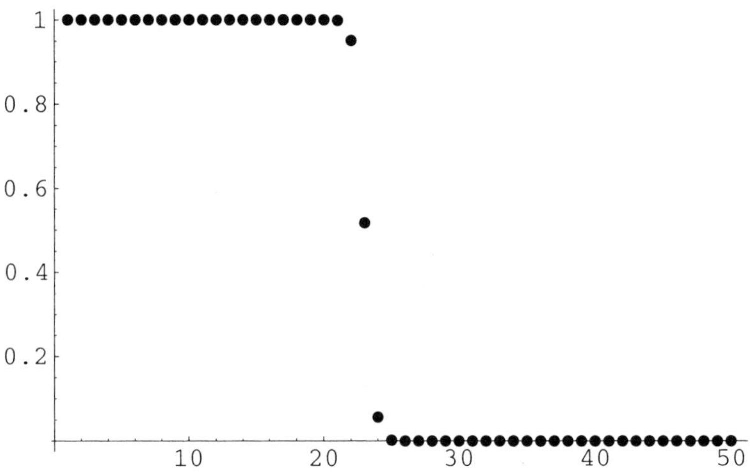

Figure 2   Eigenvalues of $\widetilde{\mathbf{C}}$ with $M = 50$ and $\mu = 0.45$

eigenvalues, with all except a small number of eigenvalues being either close to 1 or close to 0. We also suppose, without loss of generality, that $\nu_n \leq \nu_{n-1}$ and that m is such that $\nu_m \geq \frac{1}{2}$ and $\nu_{m+1} < \frac{1}{2}$. We may then partition the set of eigenvectors into the sets

$$J = \{\mathbf{v}_n : 1 \leq n \leq m\} \quad \text{and} \quad K = \{\mathbf{v}_n : m < n \leq M\}. \qquad (1.21)$$

The span of K is a finite dimensional approximation to the kernel of $\mathbf{C}$ in the sense that if the coefficients β_n are such that

$$\mathbf{v} = \sum_{n=1}^{M} \beta_n \tilde{\mathbf{e}}_n \in \text{span } K \tag{1.22}$$

we have, above the barrier,

$$\sum_{n=1}^{M} \beta_n \Gamma_{n-1}(z) \simeq 0, \quad z \in (-\mu H, 0). \tag{1.23}$$

Although it is possible to construct a vector in the span of K for which the above approximate equation does not hold ($\mathbf{v} = \mathbf{v}_m$ is the obvious example), extensive numerical experiments have shown that such special cases do not arise. Similarly the span of J is an approximation to the kernel of $\mathbf{I} - \mathbf{C}$ in the sense that for a *typical* vector $\mathbf{v} \in \text{span } J$ given by

$$\mathbf{v} = \sum_{n=1}^{M} \beta_n \tilde{\mathbf{e}}_n \in \text{span } J, \tag{1.24}$$

say, we have

$$\sum_{n=1}^{M} \beta_n \Gamma_{n-1}(z) \simeq 0 \quad \text{for} \quad z \in (-H, -\mu H). \tag{1.25}$$

Thus, if we wish to solve approximately (1.17) we may write

$$\tilde{\mathbf{a}} = (\mathbf{S} - \mathbf{L})^{-1} \sum_{n=m+1}^{M} \alpha_n \mathbf{v}_n, \tag{1.26}$$

where the coefficients α_m are to be determined. Substituting this form into the impermeability condition in (1.15) we find

$$(\widetilde{\mathbf{C}} - \mathbf{I}) \sum_{n=m+1}^{M} \alpha_n (\mathbf{S} - \mathbf{L})^{-1} \mathbf{v}_n = (\mathbf{I} - \widetilde{\mathbf{C}}) \tilde{\mathbf{e}}_2. \tag{1.27}$$

From the symmetry of $\mathbf{C}$ we may diagonalise the matrix $\mathbf{I} - \widetilde{\mathbf{C}}$ and write it in the form $\mathbf{I} - \widetilde{\mathbf{C}} = \mathbf{U}\mathbf{D}\mathbf{U}^T$, where the matrix $\mathbf{U}$ is orthogonal and the diagonal matrix $\mathbf{D}$ is defined by

$$D_{ij} = \begin{cases} 1 - \nu_i, & i = j, \\ 0, & i \neq j. \end{cases} \tag{1.28}$$

Upon pre-multiplication by $\mathbf{U}^{-1}$, (1.27) becomes

$$\mathbf{D}\mathbf{U}^T(\mathbf{L}-\mathbf{S})^{-1}\sum_{n=m+1}^{M}\alpha_n\mathbf{v}_n = \mathbf{D}\mathbf{U}^T\tilde{\mathbf{e}}_2. \qquad (1.29)$$

It may be seen that whenever the eigenvalue $\nu_i \simeq 1$ the matrix element D_{ii} is negligible and the constraint imposed by the ith row of the vector equation can be considered to be trivially satisfied for all α_n. Thus, if we neglect the first m rows of (1.29) we are left with a system of $M-m$ equations for the $M-m$ unknowns, $\alpha_{m+1},\ldots,\alpha_M$, which may be solved.

3.   NUMERICAL RESULTS

Having found the unknown coefficients $\alpha_{m+1},\ldots,\alpha_M$ we may reconstruct the vertical structure of the total pressure field above the barrier. Unless otherwise stated we shall use the parameters $H = 4000\text{m}$, $N = 0.02\text{s}^{-1}$, $\theta = 45°$, $f_0 = 1.03\times10^{-4}\text{s}^{-1}$ and $\beta = 1.62\times10^{-11}\text{s}^{-1}\text{m}^{-1}$, which corresponds to the origin of the β-plane at $45°$N, and $\omega = 4.04\times10^{-7}\text{s}^{-1}$, which represents a period of 180 days. For these parameters there are 5 propagating modes. Fig. 3 shows the total pressure field structure above

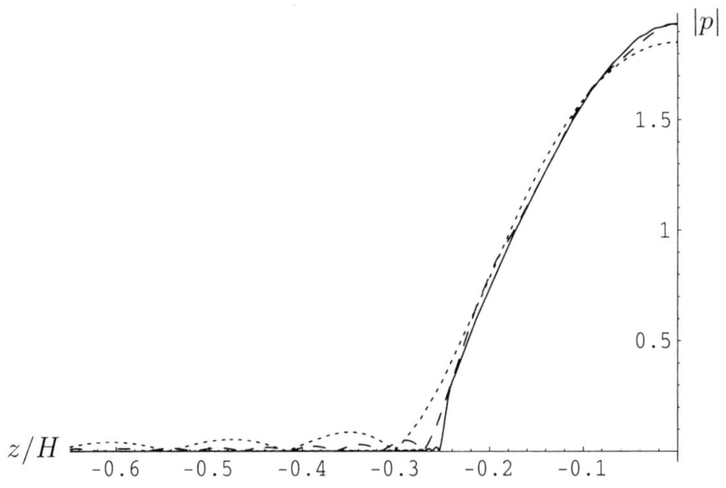

Figure 3   Pressure field above a barrier of height $0.75H$

a barrier of height $0.75H$. The dotted, dashed and solid lines denote the solution generated by truncation at 8, 30 and 120 terms respectively. We see a good degree of convergence to a solution with zero pressure on the barrier (as required by quasi-geostrophy) as the number of terms increases. At the barrier tip, the pressure increases rapidly and this is in accordance with the singularity in the analytic solution which may be found by considering the fluid flow near this point.

Similarly, Fig. 4 shows the total pressure field structure above a barrier of height $0.25H$. This also exhibits good convergence as the size of the truncated system increases as well as the same qualitative behaviour at the barrier tip. However, the pressure field is no longer monotonic and there is a local minimum located at a height comparable to the nodal point in the incident pressure field. We may also contrast the reflection

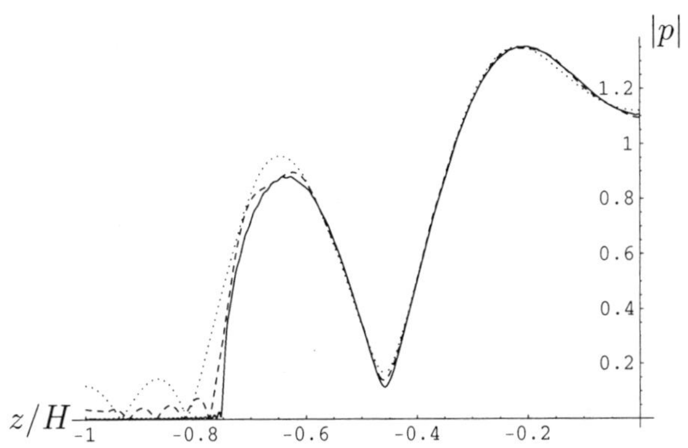

Figure 4   Pressure field above a barrier of height $0.25H$

coefficients for the modes other than the incident mode. Fig. 5 shows the reflection coefficients a_n for the three propagating modes when $\theta = 30°$. It is clear from this figure that it is quite possible for modes other than the incident mode to be the most energetic scattered mode. Finally,

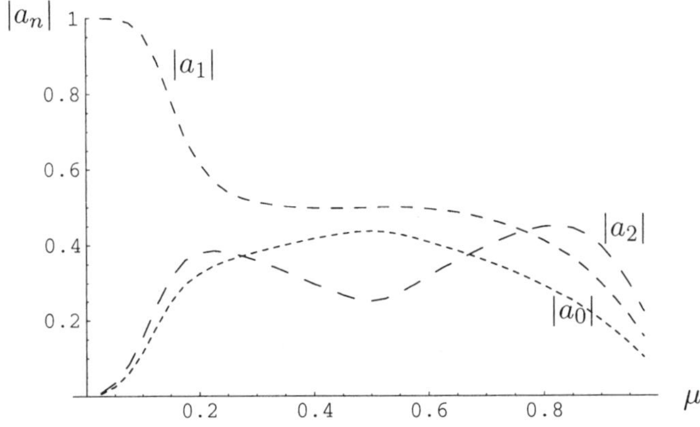

Figure 5   Reflection coefficients for barotropic and baroclinic modes with $\theta = 30°$

Fig. 6 shows the reflection coefficient of the incident mode for a complete range of barrier orientation, θ, and non-dimensional gap height, μ. The rapid variation visible near $\theta = 20°$ and, to a lesser extent $\theta = 35°$ corresponds to the angles at which the higher propagating modes cut off i.e. the wavenumber becomes complex.

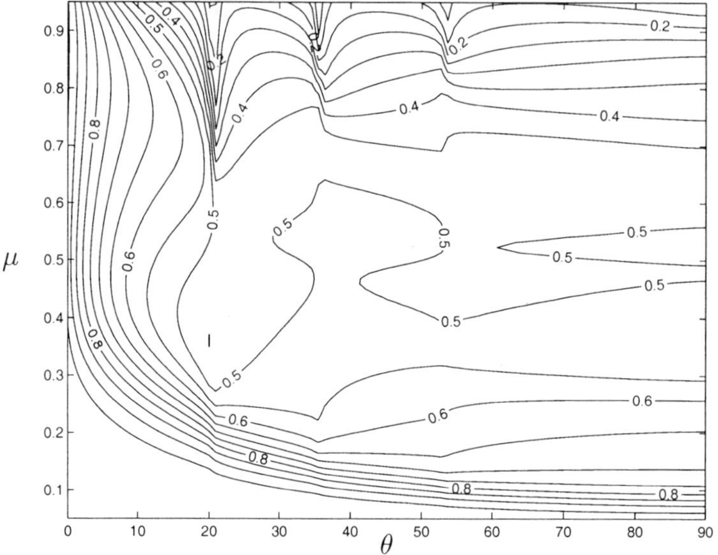

Figure 6   Contour plot of reflection coefficient $|a_1|$

References

[1] Cushman-Roisin, B (1994) *Introduction to Geophysical Fluid Dynamics*, Prentice Hall.

[2] Pedlosky, J (1979) *Geophysical Fluid Dynamics*, Springer-Verlag.

[3] Mysak, L A and LeBlond, P H (1972) *The scattering of Rossby waves by a semi-infinite barrier*, J. Phys. Oceanog. **2**, 108–114.

[4] Huthnance, J M (1981) *A note on baroclinic Rossby-wave reflection at sea-floor scarps*, Deep-Sea Res. A **28**, 83–91.

[5] Wang L P and Koblinsky C J (1994) *Influence of mid-ocean ridges on Rossby waves*, J. Geophys. Res. – Oceans **99**, 25143–25153.

[6] Schmidt, G A and Johnson, E R (1997) *The scattering of stratified topographic Rossby waves by seafloor ridges*, Geophys. Astrophys. Fluid Dynamics **84**, 29–52.

WAVE DIFFRACTION THROUGH A GAP IN A BREAKWATER OF NON-ZERO THICKNESS

N. R. T. Biggs

Department of Mathematics

Keele University, Keele, Staffordshire, ST5 5BG, UK

n.r.t.biggs@maths.keele.ac.uk

D. Porter

Department of Mathematics, University of Reading

PO Box 220, Whiteknights

Reading, RG6 7AX, UK

D.Porter@reading.ac.uk

Abstract We consider the diffraction of a prescribed train of monochromatic plane surface water waves incident on a vertical-sided, perfectly reflecting thick breakwater standing on a horizontal bed in water of uniform undisturbed depth, and containing a single gap. The corresponding linearised boundary value problem is reduced to a pair of uncoupled first kind integral equations which display a particular structure; embedding formulae are then derived for a general integral equation of the type encountered. Within the context of the diffraction problem, the embedding result gives the solution for any incident wave angle explicitly in terms of the solutions for any two other distinct angles.

1.   INTRODUCTION

In this paper, we consider a particular problem of water wave diffraction, reducing the corresponding boundary value problem to a pair of uncoupled integral equations which are characterised by a combination of sum and difference kernels. Embedding formulae are then derived for a generalised such equation, which relate solutions corresponding to different free terms. Formulae of this type applicable to the simpler case of a pure difference kernel have been derived previously by, amongst others, Porter [1], Sakhnovich [2] and Biggs et al [3].

I.D. Abrahams et al. (eds.),

IUTAM Symposium on Diffraction and Scattering in Fluid Mechanics and Elasticity, 13–21.

© 2002 *Kluwer Academic Publishers. Printed in the Netherlands.*

Space considerations dictate that only diffraction through a single gap in an otherwise infinite barrier is considered here; a treatment of diffraction through a barrier punctured by an arbitrary (but finite) number of gaps is to be found in [4], as is the associated embedding theory and a more complete list of references. We note, however, that the derivation of the embedding formulae presented below is quite different from that found in [4].

2.   THE DIFFRACTION PROBLEM

We consider the diffraction of a prescribed train of monochromatic plane waves by an array of vertical-sided, perfectly reflecting breakwaters standing on an impermeable, horizontal bed in water of uniform quiescent depth h. Using the standard assumptions of linearised water wave theory, the velocity potential Ψ representing the fluid motion can therefore be written as

$$\Psi = \mathrm{Re}\left\{ \frac{g}{i\sigma} \frac{\cosh \kappa(z+h)}{\cosh(\kappa h)} \psi(x,y) e^{-i\sigma t} \right\}.$$

The Cartesian coordinates introduced here are arranged so that z is measured vertically upwards, $z = 0$ coinciding with the undisturbed free surface and $z = -h$ with the fluid bed. The angular wave frequency σ is assumed to be given and κ, the wavenumber, is the real, positive root of the dispersion relation $\sigma^2 = g\kappa \tanh(\kappa h)$ which arises from the linearised conditions at the free surface, whose elevation is then $\eta(x,y,t) = \mathrm{Re}\left\{ \psi(x,y) e^{-i\sigma t} \right\}$. The breakwater array is aligned so that it is bounded by $y = \pm d$ and contains a single gap, situated for convenience on the unit interval $L = (0,1)$. The reduced potential $\psi(x,y)$ then satisfies the Helmholtz equation

$$\psi_{xx} + \psi_{yy} + \kappa^2 \psi = 0, \tag{2.1}$$

at all points in the (x,y) plane, excepting the domain $\mathcal{B} = (\mathbb{R}\backslash L) \times [-d,d]$.

To continue the specification of ψ, we require the normal derivative $\partial\psi/\partial n$ to vanish on each breakwater face to there represent perfect reflection. We introduce the incident plane wave in $y > d$ represented by

$$\psi_\mathrm{i}(x,y) = A e^{-i\kappa(x\cos\theta' + (y-d)\sin\theta')},$$

$A \in \mathbb{R}$ being a prescribed amplitude and $\theta' \in [0,\pi]$ the incident angle measured counter-clockwise from the line $y = d$, $x > 0$. Thus ψ may be decomposed in the form

$$\psi(x,y) = \begin{cases} \psi_\mathrm{i}(x,y) + \psi_\mathrm{r}(x,y) + \psi_\mathrm{d}(x,y), & y > d,\ x \in (-\infty,\infty), \\ \psi_\mathrm{d}(x,y), & |y| < d,\ x \in L \text{ and } y < -d,\ x \in (-\infty,\infty), \end{cases}$$

where $\psi_{\mathrm{r}}(x, y) = \psi_{\mathrm{i}}(x, -y + 2d)$ is the reflected plane wave, and $\psi_{\mathrm{d}}(x, y)$ is that part of the reduced potential which represents the diffraction process and must satisfy the Sommerfeld radiation condition.

It is convenient to now write ψ_{d} as $\psi_{\mathrm{d}} = \psi_{\mathrm{s}} + \psi_{\mathrm{a}}$, where $\psi_{\mathrm{s}}(x, y) = \frac{1}{2}\{\psi_{\mathrm{d}}(x, y) + \psi_{\mathrm{d}}(x, -y)\}$ and $\psi_{\mathrm{a}}(x, y) = \frac{1}{2}\{\psi_{\mathrm{d}}(x, y) - \psi_{\mathrm{d}}(x, -y)\}$ are the symmetric and anti-symmetric parts in y of ψ_{d} respectively. Determining $\psi_{\mathrm{s,a}}$ for $y > 0$ (say) is now sufficient to construct ψ_{d} and thus the whole solution to the diffraction problem.

Using standard Green's function techniques, we obtain the integral representations

$$\psi_{\mathrm{s,a}}(x, y) = -\frac{1}{2}i \int_{L} H_0^{(1)}\left[\kappa\sqrt{(x - x_0)^2 + (y - d)^2}\right] \frac{\partial \psi_{\mathrm{s,a}}}{\partial y_0}(x_0, d^+) \, dx_0,$$

$$(2.2)$$

in $y > d$, where $H_0^{(1)}$ denotes the Hankel function of the first kind of order zero. The corresponding representations in the breakwater gap are obtained by separation of variables. Thus, for $x \in L$, $0 < y < d$,

$$\left.\begin{aligned}
\psi_{\mathrm{s}}(x, y) &= \sum_{n=0}^{\infty} \frac{g_n(x)\cosh(\alpha_n y)}{\alpha_n \sinh(\alpha_n d)} \int_{L} \frac{\partial \psi_{\mathrm{s}}}{\partial y_0}(x_0, d^-)g_n(x_0)\, dx_0, \\
\psi_{\mathrm{a}}(x, y) &= \sum_{n=0}^{\infty} \frac{g_n(x)\sinh(\alpha_n y)}{\alpha_n \cosh(\alpha_n d)} \int_{L} \frac{\partial \psi_{\mathrm{a}}}{\partial y_0}(x_0, d^-)g_n(x_0)\, dx_0,
\end{aligned}\right\} \quad (2.3)$$

where

$$\begin{aligned}
\alpha_n &= \sqrt{(n\pi)^2 - \kappa^2} & \text{if } n\pi > \kappa \\
&\equiv i\beta_n = i\sqrt{\kappa^2 - (n\pi)^2} & \text{if } n\pi < \kappa
\end{aligned}\right\}$$

and, in particular, $\alpha_0 \equiv i\beta_0 = i\kappa$. Also, $g_n(x) = \sqrt{\varepsilon_n}\cos(n\pi x)$, where $\varepsilon_0 = 1$, $\varepsilon_n = 2$, $n \geq 1$. Equation (2.3) is clearly invalid if $\alpha_n = 0$ for some n, but a modified version is easily derived. For details of these exceptional cases, see [4].

Now matching (2.2) and (2.3) across the line $x \in L$, $y = d$, we obtain the integral equations

$$\int_{L} \left[m_{\mathrm{s,a}}^{(1)}(|x - x_0|) + m_{\mathrm{s,a}}^{(2)}(x + x_0)\right] \frac{\partial \psi_{\mathrm{s,a}}}{\partial y_0}(x_0, d)\, dx_0 = A e^{-i\kappa x \cos\theta'},$$

$$(2.4)$$

for $x \in L$, for $\partial \psi_{\mathrm{s,a}}/\partial y(x, d) = \partial \psi_{\mathrm{s,a}}/\partial y(x, d^{\pm})$. Here, to reveal their structure the kernels have been arranged as

$$m_{\mathrm{s}}^{(1)}(|x - x_0|) = \frac{1}{2}iH_0^{(1)}(\kappa|x - x_0|) + \sum_{n=1}^{\infty} \frac{\cos\left[n\pi|x - x_0|\right]}{\alpha_n \tanh(\alpha_n d)},$$

$$m_{\mathrm{s}}^{(2)}(x + x_0) = -[\kappa \tan(\kappa d)]^{-1} + \sum_{n=1}^{\infty} \frac{\cos\left[n\pi(x + x_0)\right]}{\alpha_n \tanh(\alpha_n d)},$$

$$m_{\mathrm{a}}^{(1)}(|x - x_0|) = \frac{1}{2} i H_0^{(1)}(\kappa|x - x_0|) + \sum_{n=1}^{\infty} \frac{\cos\left[n\pi|x - x_0|\right]}{\alpha_n \coth(\alpha_n d)},$$

$$m_{\mathrm{a}}^{(2)}(x + x_0) = [\kappa \cot(\kappa d)]^{-1} + \sum_{n=1}^{\infty} \frac{\cos\left[n\pi(x + x_0)\right]}{\alpha_n \coth(\alpha_n d)}.$$

The diffraction problem is thus reduced to the concise form (2.4), the representations (2.2) and (2.3) followed by the construction of ψ_{d} and ψ providing the extension to the whole flow domain.

One particular property of the solution of the diffraction problem which proves useful in the more general investigation of embedding is its far-field behaviour. This can be deduced by using the standard expansion of the Hankel function for large argument in equation (2.2). With $x = r \cos \alpha'$, $y - d = r \sin \alpha'$, and $\alpha' \in [0, \pi]$, we find that

$$\psi_{\mathrm{s,a}} \sim \frac{A e^{i(\kappa r - 3\pi/4)}}{\sqrt{2\pi\kappa r}} F_{\mathrm{s,a}}(\theta', \alpha'), \quad y > d,$$

where

$$F_{\mathrm{s,a}}(\theta', \alpha') = \frac{1}{A} \int_L \frac{\partial \psi_{\mathrm{s,a}}}{\partial y_0}(x_0, d)\, e^{-i\kappa x_0 \cos \alpha'}\, dx_0.$$

Thus the leading term in the far-field is given by

$$\psi_{\mathrm{d}} \sim \frac{A e^{i(\kappa r - 3\pi/4)}}{\sqrt{2\pi\kappa r}} F_+(\theta', \alpha'), \quad y > d, \qquad (2.5)$$

where $F_+(\theta', \alpha') = F_{\mathrm{s}}(\theta', \alpha') + F_{\mathrm{a}}(\theta', \alpha')$. For $y < -d$, the counterpart of (2.5) is

$$\psi_{\mathrm{d}} \sim \frac{A e^{i(\kappa r - 3\pi/4)}}{\sqrt{2\pi\kappa r}} F_-(\theta', \alpha'), \quad y < -d, \qquad (2.6)$$

where $F_-(\theta', \alpha') = F_{\mathrm{s}}(\theta', \alpha') - F_{\mathrm{a}}(\theta', \alpha')$, r is now the distance of the point (x, y) from $(0, -d)$, and $\alpha' \in [0, \pi]$ is measured clockwise from the new polar line $y = -d$, $x > 0$. The integrals $F_\pm(\theta', \alpha')$ are sometimes referred to as 'far-field diffraction coefficients' and determine the essential character of the far scattered wave field.

3.   EMBEDDING FORMULAE

Guided by the form of the kernels in (2.4), we derive embedding formulae for the solution of the integral equation

$$\int_0^1 \{k(|x - x_0|) + l(x + x_0)\}\, \phi_\theta(x_0)\, dx_0 = e^{-i\theta x}, \quad 0 \le x \le 1. \qquad (2.7)$$

Here θ denotes a real parameter which we shall later identify with $\kappa \cos \theta'$ to apply the results to (2.4). However, the embedding properties we derive in this section apply to any integral equation of the form (2.7).

We abbreviate (2.7) by writing it as

$$(K\phi_\theta)(x) = f_\theta(x), \quad 0 \le x \le 1, \tag{2.8}$$

where the operator K is defined by

$$(K\phi_\theta)(x) = \int_0^1 \left\{ k(|x - x_0|) + l(x + x_0) \right\} \phi_\theta(x_0) \, dx_0, \quad 0 \le x \le 1,$$

and $f_\theta(x) = e^{-i\theta x}$. We do not rely on the technicalities of operator theory in what follows, but it may be remarked at this point that we can regard K as a bounded compact operator on the Hilbert space $L_2(0, 1)$, which is the case if the kernel $k(|x - x_0|) + l(x + x_0)$ is square-integrable, for example. Since K is not invertible, we must then assume that (2.8) has a solution in $L_2(0, 1)$ and that $K\phi = 0$ implies $\phi = 0$, which guarantees that the solution is unique; these assumptions hold for examples of (2.8) arising from boundary value problems, and (2.4) in particular.

Guided by the diffraction problem (2.4), we introduce the quantity $\int_L \phi_\theta(x) e^{-i\alpha x} \, dx$ in relation to (2.8) and refer to it as the far-field diffraction coefficient. In terms of the inner product $(f, g) = \int_L f(x) \overline{g}(x) \, dx$, this coefficient can be denoted by

$$G(\theta, \alpha) = (\phi_\theta, f_{-\alpha}), \quad \alpha, \theta \in \mathbb{R}. \tag{2.9}$$

The reciprocity principle $G(\theta, \alpha) = G(\alpha, \theta)$ follows at once from (2.8) and (2.9). In the case of the diffraction problem this principle shows that the dominant contribution to the far wave-field is unaltered if the incident angle and the observation angle are interchanged. For convenience in the subsequent work we also introduce the further quantity

$$H(\theta, \alpha) = (\theta^2 - \alpha^2) G(\theta, \alpha), \tag{2.10}$$

for which the reciprocity principle takes the form

$$H(\theta, \alpha) = -H(\alpha, \theta). \tag{2.11}$$

A final piece of notation requires introduction at this point. In the following work, we make frequent use of the operator V_α and its adjoint V_α^*, which for $\alpha \in \mathbb{R}$ and $x \in [0, 1]$ satisfy

$$(V_\alpha \phi)(x) = \int_0^x \phi(\xi) e^{-i\alpha(x-\xi)} \, d\xi, \quad (V_\alpha^* \phi)(x) = \int_x^1 \phi(\xi) e^{-i\alpha(x-\xi)} \, d\xi, \tag{2.12}$$

and which are related through $(V_\alpha \phi, \psi) = (\phi, V_\alpha^* \psi)$. The identities

$$i(\alpha - \beta)V_\alpha^* V_\beta^* = V_\alpha^* - V_\beta^*, \quad i(\alpha - \beta)V_\alpha f_\beta = f_\beta - f_\alpha, \qquad (2.13)$$

are readily established.

We turn our attention to the derivation of the embedding formula itself, using an extension of a method given in Porter [1]. The key ingredient of the technique is the relationship

$$[V_\alpha - V_{-\alpha}] K\phi + K [V_\alpha - V_{-\alpha}]^* \phi = (\phi, V_\alpha \overline{k} - V_{-\alpha}\overline{l}) f_\alpha$$
$$-(\phi, V_{-\alpha}\overline{k} - V_\alpha \overline{l}) f_{-\alpha} + (\phi, f_\alpha) [V_\alpha k - V_{-\alpha} l] - (\phi, f_{-\alpha}) [V_{-\alpha} k - V_\alpha l],$$
$$(2.14)$$

for $\alpha \in \mathbb{R}$, which is constructed through a direct calculation. Now, let Φ satisfy

$$\Phi = (\theta^2 - \alpha^2)\phi_\theta - (\beta^2 - \alpha^2) [b_+ \phi_\beta + b_- \phi_{-\beta}], \qquad (2.15)$$

where θ, α and β are real parameters with $|\theta|, |\alpha|, |\beta|$ distinct from each other and zero, and the constants $b_\pm$ have been chosen to ensure that

$$(\Phi, f_{\pm \alpha}) = 0. \qquad (2.16)$$

Bearing in mind (2.16), application of (2.14) to Φ results in

$$[V_\alpha - V_{-\alpha}] K\Phi + K [V_\alpha - V_{-\alpha}]^* \Phi =$$
$$(\Phi, V_\alpha \overline{k} - V_{-\alpha}\overline{l}) f_\alpha - (\Phi, V_{-\alpha}\overline{k} - V_\alpha \overline{l}) f_{-\alpha}. \quad (2.17)$$

This equation can be simplified, for use of (2.8) and (2.15) shows that

$$K\Phi = (\theta^2 - \alpha^2) f_\theta - (\beta^2 - \alpha^2) [b_+ f_\beta + b_- f_{-\beta}],$$

and thus, by virtue of the second of (2.13), (2.17) can be rearranged as

$$(2i\alpha)^{-1} K [V_\alpha - V_{-\alpha}]^* \Phi = [a_+ f_\alpha + a_- f_{-\alpha}] + [b_+ f_\beta + b_- f_{-\beta}] - f_\theta,$$
$$(2.18)$$

for some constants $a_\pm$. The injectivity of the operator K now allows us to deduce from (2.18) that

$$(2i\alpha)^{-1} [V_\alpha - V_{-\alpha}]^* \Phi = [a_+ \phi_\alpha + a_- \phi_{-\alpha}] + [b_+ \phi_\beta + b_- \phi_{-\beta}] - \phi_\theta,$$

which, using (2.15), can be arranged as

$$W_{\theta,\alpha}^* \phi_\theta = [a_+ \phi_\alpha + a_- \phi_{-\alpha}] + W_{\beta,\alpha}^* [b_+ \phi_\beta + b_- \phi_{-\beta}], \qquad (2.19)$$

where for convenience we have introduced the operator

$$W_{\theta,\alpha}^* = I + (2i\alpha)^{-1}(\theta^2 - \alpha^2) [V_\alpha - V_{-\alpha}]^*, \quad \alpha, \theta \in \mathbb{R}, \ \alpha \neq 0,$$

and I denotes the identity operator. The relations

$$W^*_{\alpha,\theta} W^*_{\theta,\beta} = W^*_{\alpha,\beta}, \quad W^*_{\alpha,\theta} W^*_{\theta,\alpha} = I,$$

are consequences of the first of (2.13), and allow (2.19) to be solved for ϕ_θ as

$$\phi_\theta = W^*_{\alpha,\theta} \left[a_+ \phi_\alpha + a_- \phi_{-\alpha} \right] + W^*_{\beta,\theta} \left[b_+ \phi_\beta + b_- \phi_{-\beta} \right]. \tag{2.20}$$

Our remaining task is to determine the constants $a_\pm, b_\pm$, and this is easily achieved. Multiplying equation (2.19) by f_γ and integrating gives

$$(W^*_{\theta,\alpha} \phi_\theta, f_{-\gamma}) = \left(\left[a_+ \phi_\alpha + a_- \phi_{-\alpha} \right], f_{-\gamma} \right) + \left(W^*_{\beta,\alpha} \left[b_+ \phi_\beta + b_- \phi_{-\beta} \right], f_{-\gamma} \right);$$

calculations notable only for their tedium in conjunction with the condition (2.16) simplify this to

$$H(\theta, \gamma) = a_+ H(\alpha, \gamma) + a_- H(-\alpha, \gamma) + b_+ H(\beta, \gamma) + b_- H(-\beta, \gamma). \tag{2.21}$$

Now choosing $\gamma = \pm\beta$ gives

$$H(\theta, \pm\beta) = a_+ H(\alpha, \pm\beta) + a_- H(-\alpha, \pm\beta), \tag{2.22}$$

whilst the alternative choice $\gamma = \pm\alpha$ yields

$$H(\theta, \pm\alpha) = b_+ H(\beta, \pm\alpha) + b_- H(-\beta, \pm\alpha), \tag{2.23}$$

a condition which could also be found by pursuing (2.16). The reciprocity relation (2.11) allows (2.22) and (2.23) to be rewritten as

$$\left. \begin{array}{l} a_+ H(\alpha, \pm\beta) + a_- H(-\alpha, \pm\beta) = -H(\pm\beta, \theta) \\ b_+ H(\beta, \pm\alpha) + b_- H(-\beta, \pm\alpha) = -H(\pm\alpha, \theta) \end{array} \right\} \tag{2.24}$$

from which the constants $a_\pm$ and $b_\pm$ can be determined in terms of just $\phi_{\pm\alpha}$ and $\phi_{\pm\beta}$. Thus, the solution ϕ_θ to (2.8) is completely determined by (2.20) in conjunction with (2.24), from knowledge of only $\phi_{\pm\alpha}$ and $\phi_{\pm\beta}$. The corresponding formula for the diffraction coefficients is (2.21), again in conjunction with (2.24).

Our embedding results can be considerably simplified if the function l in the kernel of (2.7) is such that $l(1 + x) = l(1 - x)$, $-1 \leq x \leq 1$, a condition which is, in particular, obeyed by the integral equations of the diffraction problem (2.4). In this case, we see from (2.8) that

$$\phi_{-\theta}(x) = e^{i\theta} \phi_\theta(1 - x), \quad 0 \leq x \leq 1, \tag{2.25}$$

and, further, that

$$H(-\theta, -\alpha) = e^{i(\alpha+\theta)} H(\theta, \alpha). \tag{2.26}$$

The embedding formula (2.20) then becomes

$$
\begin{aligned}
\phi_\theta(x) = a_+(W^*_{\alpha,\theta}\phi_\alpha)(x) + e^{i\alpha}a_-(W^*_{\alpha,\theta}\phi_\alpha)(1-x) \\
+ b_+(W^*_{\beta,\theta}\phi_\beta)(x) + e^{i\beta}b_-(W^*_{\beta,\theta}\phi_\beta)(1-x), \quad (2.27)
\end{aligned}
$$

for $0 \le x \le 1$, and where $a_\pm, b_\pm$ are now determined by (2.24) appropriately modified by (2.26). In this case then, the solution ϕ_θ to (2.8) is completely determined by (2.27) from knowledge of only the two solutions ϕ_α and ϕ_β. The embedding formula for the diffraction coefficients (2.21) is also adjusted to

$$
\begin{aligned}
H(\theta,\gamma) = a_+H(\alpha,\gamma) + e^{i(\alpha-\gamma)}a_-H(\alpha,-\gamma) \\
+ b_+H(\beta,\gamma) + e^{i(\beta-\gamma)}b_-H(\beta,-\gamma). \quad (2.28)
\end{aligned}
$$

4.   NUMERICAL RESULTS

The structural results of the preceding section can be applied to each of the integral equations (2.4) of Section 2 which describe the diffraction problem, although a connection must first be made between the parameters in the free terms of (2.4) and (2.7). To that end, with κ fixed we set $\theta = \kappa\cos\theta'$ so that $\theta' \in [0,\pi]$ implies $-\kappa \le \theta \le \kappa$; other angles are dealt with in a similar fashion, whilst if θ is replaced by $-\theta$, the angle θ' is replaced by $\pi - \theta'$, and so on. Reference to (2.28) then shows that we require only two solutions of each of the two integral equations (for $\phi_\theta(x) \equiv \partial\psi_{\rm s}/\partial y(x,d)$ and $\partial\psi_{\rm a}/\partial y(x,d)$) in order to implement the embedding formulae. As the effort required to implement the embedding formula (2.28) is negligible, the overall computational cost is effectively reduced to the determination of four individual integral equation solutions.

To generate the minimum number of solutions required, a variational principle is used in conjunction with an adroit choice of trial-space. In particular, it behoves us to pick trial-functions which display cube-root singular behaviour at the ends of the interval L, as a local analysis of the fluid flow in the vicinity of a breakwater corner demonstrates that the unknowns in the integral equations (2.4) behave in such a manner. For reasons of space, details of the numerical procedure are omitted; for a detailed description, we refer the reader to [4].

Figure 1 shows far-field diffraction coefficients $|F_+(\theta',\alpha')|$ as a function of incident wave angle θ', and for four different observation angles α'. Here $\kappa d = 7$, and for convenience L has been shifted onto $[-\frac{1}{2},\frac{1}{2}]$. Numerical solutions have been calculated for each of the two integral equations (2.4) for only two incident wave angles; all those diffraction

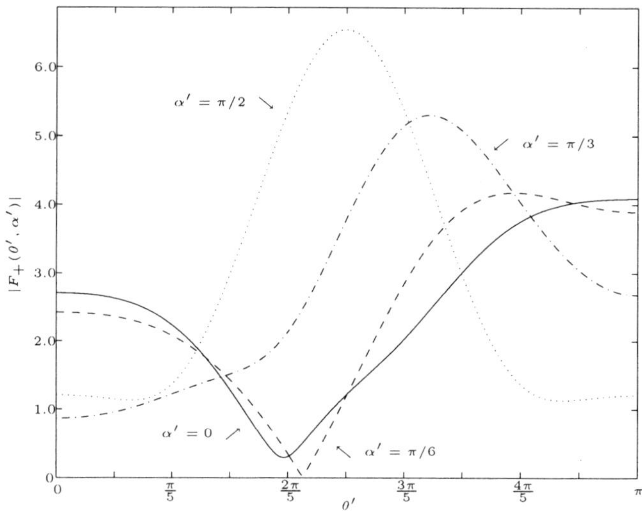

Figure 1   Far-field diffraction coefficients $|F_+(\theta',\alpha')|$.

coefficients remaining have been determined from the embedding formula (2.28), a procedure which involves minimal computational effort.

5.   SUMMARY

Embedding formulae have been derived for integral equations containing a combination of sum and difference kernels, a class hitherto unconsidered in this context. Particular attention has been paid to an example arising from a certain wave diffraction problem; the formulae in this case express the solution for an arbitrary incident wave angle in terms of those corresponding to any two other distinct incident angles.

References

[1] Porter, D (1991) *The solution of integral equations with difference kernels*, J. Int. Eqns. Appl. **3**, 429–454.

[2] Sakhnovich, L A (1996) *Integral Equations with Difference Kernels on Finite Intervals*, Basel: Birkhäuser.

[3] Biggs, N R T, Porter, D and Stirling, D S G (2000) *Wave diffraction through a perforated breakwater*, Quart. J. Mech. Appl. Math. **53**, 375–391.

[4] Biggs, N R T and Porter, D (2001) *Wave diffraction through a perforated breakwater of non-zero thickness*, Quart. J. Mech. Appl. Math. **54**, 523–547.

NONLINEAR EFFECTS IN WAVE SCATTERING AND GENERATION

Flow interaction with topography

R. H. J. Grimshaw

Department of Mathematical Sciences

Loughborough University, Leicestershire LE11 3TU, UK

R.H.J.Grimshaw@lboro.ac.uk

Abstract
When a fluid flow interacts with a topographic feature, and the fluid can support wave propagation, there is the potential for waves to be generated upstream and/or downstream. In many cases when the topographic feature has a small amplitude the situation can be successfully described using a linearised theory, and any nonlinear effects are determined as a small perturbation on the linear theory. However, when the flow is critical, that is, the system supports a long wave whose group velocity is zero in the reference frame of the topographic feature, then typically the linear theory fails and it is necessary to develop an intrinsically nonlinear theory. It is now known that in many cases such a transcritical, weakly nonlinear and weakly dispersive theory leads to a forced Korteweg-de Vries (fKdV) equation. In canonical form, this is

$$-u_t - \Delta u_x + 6uu_x + u_{xxx} + f_x = 0,$$

where $u(x,t)$ is the amplitude of the critical mode, t is the time coordinate, x is the spatial coordinate, Δ is the phase speed of the critical mode, and $f(x)$ is a representation of the topographic feature.

In this article we shall sketch the contexts where the fKdV equation is applicable, and describe some of the most relevant solutions. There are two main classes of solutions. In the first, the initial condition for the fKdV equation is $u(x,0) = 0$ so that the waves are generated directly by the flow interaction with the topography. In this case the solutions are characterised by the generation of upstream solitary waves and an oscillatory downstream wavetrain, with the detailed structure being determined by Δ and the polarity of the topographic forcing term $f(x)$. In the second class a solitary wave is incident on the topography, and depending on the system parameters may be repelled with a significant amplitude change, trapped with a change in amplitude, or allowed to pass by the topography with only a small change in amplitude.

I.D. Abrahams et al. (eds.),

IUTAM Symposium on Diffraction and Scattering in Fluid Mechanics and Elasticity, 23–34.

© 2002 *Kluwer Academic Publishers. Printed in the Netherlands.*

1.   INTRODUCTION

Scattering of waves by obstacles, that is, inhomogeneous features in an otherwise homogeneous waveguide, is a much-studied problem with a long history. For the most part, linear theories are appropriate and have been very successful, in spite of the often very complicated wave patterns that can be formed. In this context nonlinear effects are then calculated as a second-order perturbation, and although such calculations, often numerical, can reveal new physical effects such as wave-induced mean flows or long-time wave-packet modulation, it remains the case that the first-order understanding arises from linear theory. However, there exist special but important circumstances when linear theory fails and nonlinearity enters at the first order. Two situations can be identified when this occurs. The first is when the incident wave field is itself nonlinear; an extreme instance would be when the incident wave is a shock wave. The second case is the subject of this article. Consider the situation when the waveguide can support a wave mode with zero group velocity in the frame of reference of the obstacle. In this situation, wave energy cannot readily escape from the obstacle and the wave amplitude in a linear theory grows indefinitely with time. An intrinsically nonlinear theory is then needed, even when the incident wave has ony a small amplitude. In this article we will sketch some typical circumstances where this arises, and the emergence of the forced Korteweg-de Vries equation, or related equations, as a paradigm for nonlinear scattering of long waves.

Consider then a fluid flow interacting with a topographic feature, in circumstances where the fluid supports wave propagation. A typical example is water in a horizontal channel; other examples are provided by the stable density stratification of shallow coastal seas, or the boundary layer in the lower atmosphere. For simplicity, we shall discuss here mainly the important case when the waveguide has only one active spatial dimension, and the waves are long waves, that is their wavelengths are much greater than a typical transverse length scale associated with the waveguide. There is then the potential for waves to be generated upstream and/or downstream. Linear theory can successfully be used here when the topographic feature has a small amplitude and all waves in the system have finite, non-zero group velocity in the frame of reference of the obstacle. However, when the flow is critical, that is, the system supports a long wave whose group velocity is close to zero in the reference frame of the obstacle, then it is necessary to develop an intrinsically nonlinear theory. It is now known that in many cases such a transcritical, weakly nonlinear and weakly dispersive theory leads to a forced Korteweg-de Vries equation. In canonical non-dimensional form, this is given by

$$-u_t - \Delta u_x + 6uu_x + u_{xxx} + f_x(x) = 0, \tag{3.1}$$

where $u(x,t)$ is the amplitude of the critical mode, t is the time coordinate, x is the spatial coordinate, Δ is the long-wave speed of the critical mode, and $f(x)$ is a representation of the topographic obstacle.

Explicit asymptotic derivations of the fKdV equation (3.1) have been carried out for water waves in a channel [1, 2, 3, 4], internal waves in a shallow fluid [5, 6, 7, 8], inertial waves in a narrow tube [9] and topographic waves in an oceanic coastal waveguide [10, 11], along with analogous derivations in several other physical contexts. Here we sketch a schematic derivation which indicates how a more formal derivation might proceed. Suppose that in linear long wave theory there is a wave mode whose group velocity Δ is close to zero in the reference frame of the obstacle. Then using a standard linear long wave theory, the amplitude $u(x,t)$ of this mode can be shown to satisfy the forced first-order wave equation

$$u_t + \Delta u_x = f_x(x), \tag{3.2}$$

where $f(x)$ is a typically a projection of the actual obstacle profile onto the wave mode in question. The solution of equation (3.2) for the zero initial condition $u(x,0) = 0$ is given by

$$u = \frac{f(x) - f(x - \Delta t)}{\Delta}, \text{ if } \Delta \neq 0, \text{ or } u = tf_x(x) \text{ if } \Delta = 0. \tag{3.3}$$

Clearly this solution fails as $\Delta \to 0$. In this case it is necessary to add a nonlinear term, which is generically quadratic and of the form given in (3.1); further it is now well-understood that when adding a weakly nonlinear correction it is necessary to add a balancing weakly linear dispersive term, which typically has the form given in (3.1). Formally this is achieved by using a multi-scale asymptotic expansion where $\partial/\partial x \sim \epsilon$, $\partial/\partial t \sim \epsilon^3$, $u \sim \epsilon^2$, $\Delta \sim \epsilon^2$ and $f \sim \epsilon^4$. There are plenty of illustrations in the afore-mentioned references to explicit derivations. One of the most important features of this fKdV model emerges already from this scaling, in that the response to a small forcing of $O(f)$ is $O(f^{1/2})$.

Before proceeding we need to describe the steady-state solutions of the unforced KdV equation (i.e. (3.1) with $f \equiv 0$). First there is the well-known solitary wave,

$$u = a\,\mathrm{sech}^2\beta(x - Vt),$$
$$\text{where} \qquad \Delta - V = 2a = 4\beta^2. \tag{3.4}$$

Next, there are the periodic solutions, namely the so-called "cnoidal" waves, given by

$$u = a\left\{b(m) + \mathrm{cn}^2\beta(x - Vt)\right\} + d,$$

where

$$b = \frac{1 - m}{m} - \frac{E(m)}{mK(m)},$$

$$\Delta - V = 6d + 2a\left\{\frac{2 - m}{m} - \frac{3E(m)}{mK(m)}\right\},$$

and

$$a = 2m\beta^2. \tag{3.5}$$

Here $\mathrm{cn}(x)$ is the Jacobian elliptic function of modulus m, while $K(m)$ and $E(m)$ are the complete elliptic integrals of the first and second kind respectively. The mean value of A over one period is d, while the spatial period is $2K(m)/\beta$. As $m \to 1$, $\mathrm{cn}^2(x) \to \mathrm{sech}^2(x)$, $b(m) \to 0$, and then (3.5) becomes solitary wave (3.4), relative to the level d. As $m \to 0$, $b + \mathrm{cn}^2(x) \to \cos 2(x)$, $a \to 0$ and $V \to \Delta - 6d + 4\beta^2$; this is just a sinusoidal wave train relative to the level d.

In this article we shall describe some of the most relevant solutions of equation (3.1). There are two main classes of solutions. In the first, discussed in Section 2, the initial condition for (3.1) is $u(x,0) = 0$ so that the waves are generated directly by the flow interaction with the topography. In this case the solutions are characterised by the generation of upstream solitary waves and an oscillatory downstream wave train, with the detailed structure being determined by Δ and the polarity of the topographic forcing term $f(x)$. In the second class, discussed in Section 3, a solitary wave is incident on the topography, and depending on the system-parameters may be repelled with a significant amplitude change, trapped with a change in amplitude, or allowed to pass by the topography with only a small change in amplitude.

2.   GENERATION OF UPSTREAM AND DOWNSTREAM WAVES

Here we consider the case when the waves are generated directly by the topography, when the appropriate initial condition for (3.1) is that $u(x,0) = 0$. A set of typical solutions of are shown in Fig. 1 for the case when the forcing term is positive and isolated. That is $f(x)$ is positive, and non-zero only in a vicinity of $x = 0$, with a maximum value of f_M. For exact criticality, when $\Delta = 0$ (Fig. 1(a)) the solution is characterised by upstream and downstream wavetrains connected by a locally steady solution over the obstacle. When the oncoming flow is subcritical, so that $\Delta < 0$ (Fig. 1(b)), the upstream wavetrain weakens, and for sufficiently large $|\Delta|$ detaches from the obstacle, while the downstream wavetrain intensifies and for sufficiently large $|\Delta|$ forms a stationary lee wave field. When the oncoming flow is supercritical, so that $\Delta > 0$ (Fig. 1(c)) the upstream wavetrain develops into well-separated solitary waves, while the downstream wavetrain weakens and moves further downstream. For more details, see [5, 12].

The origin of the upstream and downstream wavetrains can be found in the structure of the locally steady solution over the obstacle. In the transcritical régime this is characterised by a transition from a constant state U_- upstream of the obstacle to a constant state U_+ downstream of the obstacle, where $U_- > 0$ and $U_+ < 0$. It is readily shown that $\Delta = 3(U_+ + U_-)$ independently of the details of the forcing term $f(x)$. Explicit determination of U_+ and U_- requires some knowledge of the forcing term $f(x)$. However, in the "hydraulic" limit when the linear dispersive term in (3.1) can be neglected, it is readily shown that

$$6U_\pm = \Delta \mp (12f_M)^{1/2}. \tag{3.6}$$

This expression also serves to define the transcritical régime, which is

$$|\Delta| < (12f_M)^{1/2}. \tag{3.7}$$

Thus upstream of the obstacle there is a transition from the zero state to U_-, while downstream the transition is from U_+ to 0; each transition is effectively generated at $x = 0$.

Both transitions are resolved by "undular bore" solutions. That is, we use the Whitham modulation theory ([13, 14]) in which the periodic wavetrain (3.5) is allowed to have a slowly-varying amplitude a, modulus m and mean depth d. For the undular bore solution, these are functions of the similarity variable x/t. The downstream "undular bore" is then given by

$$\Delta - \frac{x}{t} = 2U_+ \left\{ 2 - m + \frac{2m(1-m)(K(m)}{E(m) - (1-m)K(m)} \right\}$$

for $\max\{0, \Delta - 2U_+\} < (x/t) < \Delta - 12U_+,$

$$a = -2U_+ m, \quad \text{and} \quad d = U_+ \left\{ 2 - m - \frac{2E(m)}{K(m)} \right\}. \tag{3.8}$$

Ahead of the wavetrain where $x/t \approx \Delta - 12U_+$, $m \to 0$ and the waves are approximately sinusoidal. Behind the wavetrain where $x/t \approx \Delta - 2U_+$, $m \to 1$ and the waves are approximately solitary waves of amplitude $-2U_+$ relative to the mean level of U_+. Further, it can be shown that on any individual crest in the wavetrain, $m \to 1$ as $t \to \infty$. In this sense,

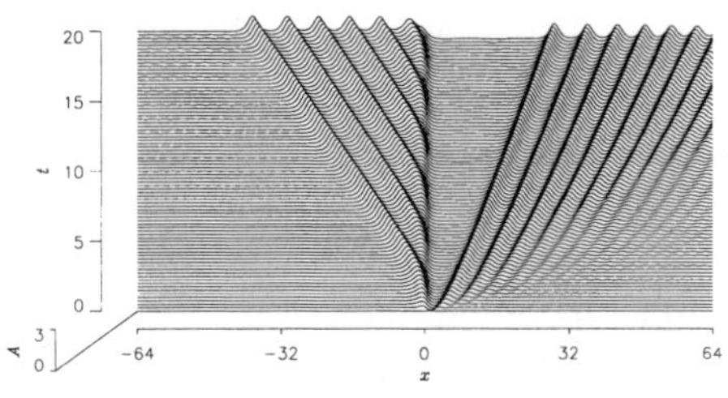

(a) $\Delta = 0$

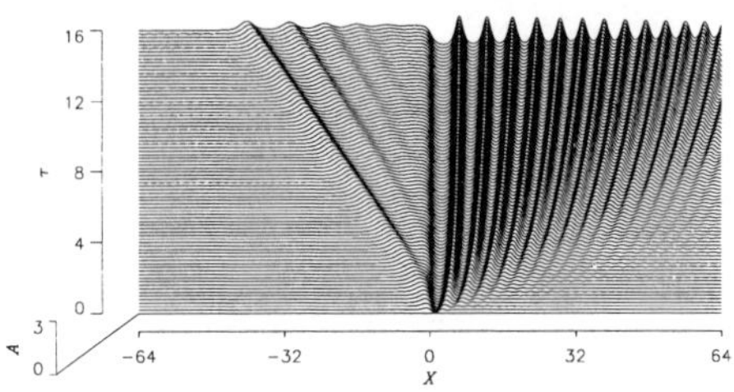

(b) $\Delta < 0$

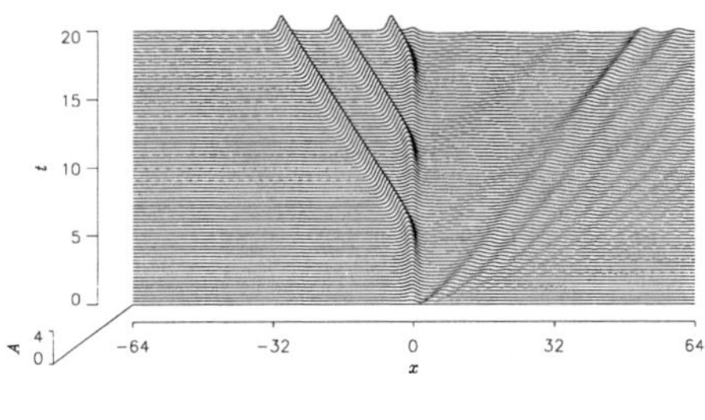

(c) $\Delta > 0$

Figure 1: Typical solutions of the forced KdV equation (1.1) for (a) $\Delta = 0$, (b) $\Delta < 0$ and (c) $\Delta > 0$.

the undular bore evolves into a train of solitary waves. This downstream wavetrain is constrained to lie in $x > 0$, and hence is only fully realised if $\Delta > 2U_+$. Combining this criterion with (3.6) and (3.7) then leads to the régime,

$$-\frac{1}{2}(12f_M)^{1/2} < \Delta < (12f_M)^{1/2}. \tag{3.9}$$

On the other hand, when $\Delta < 2U_+$, we obtain the régime,

$$-(12f_M)^{1/2} < \Delta < -\frac{1}{2}(12f_M)^{1/2}. \tag{3.10}$$

Now the downstream undular bore is attached to the obstacle, with a modulus $m_s(< 1)$ at the obstacle, where m_s can be found by putting $x = 0$ in (3.8). Indeed, a stationary lee wavetrain develops just behind the obstacle (for further details, see [12]).

The upstream "undular bore" is given by

$$\Delta - \frac{x}{t} = 2U_- \left\{ 1 + m - \frac{2m(1 - m)(K(m)}{E(m) - (1 - m)K(m)} \right\}$$

for $\max\{0, \Delta - 4U_-\} < (x/t) < \Delta + 6U_-$,

$$a = 2U_- m, \quad \text{and} \quad d = U_- \left\{ m - 1 + \frac{2E(m)}{K(m)} \right\}. \tag{3.11}$$

Ahead of the wavetrain where $x/t \approx \Delta - 4U_-$, $m \to 1$ and the waves are approximately solitary waves of amplitude $2U_-$ where the leading wave has an amplitude of $2U_-$. Behind the wavetrain where $x/t \approx \Delta - 6U_-$, $m \to 0$ and the waves are approximately sinusoidal. Further, as before, it can be shown that on any individual crest in the wavetrain, $m \to 1$ as $t \to \infty$, so that in this sense, the undular bore evolves into a train of solitary waves. This upstream wavetrain is constrained to lie in $x < 0$, and hence is only fully realised if $\Delta < -6U_-$. Combining this criterion with (3.6) and (3.7) then leads to the régime (3.10). Thus a fully detached upstream "undular bore" coincides with the case when the downstream "undular bore" is attached. On the other hand, when $\Delta > -6U_-$, we obtain the régime (3.9), and so a fully detached downstream "undular bore" coincides with the case when the upstream "undular bore" is attached. It has a modulus m_0 at the obstacle, where m_0 is determined by putting $x = 0$ in (3.11), see [12] for further details.

For the case when the obstacle provides a negative, but still isolated, forcing term (i.e. $f(x)$ is negative, and non-zero only in the vicinity of $x = 0$), the upstream and downstream solutions are qualitatively similar to those described above for positive forcing. However, the solution in the vicinity of the obstacle remains transient, and this causes a modulation of the "undular bore" solutions.

3.   INTERACTION OF A SOLITARY WAVE WITH AN OBSTACLE

Here we consider the situation when a solitary wave is incident on the topographic obstacle. As in the previous section, we shall suppose that the obstacle is isolated, but may have either polarity. The initial condition is a free solitary wave located far from the obstacle. That is, $u(x,0)$ is given by (3.4) with $t = 0$ and x replaced by $x - x_0$ where x_0 is such that $f(x_0) \approx 0$. As the solitary wave propagates towards the obstacle its speed V will adjust, and we would expect either *trapping* if $V \to 0$ as $t \to \infty$, or *repulsion* if V passes through zero and changes sign. Both these cases clearly require that $\Delta > 0$, i.e. the flow is supercritical. Otherwise, if V retains the same sign throughout the interaction, then we have *passage*. Clearly, this will always be the case if $\Delta < 0$, i.e. the flow is subcritical, since then V remains negative throughout the interaction. The *passage* regime may also ocur if $\Delta > 0$ provided that either the solitary wave amplitude a remains large so that V remains negative, or a remains small, so that V remains positive.

These notions can be quantified by assuming that the obstacle provides only a small, and slowly-varying, effect on the solitary wave. Thus, we assume that locally the solitary wave retains its shape, but its amplitude and speed vary slowly. The derivation of the equations governing the amplitude is a multi-scale asymptotic procedure. As this is well-known (see, for instance, [15]), and has been described for the present circumstances in [16], we shall omit all details here. The outcome is that the solitary wave is described by,

$$u = a \operatorname{sech}^2(\beta(t)\Phi),$$

$$\text{where} \quad \Phi(t) = x - \Psi(t), \quad \Psi(t) = \int_0^t V(t')\,dt',$$

$$\text{and} \quad \Delta - V = 2a = 4\beta^2. \tag{3.12}$$

The evolution of the amplitude is determined essentially by an "energy" equation, which here takes the form,

$$\frac{da}{dt} = \beta \int_{-\infty}^{\infty} \operatorname{sech}^2(\beta\Phi)\frac{\partial f}{\partial \Phi}(\Phi + \Psi)\,d\Phi. \tag{3.13}$$

A second equation is obtained from the leading order relationship between the speed and the amplitude in (3.12), which is

$$V = \frac{d\Psi}{dt} = \Delta - 2a. \tag{3.14}$$

Together, (3.13) and (3.14) form a coupled system of ordinary differential equations for the amplitude a and position Ψ of the solitary wave. This system has been discussed in detail by [16] including the effects of higher-order corrections to the "speed" equation (3.14). The effects of allowing Δ to vary slowly in time have been discussed in [17], while the effects of including some dissipation were discussed in [18].

In order to obtain some simple formulas, we shall here make the further approximation that the solitary wave is much narrower than the obstacle. This is the so-called "broad" forcing case of [16], which was shown there to be representative. In this limit, (3.13) becomes

$$\frac{da}{dt} = 2\frac{\partial f}{\partial \Phi}(\Phi). \tag{3.15}$$

The system consisting of (3.15) and (3.14) is Hamiltonian with the Hamiltonian

$$H = \Delta a - a^2 - 2f(\Psi). \tag{3.16}$$

H is an invariant, and the orbits are given by $H = $ constant. We will consider here symmetric and isolated functions $f(\Psi)$ with a single stationary value at $\Psi = 0$ of f_M which is a maximum (minimum) for $f_M > 0$ (< 0). Then the system (3.15), (3.14) has a single critical point at $\Psi = 0$ and $a = a_c = \Delta/2$. Since a is constrained to remain positive, this critical point does not exist if $\Delta < 0$. In this latter case all the orbits pass from $\Psi = \infty$ to $\Psi = -\infty$ as t increases, which corresponds to a *passage* regime. Otherwise, when $\Delta > 0$ the critical point is a centre when $f_M < 0$ and so the obstacle is positive (that is, it has the same polarity as the solitary wave), while the critical point is a saddle point when $f_M > 0$ and so the obstacle is negative (that is, it has the opposite polarity to the solitary wave).

In the case of a centre, there is a family of periodic orbits surrounding the critical point, corresponding to the *trapping* regime. The critical point represents a stationary solitary wave trapped at the obstacle location. This family of periodic orbits is contained within a pair of "homoclinic" orbits, which pass to infinity at the value $a = a_\infty = a_c$ (that is, at the same amplitude as that at the critical point), and reach maximum (minimum) amplitudes of $a_c \pm (-2f_M)^{1/2}$ at $\Psi = 0$ (that is, at the location of the obstacle maximum). Note that if $\Delta^2 < -8f_M$, then the minimum amplitude is less than zero, and so cannot be realised (but see [16]) for the resolution of this dilemma). All orbits outside these "homoclinic" orbits correspond to the *passage* regime, with those orbits whose amplitude at infinity is greater (less) than a_c passing from right to left (left to right).

For the case of a saddle point, there are "homoclinic" orbits emanating from the saddle point, these being two unstable manifolds and two stable manifolds. These pass to infinity with an amplitude $a = a_\infty = a_c \pm (2f_M)^{1/2}$. Note that if $\Delta^2 < 8f_M$, then the smaller of these amplitudes is less than zero, and so cannot be realised (again, see [16] for the resolution of this dilemma). All orbits which at infinity have an amplitude lying between these two values correspond to the *repulsion* regime. Those orbits starting from the left (right) show an increase (decrease) in amplitude. All orbits whose amplitude at infinity lies outside this range correspond to a *passage* regime, and those orbits with the smaller (larger) amplitude pass from left to right (right to left) in the sense of increasing Ψ.

These typical scenarios have been confirmed by numerical simulations of the fKdv equation (3.1) reported in [16]. Here we show two typical results in Fig. 2, these being (a) for a *trapping* regime, and (b) for a *repulsion* regime. Other similar results are shown in [16].

References

[1] Akylas, T R (1984) *On the excitation of long nonlinear water waves by a moving pressure distribution*, J. Fluid Mech. **141**, 455–466.

[2] Cole, S L (1985) *Transient waves produced by flow past a bump*, Wave Motion **7**, 579–587.

[3] Lee, S J, Yates, G T and Wu, T Y (1989) *Experiments and analyses of upstream-advancing solitary waves generated by moving disturbances*, J. Fluid Mech. **199**, 569–593.

[4] Wu, T Y (1987) *Generation of upstream advancing solitons by moving disturbances*, J. Fluid Mech. **184**, 75–99.

[5] Grimshaw, R and Smyth N (1986) *Resonant flow of a stratified fluid over topography*, J. Fluid Mech. **169**, 429–464.

[6] Melville, W K and Helfrich, K R (1987) *Transcritical two-layer flow over topography*, J. Fluid Mech. **178**, 31–52.

[7] Clarke, S R and Grimshaw, R H J (1994) *Resonantly generated internal wave in a contraction*, J. Fluid Mech. **274**, 139–161.

[8] Clarke, S R and Grimshaw, R H J (2000) *Weakly-nonlinear internal wave fronts trapped in contractions*, J. Fluid Mech. **415**, 323–345.

[9] Grimshaw, R (1990) *Resonant flow of a rotating fluid past an obstacle: the general case*, Stud. Appl. Math. **83**, 249–269.

[10] Grimshaw, R (1987) *Resonant forcing of barotropic coastally trapped waves*, J. Phys. Oceanogr. **17**, 53–65.

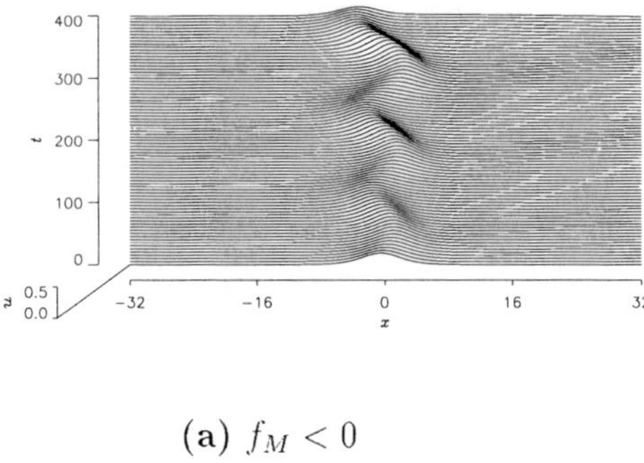

(a) $f_M < 0$

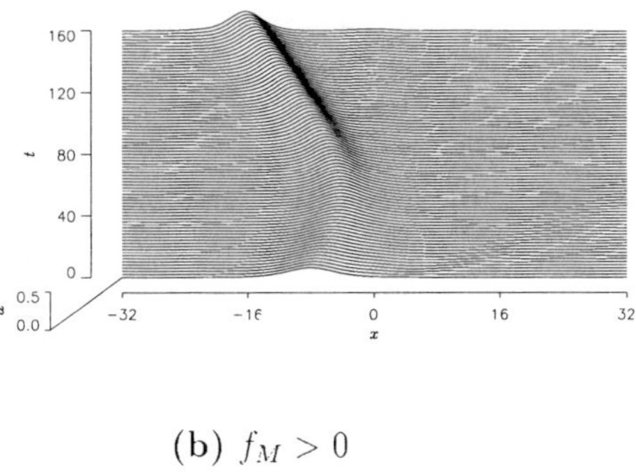

(b) $f_M > 0$

Figure 2: Typical solutions of the forced KdV equation (1.1) for the case of a solitary wave incident on an obstacle; (a) a *trapping* regime when $f_M < 0$, a *repulsion* regime when $f_M > 0$.

[11] Mitsudera, H and Grimshaw, R (1990) *Resonant forcing of coastally trapped waves in a continuously stratified ocean*, Pure & Appl. Geophys. **133**, 635–644.

[12] Smyth, N (1987) *Modulation theory solution for resonant flow over topography*, Proc. Roy. Soc. London A **409**, 79–97.

[13] Whitham, G B (1974) *Linear and Nonlinear Waves*, New York: Wiley.

[14] Gurevich, A V and Pitaevskii, L P (1974) *Nonstationary structure of a collionsless shock wave*, Sov. Phys. JETP **38**, 291–297.

[15] Grimshaw, R and Mitsudera, H (1993) *Slowly-varying solitary wave solutions of the perturbed Korteweg-de Vries equation revisited*, Stud. Appl. Math. **90**, 75–86.

[16] Grimshaw, R, Pelinovsky, E and Tian, X (1994) *Interaction of a solitary wave with an external force*, Physica D **77**, 405–433.

[17] Grimshaw, R, Pelinovsky, E and Sakov, P (1996) *Interaction of a solitary wave with an external force moving with variable speed*, Stud. Appl. Math. **97**, 235–276.

[18] Grimshaw, R, Pelinovsky, E and Bezen, A (1997) *Hysteresis phenomena in the interaction of a damped solitary wave with an external force*, Wave Motion **26**, 253–274.

REFLECTION BY SLOTTED THICK BARRIERS UNDER OBLIQUE WAVE ATTACK: A NUMERICAL STUDY

M. Kanoria

Department of Applied Mathematics, Calcutta University
92 A.P.C. Road, Calcutta - 700 009, India

mridulak@cubmb.ernet.in

B. N. Mandal

Physics and Applied Mathematics Unit
Indian Statistical Institute
203 B. T. Road, Calcutta - 700 035, India

biren@isical.ac.in

Abstract   This paper is concerned with scattering of an obliquely incident surface water wave by a thick vertical slotted barrier of rectangular cross section with an arbitrary number of slots of unequal lengths along the vertical direction, and present in finite depth water. Four different geometrical configurations of the slotted barrier are considered. The problem is reduced to solving first kind integral equations valid in the union of several disjoint intervals. Galerkin approximations involving ultraspherical Gegenbauer polynomials are utilized to solve these integral equations to obtain very accurate numerical estimates for the reflection coefficient, which are depicted graphically against the wave number for each configuration of the slotted barrier. Numerical codes prepared for this problem are valid for an arbitrary number of slots, the lengths of the slots as well as the wetted portions of the barrier for each configuration being unequal. Some results in the limiting cases have been compared with known results and good agreement is seen to have been achieved.

1.   INTRODUCTION

Breakwaters are used to protect harbours and marinas from the waves. Water wave scattering problems involving various types of breakwaters have been investigated in the literature for quite some time. Slotted breakwaters have a number of desirable features that have encouraged their use within harbours. Problems on water wave scattering by a *thin*

I.D. Abrahams et al. (eds.),
IUTAM Symposium on Diffraction and Scattering in Fluid Mechanics and Elasticity, 35–44.
© 2002 *Kluwer Academic Publishers. Printed in the Netherlands.*

slotted vertical barrier in *deep* water or in *finite depth* water were well studied in the literature. Mei and Black [1] have studied the water wave scattering by a surface piercing or bottom standing *thick* rectangular barrier in finite depth water by employing a variational formulation to compute numerically the reflection and transmission coefficients with an accuracy within one percent. Bai [2] later investigated the problem of oblique wave scattering by a surface piercing long cylinder of rectangular cross section by employing the finite element technique. When the obstacle is in the form of a thick wall with a submerged *narrow* gap, Liu and Wu [3] used the method of matched asymptotic expansion to obtain an approximate expression for the transmission coefficient assuming the thickness of the wall to be of the same magnitude as the wave length. However, the results are not valid if the gap is not narrow.

Recently, several water wave scattering problems involving a thick rectangular barrier have been investigated by Kanoria *et al.* [4] for normal incidence of the wave train and by Mandal and Kanoria [5] for oblique incidence wherein the barrier has four types of configurations, namely, surface piercing (Type I), bottom standing (Type II), in the form of a submerged block (Type III) or a thick wall with a submerged gap (Type IV). Also, Kanoria [6] considered the case when the barrier is in the form of a submerged thick wall with a gap for both normally and obliquely incident waves. In all these problems the technique of Galerkin approximations involving ultra-spherical Gegenbauer polynomials have been utilized for solving some first kind integral equations arising in the mathematical analysis to obtain very accurate numerical estimates for the reflection coefficient.

In the present paper we have considered scattering of surface water waves obliquely incident on a *thick slotted* barrier of rectangular cross section. The barrier has the above four different types of geometrical configurations. The boundary value problem corresponding to each type of barrier is solved in a manner described in [4].

Fairly accurate numerical estimates for the reflection coefficient for each type of barriers are obtained and are depicted graphically against the wave number for various values of the different parameters. Numerical codes prepared for each type of barriers are valid for *any* number of slots. Some results obtained here in the limiting cases are compared with [5] for type-I and type-III barriers with no slot and type-IV barrier with one slot. Good agreement is achieved. For type-II barrier with one slot the numerical results obtained here are compared with [6] and here also the agreement is very good. These perhaps provide some partial checks on the correctness of the numerical results obtained here.

It is expected that the increase in the number of slots should decrease the reflection coefficent for all wave numbers. While this is true for the type-IV barrier for all wave numbers, this is not true for the other type of barriers for some wave numbers. This may be due to some sort of interaction in the flow field within the gaps such as multiple reflection by the submerged edges of the slots. Also, in the long wave limit, the reflection coefficient for any barrier configuration with any number of slots, is seen to tend to zero. This is in agreement with the general result confirmed earlier for any obstacle by Martin and Dalrymple [7] and McIver [8] by using the method of matched asymptotic expansion. This can also be regarded as another check on the correctness of the numerical results and the mathematical analysis followed here.

The results obtained here may be useful in the modelling of efficient breakwaters which are constructed to protect a sheltered area on the coast from the rough sea.

2.   FORMULATION OF THE PROBLEM

We consider a long thick horizontal barrier with $q - 1$ $(q \geq 2)$ slots of width $2b$ present in water of uniform finite depth h, and choose the y-axis vertically downwards along the vertical line of symmetry of the barrier. The xz-plane coincides with the rest position of the free surface. The wetted portion of the barriers occupy the region $-b \leq x \leq b$, $y \in L = L_j (j = 1, 2, 3, 4)$. Here

$$L_1 = (0, a_1^{(1)}) \bigcup_{l=1}^{q-1} (a_{2l}^{(1)}, a_{2l+1}^{(1)}) \quad (0 < a_1^{(1)} < \cdots < a_{2q-1}^{(1)} < h),$$

$$L_2 = \bigcup_{l=0}^{q-1} (a_{2l+1}^{(2)}, a_{2l+2}^{(2)}) \quad (0 < a_1^{(2)} < \cdots < a_{2q}^{(2)} = h),$$

$$L_3 = \bigcup_{l=0}^{q-1} (a_{2l+1}^{(3)}, a_{2l+2}^{(3)}) \quad (0 < a_1^{(3)} < .... < a_{2q}^{(3)} < h),$$

$$L_4 = (0, a_1^{(4)}) \bigcup_{l=0}^{q-2} (a_{2l+2}^{(4)}, a_{2l+3}^{(4)}) \quad (0 < a_1^{(4)} < \cdots < a_{2q-1}^{(4)} = h),$$

L_1, L_2, L_3 and L_4 correspond respectively to type-I, type-II, type-III and type-IV barrier configurations.

Assuming linear theory, a surface wave train *obliquely* incident on a particular thick-barrier configuration from a large distance on its right, is represented by velocity potential $Re\{\phi^{inc}(x, y)e^{i\nu z - i\sigma t}\}$, where

$$\phi^{inc}(x, y) = 2\cosh k_0(h - y)e^{-i\mu(x-b)}/\cosh k_0 h,$$

k_0 being the unique real positive root of the transcendental equation $k \tanh kh = K$ with $K = \sigma^2/g$, σ being the circular frequency of the incoming wave train, g being the acceleration due to gravity, $\mu = k_0 \cos \alpha, \nu = k_0 \sin \alpha$, α being the angle of incidence of the wave train. Due to geometrical symmetry of the thick barrier, the z-dependence can be eliminated by assuming the velocity potential describing the motion in the fluid to be of the form $Re\{\phi(x,y)e^{i\nu z - i\sigma t}\}$. Then $\phi(x,y)$ satisfies

$$(\nabla^2 - \nu^2)\phi = 0 \quad \text{in the fluid region}$$

$$K\phi + \phi_y = 0 \quad \text{on } y = 0, \begin{cases} |x| > b & \text{for type I, IV barrier,} \\ |x| < \infty & \text{for type II, III barrier,} \end{cases}$$

$$\phi_x = 0 \quad \text{on } x = \pm b, y \in L,$$

$$r^{1/3}\nabla\phi \quad \text{is bounded as } r \to 0$$

where r is the distance from any submerged edge of the thick barrier,

$$\phi_y = 0 \quad \text{on } y = a_i^{(j)}, \ |x| < b,$$

for $i = 1, 2, \ldots, m_j$ and $j = 1, 2, 3, 4$, where $m_1 = m_2 = 2q - 1$, $m_3 = 2q$ and $m_4 = 2q - 2$,

$$\phi_y = 0 \quad \text{on } y = h, \begin{cases} |x| < \infty & \text{for type I, III barrier} \\ |x| > b & \text{for type II, IV barrier,} \end{cases}$$

$$\phi(x,y) \to \begin{cases} \phi^{inc}(x,y) + R\phi^{inc}(-x,y) & \text{as } x \to \infty \\ T\phi^{inc}(x,y) & \text{as } x \to -\infty \end{cases}$$

where R and T denote respectively the reflection and transmission coefficients (complex) and are to be determined.

The details of the method of solution are omitted here. Only the numerical results are presented in the next section.

3.   NUMERICAL RESULTS

For a *surface-piercing thick slotted barrier* (type I) the numerical estimates for $|R|$ are depicted graphically in Figs 1a–d against the wave number Kh. Fig. 1a displays the behaviour of $|R|$ against Kh for $a_1^{(1)}/h = 0.1$, $a_2^{(1)}/h = 0.15$, $a_3^{(1)}/h = 0.2$, $a_4^{(1)}/h = 0.25$, $a_5^{(1)}/h = 0.3$, $b/h = 0.1$ and $\alpha = 30°, 60°$. From this figure it is observed that $|R|$ decreases as α increases for fixed wave number. Fig. 1b depicts $|R|$ against Kh for $b/h = 0.01, 0.1, 1.0$ and $a_i^{(1)}$ ($i = 1, 2, 3, 4, 5$) having the same values as in Fig. 1a and $\alpha = 30°$. It is observed from this figure that $|R|$ increases with the increase of the thickness parameter b/h. However, when the barrier is comparatively thin $(b/h = 0.01)$, $|R|$ vanishes at a certain wave number. This behaviour may be due to some interaction of flow between the gaps. In Fig. 1c, $|R|$ is plotted against Kh when the number of slots in the barrier is one $(q = 2)$ and two $(q = 3)$. For $q = 2$, we have taken $a_1^{(1)}/h = 0.1$, $a_2^{(1)}/h = 0.15$ and $a_3^{(1)}/h = 0.3$. For $q = 3$, we have taken $a_1^{(1)}/h = 0.1$, $a_2^{(1)}/h = 0.15$, $a_3^{(1)}/h = 0.2$, $a_4^{(1)}/h = 0.25$ and $a_5^{(1)}/h = 0.3$. For $0 < Kh < 1.7$, the single-slot curve is above the two-slot curve. Thus $|R|$ decreases with the increase of the number of slots so long as the wave number remains moderate (here less than 1.7), which appears to be quite natural. However, as the wave number further increases the incident surface wave train may not feel the presence of lower gap or gaps and thus their effect on $|R|$ may not be appreciable.

In the limiting case when the slot length is made very small $(a_1^{(1)}/h = 0.2998$, $a_2^{(1)}/h = 0.2999$ and $a_3^{(1)}/h = 0.3)$ so that the surface piercing barrier with one slot almost assumes the form of a similar barrier without any slot, the corresponding curve for $|R|$ (denoted by solid line) is plotted in Fig. 1d. It is observed that this curve almost coincides with the one given in [5] for the surface piercing barrier also plotted in the same figure (denoted by cross mark) by taking $a/h = 0.3$.

For a *bottom standing thick slotted barrier* (type II) $|R|$ is depicted in Figs. 2a–d against the wave number Kh. In Fig. 2a, we have taken $a_1^{(2)}/h = 0.3$, $a_2^{(2)}/h = 0.45$, $a_3^{(2)}/h = 0.6$, $a_4^{(2)}/h = 0.7$ and $a_5^{(2)}/h = 0.8$ (two slots). In Fig. 2b, the same positions of the slots and slot lengths are used but the thickness parameter b/h has three different values. Occurrence of zeros of $|R|$ as a function of the wave number is observed. As the thickness increases the number of zeros of $|R|$ also increases (Fig. 2b). This is consistent with a similar phenomenon observed in [5] for *oblique* wave scattering by a bottom-standing thick barrier without slots. Also the number of zeros of $|R|$ increases with the decrease of the angle of incidence (Fig. 2a). This is also consistent with the observation in [5]. In Fig. 2c, $|R|$ is depicted for one, two and three slots of the bottom standing barrier. For one slot, we have taken $a_1^{(2)}/h = 0.3$, $a_2^{(2)}/h = 0.45$ and $a_3^{(2)}/h = 0.6$; and for two slots we have added $a_4^{(2)}/h = 0.7$ and $a_5^{(2)}/h = 0.8$ while for three slots we have further added $a_6^{(2)}/h = 0.85$ and $a_7^{(2)}/h = 0.9$. From this figure, it is observed that zeros of $|R|$ almost remain the same while $|R|$ slightly increases or decreases with the increase of number of slots. For the purpose of comparision, $|R|$ is depicted in Fig. 2d (solid line) for a bottom standing barrier with a single slot by taking $a_1^{(2)}/h = 0.3$, $a_2^{(2)}/h = 0.45$, $a_3^{(2)}/h = 0.6$ and $\alpha = 30°$

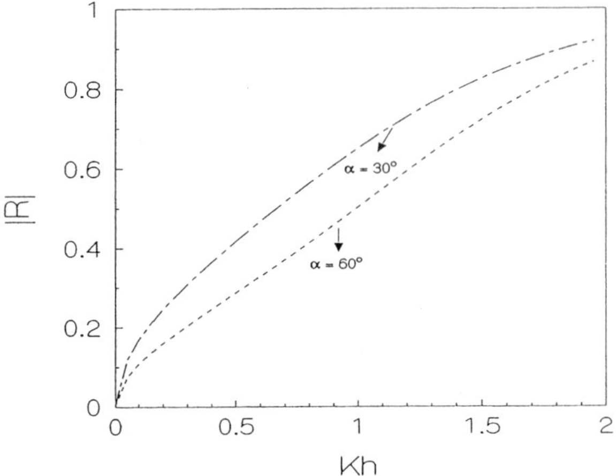

Fig. 1a: Reflection coefficient for type I barrier, b/h = .1, q = 3

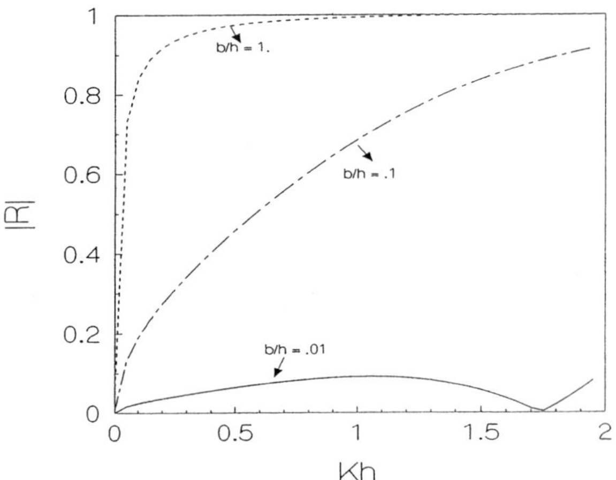

Fig.1b : Reflection coefficient for type I barrier, q = 3, $\alpha = 30°$

in the numerical procedure adopted here. It is observed that this figure almost coincides with the figure plotted here (cross marks) from the results in [6] for a submerged thick wall with a gap.

For a *submerged slotted thick block* (type III), occurrence of the zeros of $|R|$ is also observed. In Fig. 3, $|R|$ is shown against Kh for one and two slots. For the case of one slot, we have taken $a_1^{(3)}/h = 0.2$, $a_2^{(3)}/h = 0.25$, $a_3^{(3)}/h = 0.3$ and $a_4^{(3)}/h = 0.5$ while for two slots we have taken $a_i^{(3)}/h$ ($i = 1, 2, 3$) as the same together with $a_4^{(3)}/h = 0.4$, $a_5^{(3)}/h = 0.45$ and $a_6^{(3)}/h = 0.5$. As in the case of bottom standing barrier, we observe from

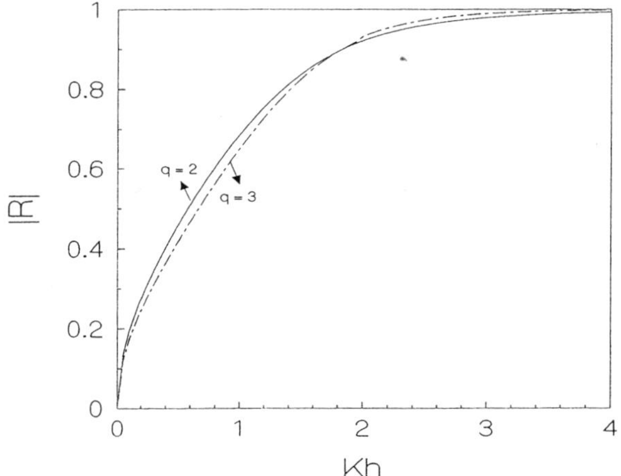

Fig 1c: Partially immersed barrier with gaps, b/h = .1, α = 30°

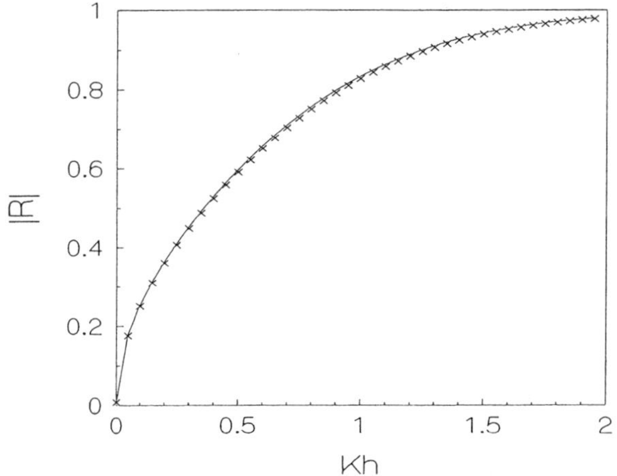

Fig.1d: Reflection coefficient for type I barrier, b/h = .1, α = 30°

Fig. 3 that the zeros of $|R|$ almost remain the same while $|R|$ slightly increases or decreases with the increase in the number of slots.

For a *thick slotted wall extending from the surface down to the bottom* (type IV), it is observed that $|R|$ decreases with the angle of incidence and with the decrease of thickness for fixed wave number. Also $|R|$ increases steadily with Kh and $|R| \to 1$ as $Kh \to \infty$. This is plausible since for large wave number, the incident wave train is confined within a thin layer below the free surface and as such most of the incident wave energy is reflected by the top most part of the barrier. This feature is also common to the surface piercing barrier. To visualize the effect of

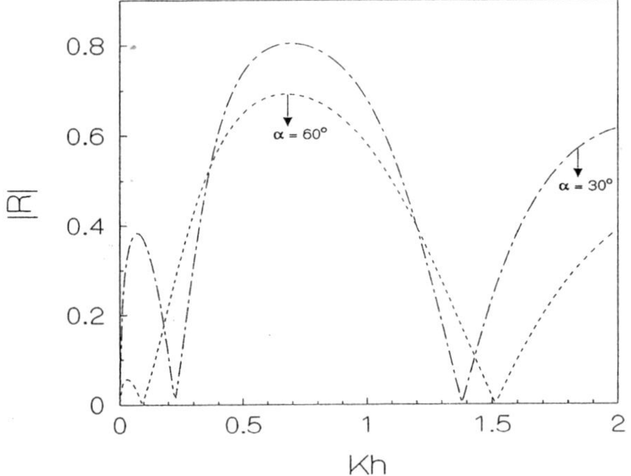

Fig.2a: Reflection coefficient for type II barrier, b/h = 1. q =3

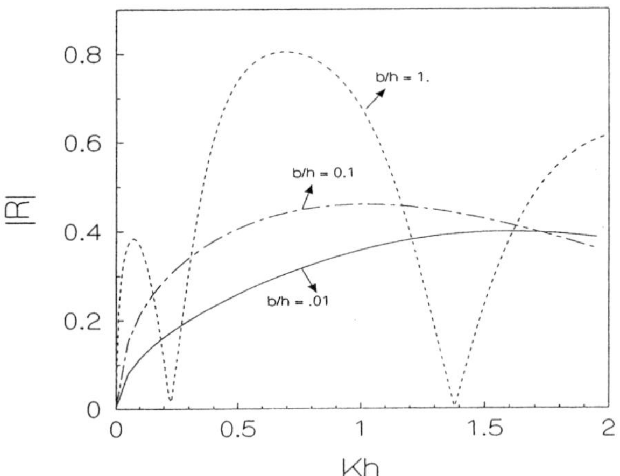

Fig.2b: Reflection coefficient for type II barrier, q = 3, α = 30°

the number of slots, $|R|$ is depicted in Fig. 4 against Kh as the number of slots in a thick wall varies from 1 to 5. The following values for $a_i^{(4)}/h$ ($i = 1,\ldots,2j-1$, $j = 1,2,\ldots,5$) are chosen in plotting these figures; $a_i^{(4)}/h = i/10$, $i = 1,2,\ldots,9$ and $a_{10}^{(4)}/h = 0.98$, so that all the slots except the one nearest to the bottom are of the same length. It is observed from Fig. 4 that $|R|$ decreases as the number of slots increases for a fixed wave number. This is a plausible result. However, as the wave number becomes large, $|R|$ for any number of slots asymptotically increases to unity.

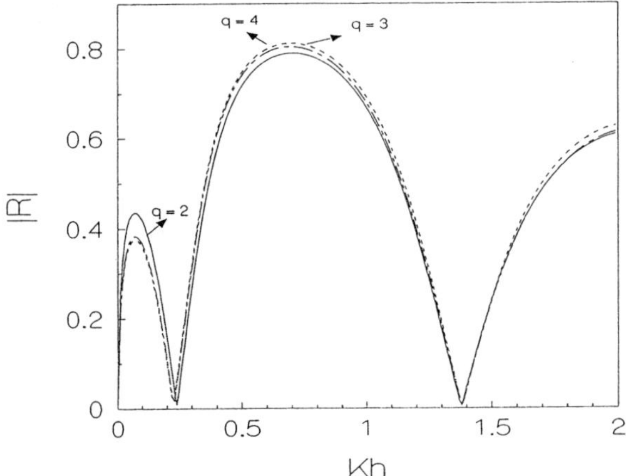

Fig.2c: Reflection coefficient for type II barrier, b/h = 1., α=30°

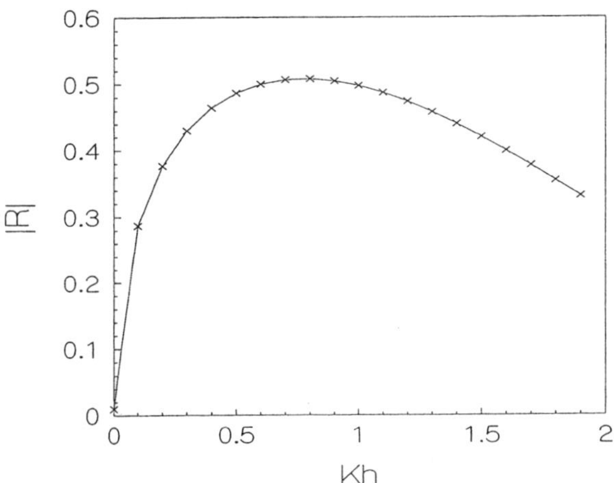

Fig. 2d : Reflection coefficient for type II barrier, b/h = .1, α = 30°

This work is partially supported by CSIR.

References

[1] Mei, C C and Black, J L (1969) J. Fluid Mech. **38**, 499–511.

[2] Bai, K J (1975) J. Fluid Mech. **68**, 513–535.

[3] Liu, P L -F and Wu, T (1987) J. Wtry., Port, Coastal & Ocean Eng. **113**, 660–671.

[4] Kanoria, M, Dolai D P and Mandal B N (1999) J. Eng. Math. **35**, 361–384.

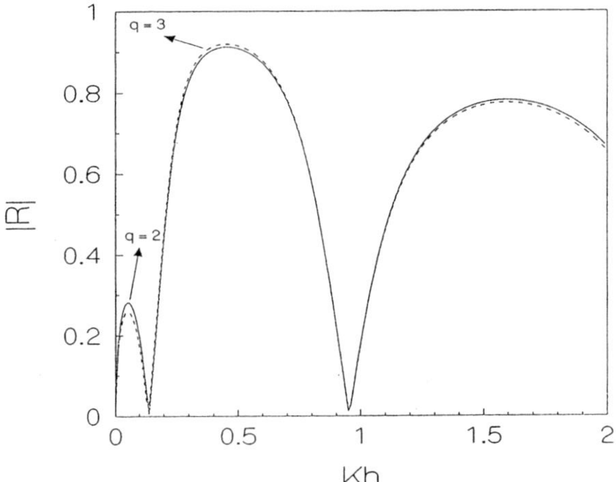

Fig.3: Reflection coefficient for type III barrier, $b/h = 1.$, $\alpha = 30^0$

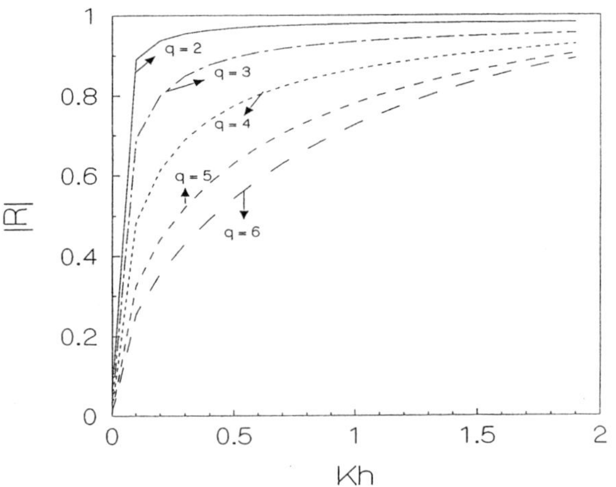

Fig.4: Reflection coefficent for type IV barrier, $b/h = .1$, $\alpha = 30°$

[5] Mandal, B N and Kanoria, M (2000) J. Offshore Mech. Arctic Eng. **122**, 100–108.

[6] Kanoria, M (1999) Appl. Ocean Res. **21**, 69–80.

[7] Martin, P A and Dalrymple, R A (1988) J. Fluid Mech. **188**, 465–490.

[8] McIver, P (1994) *Ocean Wave Engineering*, Comp. Mech. Pubs., 1–49.

THE FINITE DOCK PROBLEM

C. M. Linton

Department of Mathematical Sciences, Loughborough University

Loughborough, Leicestershire LE11 3TU, UK

C.M.Linton@lboro.ac.uk

1. INTRODUCTION

Problems concerning the interaction of water waves with a rigid plate of finite width and infinite length lying in the free surface have a long history. Such problems are interesting for many reasons. First the simple geometry allows considerable mathematical progress to be made and thus dock problems can be used as model problems against which to test new techniques or numerical results. Secondly they can be used as the first approximation in a perturbation analysis of wave interactions with shallow-draft ships.

Most work on finite dock problems has concentrated on the infinite depth case, but numerical calculations of the reflection and transmission coefficients for the scattering problem in finite depth were presented in [1]. Only normal incidence was considered, though the matched eigenfunction technique that was employed easily generalizes to the oblique incidence case. The purpose of this paper is to show that this problem can in fact be solved in an elegant manner, one which has the advantage over Mei and Black's solution procedure in that it takes into account the fact that the semi-infinite dock problem possesses an explicit solution. Another advantage of the method described below over Mei and Black's approach is that it accurately models the known singularity in the derivative of the potential at the plate edge.

The technique that we use is based on a combination of matched eigenfunction expansions and residue calculus theory. This is quite a technical procedure, the details of which are omitted, but the resulting formulas for the reflection and transmission coefficients, R and T, make it straightforward to evaluate these quantities. To obtain numerical values it is necessary to first compute the equivalent quantities for the semi-infinite dock (for which there are explicit formulas) and then to compute finite-width corrections by solving two infinite systems of real

I.D. Abrahams et al. (eds.),

IUTAM Symposium on Diffraction and Scattering in Fluid Mechanics and Elasticity, 45–51.

© 2002 *Kluwer Academic Publishers. Printed in the Netherlands.*

algebraic equations. These systems converge extremely rapidly provided the ratio of plate width to water depth $(2a/h)$ is not too small. As well as providing an efficient method for accurately computing R and T, the formulation also enables us to derive an approximation based on a 1×1 truncation of the infinite systems, an approximation which leads to simple formulas for R and T that are extremely accurate except for small values of a/h.

2.   FORMULATION

Consider the diffraction of an incident plane wave of unit amplitude, with angular frequency ω and making an angle θ_I with the positive x-axis, by a dock which occupies $z = 0$, $-a \leq x \leq a$, $-\infty < y < \infty$ in water of uniform depth h. The solutions to the dispersion relation $K + u \tan uh = 0$ ($K = \omega^2/g$) will be denoted by $u = \pm k_n$, $n \geq 0$, where $k_0 = -ik$ is purely imaginary and k_n, $n \geq 1$ are real and positive. The total velocity potential for the scattering problem can be written $\mathrm{Re}\left\{\phi(x,z)e^{i\ell y}e^{-i\omega t}\right\}$, where $\ell = k \sin \theta_I$, and then $\phi(x,z)$ satisfies

$$\nabla^2_{xz}\phi - \ell^2\phi = 0 \quad \text{in} \quad -h < z < 0, \ -\infty < x < \infty, \qquad (5.1)$$

with the boundary conditions

$$\phi_z = 0 \quad \text{on} \quad z = -h, \qquad (5.2)$$
$$\phi_z = K\phi \quad \text{on} \quad z = 0, \ |x| > a, \qquad (5.3)$$
$$\phi_z = 0 \quad \text{on} \quad z = 0, \ |x| \leq a \qquad (5.4)$$

and we choose to define the reflection and transmission coefficients through the far-field behaviour

$$\phi \sim -\frac{ig \cosh k(z+h)}{\omega \cosh kh}(e^{i\alpha(x+a)} + Re^{-i\alpha(x+a)}) \quad \text{as} \quad x \to -\infty, \qquad (5.5)$$

$$\phi \sim -\frac{ig \cosh k(z+h)}{\omega \cosh kh}Te^{i\alpha(x-a)} \quad \text{as} \quad x \to \infty, \qquad (5.6)$$

where $\alpha = k \cos \theta_I = (k^2 - \ell^2)^{1/2}$.

One final condition needs to be applied and that is a condition which specifies the nature of the solution near the plate edge, $(x,z) = (a,0)$. If we insist that ϕ is regular at this point then [2], Theorem 3.3, shows that $\phi \sim P(r, r \ln r)$ as $r = [(x-a)^2 + z^2]^{1/2} \to 0$, where P is some polynomial. It then follows that

$$\phi_r \sim A \ln r \quad \text{as} \quad r \to 0, \qquad (5.7)$$

for some constant A.

The solution to this problem can be obtained by splitting it up into parts which are symmetric and antisymmetric about $x = 0$ and then setting up eigenfunction expansions below and away from the dock. By matching these expansions across $x = a$ we can reduce the problem to the solution of the two infinite systems of complex linear algebraic equations

$$\sum_{n=0}^{\infty} V_n^{\pm} \left(\frac{1}{\alpha_n - \beta_m} \pm \frac{e^{-2\beta_m a}}{\alpha_n + \beta_m} \right) = \frac{1}{\alpha_0 + \beta_m} \pm \frac{e^{-2\beta_m a}}{\alpha_0 - \beta_m}, \qquad m \geq 0,$$

$$(5.8)$$

for the unknowns $V_n^{\pm}$, where $\alpha_n = (k_n^2 + \ell^2)^{1/2}$, $\beta_n = (\lambda_n^2 + \ell^2)^{1/2}$, $\lambda_n = n\pi/h$ and we require that $V_n^{\pm} = O(n^{-1})$ as $n \to \infty$ in order to satisfy the edge condition (5.7).

The method that we use to solve this system is the modified residue calculus technique originally devised in [3] and described in [4], §2.12, which takes advantage of the fact that the terms $\exp(-2\beta_m a)$ all tend to zero rapidly as a/h gets large. If the exponential terms were not present we would have the single system

$$\sum_{n=0}^{\infty} \frac{V_n}{\alpha_n - \beta_m} = \frac{1}{\alpha_0 + \beta_m} \qquad m \geq 0, \qquad (5.9)$$

which can be solved explicitly using the residue calculus theory (described in detail in [5]) in which a function of a complex variable $f(z)$ is constructed which has the property that when Cauchy's residue theorem is applied to certain integrals involving f, the system (5.9) results and V_n can be identified with the residues of f. The key to the modified residue calculus theory is to use a modification of the function f which solves (5.9) in the solution procedure for (5.8).

We find that the reflection and transmission coefficients for this problem are given by

$$R = \frac{1}{2}(e^{2i\delta_+} + e^{2i\delta_-})R_\infty, \qquad T = \frac{1}{2}(e^{2i\delta_+} - e^{2i\delta_-})R_\infty, \qquad (5.10)$$

where $\delta_\pm$ and R_∞ are defined as follows. First

$$R_\infty = e^{-2i\theta_I} \prod_{n=1}^{\infty} \frac{(1 - i\alpha/\alpha_n)(1 + i\alpha/\beta_n)}{(1 + i\alpha/\alpha_n)(1 - i\alpha/\beta_n)} = e^{-2i\theta_I} e^{2i\delta_\infty}, \qquad (5.11)$$

where

$$\delta_\infty = \sum_{n=1}^{\infty} \left(\tan^{-1}(\alpha/\beta_n) - \tan^{-1}(\alpha/\alpha_n) \right). \qquad (5.12)$$

This part of the solution essentially comes from solving (5.9). Secondly

$$\delta_\pm = \arg\left(1 - \sum_{n=0}^{\infty} \frac{C_n^\pm}{i\alpha + \beta_n}\right), \tag{5.13}$$

where $C_m^\pm$, $m \geq 0$, are the solutions to the infinite systems of real equations

$$C_m^\pm \pm D_m \sum_{n=0}^{\infty} \frac{C_n^\pm}{\beta_m + \beta_n} = \pm D_m \qquad m \geq 0, \tag{5.14}$$

where

$$D_0 = 2\ell e^{-2\ell a} \prod_{n=1}^{\infty} \frac{(1 - \ell/\alpha_n)(1 + \ell/\beta_n)}{(1 + \ell/\alpha_n)(1 - \ell/\beta_n)}, \tag{5.15}$$

and for $m \geq 1$,

$$D_m = \frac{2\beta_m(\ell + \beta_m)(\alpha_m - \beta_m)}{(\ell - \beta_m)(\alpha_m + \beta_m)} e^{-2\beta_m a} \prod_{\substack{n=1 \\ n \neq m}}^{\infty} \frac{(1 - \beta_m/\alpha_n)(1 + \beta_m/\beta_n)}{(1 + \beta_m/\alpha_n)(1 - \beta_m/\beta_n)}. \tag{5.16}$$

Because of the presence of the factor $e^{-2\beta_m a}$ in the expression (5.16) for D_m, the systems of equations (5.14) converge very quickly provided a/h is not too small. As $a/h \to \infty$, (5.14) shows that $C_m^\pm \to 0$, $m \geq 0$ and hence that $e^{2i\delta_\pm} \to 1$. It follows that R_∞ is the reflection coefficient for the semi-infinite dock problem. The quantities $\delta_\pm$ characterize the modifications to the reflection and transmission coefficients due to the finite length of the dock.

Results for normal incidence can be extracted from the above analysis by taking the limit as $\theta_I \to 0$ (i.e. $\ell \to 0$ for fixed k). First we note that in this limit

$$\ln \prod_{n=1}^{\infty} \frac{(1 - \ell/\alpha_n)(1 + \ell/\beta_n)}{(1 + \ell/\alpha_n)(1 - \ell/\beta_n)} \sim 2\ell\sigma + O(\ell^3), \tag{5.17}$$

where

$$\sigma = \sum_{n=1}^{\infty} \left(\frac{1}{\lambda_n} - \frac{1}{k_n}\right) \tag{5.18}$$

and hence

$$D_0 \sim 2\ell + 4\ell^2(\sigma - a) + O(\ell^3). \tag{5.19}$$

The $m = 0$ equation in (5.14) with the plus sign then becomes $C_0^+ = 0$ and we only need to solve the system for $m \geq 1$. The leading order behaviour of the $m = 0$ equation in (5.14) with the minus sign is more complicated; we find that

$$(\sigma - a)C_0^- + \sum_{n=1}^{\infty} \frac{C_n^-}{\beta_n} = 1. \tag{5.20}$$

Apart from these changes, we can simply set $\ell = 0$ in the general expressions for R and T.

It is also possible to examine the long wave limit, i.e. $Kh \to 0$, for fixed θ_{I}. In this limit $kh \sim (Kh)^{1/2}$ and an analysis of (5.14) reveals that

$$\delta^+ \sim \theta_{\mathrm{I}} - ka \sin \theta_{\mathrm{I}} \tan \theta_{\mathrm{I}}, \qquad \delta^- \sim \theta_{\mathrm{I}} - \frac{\pi}{2} + ka \cos \theta_{\mathrm{I}} \tag{5.21}$$

and hence that

$$R \sim -ika \sec \theta_{\mathrm{I}}, \qquad T \sim 1 + ika \sec \theta_{\mathrm{I}} \cos 2\theta_{\mathrm{I}}. \tag{5.22}$$

For $\theta_{\mathrm{I}} = 0$, these results agree with those in [6] after taking account the different definitions of T used in that paper.

3.   RESULTS

Equations (5.10)–(5.16) provide a numerically straightforward way of computing the reflection and transmission coefficients for the finite dock problem in finite depth. The infinite systems of equations that need to be solved converge extremely rapidly (much more rapidly than (5.8)) and the sums and products that need to be evaluated cause no difficulty. For example, the terms in the summation in the definition of δ_∞, equation (5.12), are $O(n^{-3})$ as $n \to \infty$. This is computationally acceptable, but the series is easily accelerated. By subtracting off the leading order asymptotics of the summand we can derive the expression

$$\delta_\infty = -\frac{ahKh}{\pi^3}\zeta(3) + \sum_{n=1}^{\infty}\left(\tan^{-1}(\alpha/\beta_n) - \tan^{-1}(\alpha/\alpha_n) + \frac{ahKh}{n^3\pi^3}\right), \tag{5.23}$$

in which ζ is the Riemann zeta function and the terms are $O(n^{-5})$ as $n \to \infty$. All the infinite products can be accelerated in the same way after first taking their logarithms.

To demonstrate the rapid convergence of the infinite systems of equations we can consider the case of a 1×1 truncation. If we only include one term from the summation in (5.14), solve for $C_0^{\pm}$ and substitute into (5.13) we obtain

$$\tan \delta_{\pm} \approx \frac{\pm \sin 2\theta_{\mathrm{I}}}{b^{-1} \pm \cos 2\theta_{\mathrm{I}}}, \tag{5.24}$$

where $b = D_0/2\ell$. If we substitute this expression into (5.10) and use (5.11) we obtain the approximations

$$R \approx e^{2i\delta_\infty} e^{2i\theta_I} \left(\frac{b^2 - 1}{b^2 - e^{4i\theta_I}} \right), \qquad T \approx b\, e^{2i\delta_\infty} \left(\frac{1 - e^{4i\theta_I}}{b^2 - e^{4i\theta_I}} \right). \qquad (5.25)$$

If we take the limit of (5.25) as $\theta_I \to 0$ using (5.19) we obtain the following approximations for the normal incidence case:

$$R \approx \frac{k(\sigma - a)e^{2i\delta_\infty}}{k(\sigma - a) - i}, \qquad T \approx \frac{-ie^{2i\delta_\infty}}{k(\sigma - a) - i}. \qquad (5.26)$$

The accuracy of these approximations depends strongly on the value of a/h and to a lesser extent on the value of θ_I, with larger values of either parameter resulting in greater accuracy. This is illustrated in Table 1 which shows the errors that result from using these approximations to compute $|R|$. For each value of a/h and θ_I the table gives the maximum percentage error in the computed value of $|R|$ as K varies over the entire frequency range. The table shows that 1% accuracy is achieved for all values of $a/h \geq 0.5$, with the accuracy increasing rapidly as a/h increases.

	a/h			
θ_I	0.25	0.5	0.75	1
0°	8.6	0.94	0.13	0.021
40°	5.2	0.43	0.046	0.0056
80°	0.61	0.026	0.0020	0.00020

Table 1 Maximum percentage error (over all frequencies) when computing $|R|$ from the approximation (5.25).

Table 2 shows a sample set of results for $\theta_I = 45°$ and $a/h = 1$, based on a 2×2 truncation of the system of equations (5.14). All the digits displayed are believed to be accurate.

References

[1] Mei, C C and Black, J L (1969) *Scattering of surface waves by rectangular obstacles in waters of finite depth*, J. Fluid Mech. **38**, 499–511.

[2] Wigley, N M (1964) *Asymptotic expansions at a corner of solutions of mixed boundary value problems*, J. Math. Mech., **13**, 549–576.

Kh	$|R|$	$|T|$
0.2	0.5781	0.8160
0.4	0.7506	0.6608
0.6	0.8479	0.5302
0.8	0.9070	0.4211
1.0	0.9437	0.3307
1.2	0.9665	0.2566
1.4	0.9805	0.1966
1.6	0.9889	0.1489
1.8	0.9938	0.1115
2.0	0.9966	0.0826

Table 2 Values of $|R|$ and $|T|$ for different values of Kh when $\theta_{\mathrm{I}} = 45°$ and $a/h = 1$.

[3] Mittra, R, Lee, S W and Van Blaricum, G F (1968) *A modified residue calculus technique*, Int. J. Eng. Sci. **6**, 395–408.

[4] Jones, D S (1994) *Methods in Electromagnetic Wave Propagation*, 2nd edition, Oxford: Clarendon Press.

[5] Mittra, R and Lee, S W (1971) *Analytical Techniques in the Theory of Guided Waves*, New York: Macmillan.

[6] Martin, P A and Dalrymple, R A (1988) *Scattering of long waves by cylindrical obstacles and gratings using matched asymptotic expansions*, J. Fluid Mech. **188**, 465–490.

LONG PERIODIC WAVES ON A BEACH

A. Shermenev*, M. Shermeneva
Wave Research Center, Russian Academy of Science
38 Vavilov Street, Moscow 117942, Russia
sher@orc.ru

1. INTRODUCTION

Traditionally, the motion of periodic long nonlinear waves is described by so called Boussinesq-type equations. A special feature of these equation is the possibility to reduce the dimension of the problem by expanding the velocity potential in power series in the vertical coordinate.

In 1966, Mei and Le Méhauté [1] extended these equations to an uneven bottom in one dimension using the potential at the bottom as basic variable. Later, similar equations based on the depth-averaged velocity and on the velocity at the still water level were derived; see survey in [2].

In this paper, we follow Mei and Le Méhauté and write down Boussinesq-type equations for the bottom potential.

There are two small parameters associated with Boussinesq-type equations: the ratio of amplitude to depth, ε; and the ratio of depth to wavelength, μ. The classical Boussinesq equations include terms of orders $O\left(\varepsilon\right)$ and $O\left(\mu^2\right)$, and assume that $O\left(\varepsilon\right) = O\left(\mu^2\right)$. We work to the next order using the same assumption: our equations include terms in ε^2, $\varepsilon\mu^2$, and μ^4.

We consider regular periodic wave motion over the uniform slope sx excluding the deep water region where the shallow water assumptions are violated, and the neighbourhood of $x = 0$ (shoreline) where a singularity is possible. The potential at the bottom is expanded in Fourier series

$$C^0(x) + S^1(x)\sin\left(\omega t\right) + C^1(x)\cos\left(\omega t\right) + S^2(x)\sin\left(2\omega t\right)$$
$$+ C^2(x)\cos\left(2\omega t\right) + S^3(x)\sin\left(3\omega t\right) + C^3(x)\cos\left(3\omega t\right).$$

The main result of this work consists in the explicit expressions for functions $S^m(x)$ and $C^m(x)$ up to orders $(\varepsilon^2, \varepsilon\mu^2, \mu^4)$ which are homoge-

---

*The work was supported by the RFBR Projects 96-15-96525 and 99-02-17005.

I.D. Abrahams et al. (eds.),
IUTAM Symposium on Diffraction and Scattering in Fluid Mechanics and Elasticity, 53–60.
© 2002 *Kluwer Academic Publishers. Printed in the Netherlands.*

nous polynomials in Bessel functions $Z_0(\xi)$ and $Z_1(\xi)$ whose coefficients are polynomials in $x^{-1/2}$ and $x^{1/2}$; here, $\xi = 2\omega\sqrt{x/s}$. A similar expansion for the surface elevation is also given. These expressions give periodic solutions to the Boussinesq equations within the same accuracy as the equations are derived. Therefore, the result can be interpreted as periodic solution of Euler equations (6.1)–(6.4) calculated up to orders ε^2, $\varepsilon\mu^2$, and μ^4 . (The first term of this expansion $J_0(\xi)\sin(\omega t)$ is used, for example, in the book of Mei [3]).

We believe that the derived formulas are only the lowest terms of expansion satisfying the Euler equations (6.1)–(6.4) over a sloping beach.

This paper is closely related to note [4] where the special case of standing waves is considered. For the standing waves Carrier and Greenspan [5] have given an exact solution to Airy's approximate equations which are fully nonlinear but nondispersive (terms of order ε are retained).

2.   BASIC EQUATIONS

Non-dimensional coordinates are used as follows:

$$x = \frac{x'}{l_0'}, \quad z = \frac{z'}{h_0'}, \quad t = \frac{\sqrt{gh_0'}}{l_0'}t', \quad \eta = \frac{\eta'}{a_0'}, \quad \varphi = \frac{h_0'}{a_0' l_0' \sqrt{gh_0'}}\varphi', \quad h = \frac{h'}{h_0'},$$

where the prime denotes physical variables, and a_0', l_0', and h_0' denote the characteristic wave amplitude, depth, and wavelength, respectively; g is the acceleration of gravity; x is the horizontal coordinate; z is the vertical coordinate; and t is the time. The scaled governing equation and the boundary conditions for the irrotational wave problem read

$$\mu^2\varphi_{xx} + \varphi_{zz} = 0, \quad -h(x) < z < \varepsilon\eta(x,t), \quad (6.1)$$
$$\eta_t + \varepsilon\varphi_x\eta_x - \mu^{-2}\varphi_z = 0, \quad z = \varepsilon\eta(x,t), \quad (6.2)$$
$$\varphi_t + \tfrac{1}{2}\varepsilon\left(\varphi_x^2 + \mu^{-2}\varphi_z^2\right) + \eta = 0, \quad z = \varepsilon\eta(x,t), \quad (6.3)$$
$$\varphi_z = -\mu^2 h_x\varphi_x, \quad z = -h(x), \quad (6.4)$$

where ε and μ are the measures of nonlinearity and frequency dispersion defined by $\varepsilon = a_0'/h_0'$ and $\mu = h_0'/l_0'$.

We assume that the function defining the bottom $z = -h(x)$ is linear, $h(x) \equiv sx$, and expand the potential $\varphi(x,z,t)$ in powers of vertical coordinate

$$\varphi(x,z,t) = \sum_{m=0}^{\infty}(z + h(x))^m F_m(x,t). \quad (6.5)$$

Substituting (6.5) into (6.1) and equating to zero the coefficients of each power of $z + h(x)$, we have

$$F_{m+2} = -\mu^2 \frac{2s(m+1)F_{m+1,x} + F_{m,xx}}{(m+2)(m+1)(1+s^2\mu^2)}.$$ (6.6)

The boundary condition at the bottom (6.4) gives

$$F_1 = -\mu^2 \frac{2sF_{0,x}}{1+s^2\mu^2}.$$ (6.7)

Denoting $f(x,t) \equiv F_0(x,t)$, and expanding all expressions in powers of μ, we obtain the first terms of φ,

$$\begin{aligned}
\varphi = {} & f + \left(-s + s^3\mu^2 - s^5\mu^4\right)\mu^2(z+h)f_x \\
& + \left(-\tfrac{1}{2} + \tfrac{3}{2}s^2\mu^2 - \tfrac{5}{2}s^4\mu^4\right)\mu^2(z+h)^2 f_{xx} \\
& + \left(\tfrac{1}{2}s - \tfrac{5}{3}s^3\mu^2\right)\mu^4(z+h)^3 f_{xxx} + \left(\tfrac{1}{24} - \tfrac{5}{12}s^2\mu^2\right)\mu^4(z+h)^4 f_{xxxx} \\
& - \tfrac{1}{24}s\mu^6(z+h)^5 f_{xxxxx} - \tfrac{1}{720}\mu^6(z+h)^6 f_{xxxxxx}.
\end{aligned}$$ (6.8)

Expression (6.8) satisfies (6.1) and (6.4). Substitution of (6.8) into (6.2) and (6.3) gives the Boussinesq-type equations for potential at bottom $f(x,t)$ and for surface elevation $\eta(x,t)$:

$$\begin{aligned}
& \eta_t + sf_x + sxf_{xx} + \left(-s^3 f_x - 3s^3 x f_{xx} - \tfrac{3}{2}s^3 x^2 f_{xxx} - \tfrac{1}{6}s^3 x^3 f_{xxxx}\right)\mu^2 \\
& + \left(s^5 f_x + 5s^5 x f_{xx} + 5s^5 x^2 f_{xxx} + \tfrac{5}{3}s^5 x^3 f_{xxxx} + \tfrac{5}{24}s^4 x^4 f_{xxxxx}\right. \\
& \left. + \tfrac{1}{120}s^5 x^5 f_{xxxxxx}\right)\mu^4 + \left(f_{xx}\eta + f_x\eta_x\right)\varepsilon - \left(\left(3s^2 f_{xx} + 3s^2 x f_{xxx}\right.\right. \\
& \left.\left. + \tfrac{1}{2}s^2 x^2 f_{xxxx}\right)\eta + \left(s^2 f_x + 2s^2 x f_{xx} + \tfrac{1}{2}s^2 x^2 f_{xxx}\right)\eta_x\right)\varepsilon\mu^2 = 0,
\end{aligned}$$

$$\begin{aligned}
& \eta + f_t + \tfrac{1}{2}f_x^2\varepsilon + \left(-s^2 x f_{xt} - \tfrac{1}{2}s^2 x^2 f_{xxt}\right)\mu^2 + \left(-\tfrac{1}{2}s^2 f_x^2 - sn f_{xt}\right. \\
& - s^2 x f_x f_{xx} + \tfrac{1}{2}s^2 x^2 f_{xx}^2 - sxn f_{xxt} - \left.\tfrac{1}{2}s^2 x^2 f_x f_{xxx}\right)\varepsilon\mu^2 \\
& + \left(s^4 x f_{xt} + \tfrac{3}{2}s^4 x^2 f_{xxt} + \tfrac{1}{2}s^4 x^3 f_{xxxt} + \tfrac{1}{24}s^4 x^4 f_{xxxxt}\right)\mu^4 = 0.
\end{aligned}$$

3.   PERIODIC PROBLEM

We suppose that the solution is periodic in time and can be expanded in Fourier series in an area excluding the deep water region and a neighbourhood of $x = 0$ (shore line):

$$\begin{aligned}
f(x,t) = {} & S_{10}^0\varepsilon \\
& + \left(S_{00}^1(x) + S_{20}^1(x)\varepsilon^2 + S_{02}^1(x)\mu^2 + S_{04}^1(x)\mu^4\right)\sin(\omega t) \\
& + \left(C_{00}^1(x) + C_{20}^1(x)\varepsilon^2 + C_{02}^1(x)\mu^2 + C_{04}^1(x)\mu^4\right)\cos(\omega t)
\end{aligned}$$

$$+ \left(S_{10}^2(x)\varepsilon + S_{12}^2(x)\varepsilon\mu^2\right)\sin\left(2\omega t\right)$$
$$+ \left(C_{10}^2(x)\varepsilon + C_{12}^2(x)\varepsilon\mu^2\right)\cos\left(2\omega t\right)$$
$$+ S_{20}^3(x)\varepsilon^2\sin\left(3\omega t\right) + C_{20}^3(x)\varepsilon^2\cos\left(3\omega t\right),$$

$$
\begin{aligned}
\eta(x,t) \;=\; & P_{10}^0(x)\varepsilon + P_{12}^0(x)\varepsilon\mu^2 \\
& + \left(P_{00}^1(x) + P_{20}^1(x)\varepsilon^2 + P_{02}^1(x)\mu^2 + P_{04}^1(x)\mu^4\right)\sin\left(\omega t\right) \\
& + \left(Q_{00}^1(x) + Q_{20}^1(x)\varepsilon^2 + Q_{02}^1(x)\mu^2 + Q_{04}^1(x)\mu^4\right)\cos\left(\omega t\right) \\
& + \left(P_{10}^2(x)\varepsilon + P_{12}^2(x)\varepsilon\mu^2\right)\sin\left(2\omega t\right) \\
& + \left(Q_{10}^2(x)\varepsilon + Q_{12}^2(x)\varepsilon\mu^2\right)\cos\left(2\omega t\right) \\
& + P_{20}^3(x)\varepsilon^2\sin\left(3\omega t\right) + Q_{20}^3(x)\varepsilon^2\cos\left(3\omega t\right).
\end{aligned}
$$

We denote by $S = S(x)$ and $C = C(x)$ two solutions of the equation

$$\omega^2 Z + sZ_x + sxZ_{xx} = 0,$$

and by S' and C' their derivatives. Functions $S(x)$, $C(x)$, $S'(x)$ and $C'(x)$ can be expressed in terms of Bessel functions as

$$
\begin{aligned}
S\left(x\right) &= a_{11}J_0(\xi) + a_{12}Y_0(\xi), \\
S'\left(x\right) &= -\omega(sx)^{-1/2}\left(a_{11}J_1(\xi) + a_{12}Y_1(\xi)\right), \\
C\left(x\right) &= a_{21}J_0(\xi) + a_{22}Y_0(\xi), \\
C'\left(x\right) &= -\omega(sx)^{-1/2}\left(a_{21}J_1(\xi) + a_{22}Y_1(\xi)\right),
\end{aligned}
$$

where $\xi = 2\omega\sqrt{x/s}$.

Denote by a the determinant $a_{11}a_{22} - a_{12}a_{21}$. Then the Wronskian $SC' - CS' = a/(\pi x)$.

The major findings of this paper are the following expressions of $S_{\alpha\beta}^i$, $C_{\alpha\beta}^i$ $P_{\alpha\beta}^i$, and $Q_{\alpha\beta}^i$:

Zero Harmonic:

$$
\begin{aligned}
S_{10}^0 &= -(a\omega)/(2\pi s)\,x^{-1}, \\
L_{10}^0 &= -\tfrac{1}{4}S'^2 - \tfrac{1}{4}C'^2, \\
L_{12}^0 &= \tfrac{1}{4}\omega^4(S^2 + C^2) + \tfrac{1}{18}\omega^4 x(SS' + CC') - \tfrac{7}{36}s\omega^2 x(S'^2 + C'^2).
\end{aligned}
$$

First Harmonic:

$$S_{00}^1 \;=\; S,$$

$$S_{20}^1 \;=\; -\frac{a\omega}{2sx^2}C - \frac{\omega^4}{8s^3x}S^3 - \frac{\omega^2}{8s^2x}S^2S' - \frac{\omega^2}{8s^2}SS'^2 - \frac{1}{8s}S'^3,$$

$$- \left(\frac{\omega^4}{8s^3 x} C^2 + \frac{3\omega^2}{8s^2} C'^2 \right) S + \left(\frac{1}{8s} C'^2 - \frac{\omega^2}{8s^2 x} C^2 + \frac{\omega^2}{4s^2} CC' \right) S',$$

$$S_{02}^1 = -\tfrac{2}{9} s\omega^2 x S + \tfrac{1}{9} \left(7s^2 x + s\omega^2 x^2 \right) S',$$

$$S_{04}^1 = \left(-\tfrac{479}{1350} s^3 \omega^2 x - \tfrac{29}{225} s^2 \omega^4 x^2 - \tfrac{1}{162} s\omega^6 x^3 \right) S$$
$$+ \left(-\tfrac{479}{1350} s^4 x - \tfrac{136}{675} s^3 \omega^2 x^2 - \tfrac{7}{450} s^2 \omega^4 x^3 \right) S',$$

$$C_{00}^1 = C,$$

$$C_{20}^1 = \frac{a\omega}{2sx^2} S - \frac{\omega^4}{8s^3 x} C^3 - \frac{\omega^2}{8s^2 x} C^2 C' - \frac{\omega^2}{8s^2} CC'^2 - \frac{1}{8s} C'^3$$
$$- \left(\frac{\omega^4}{8s^3 x} S^2 + \frac{3\omega^2}{8s^2} S'^2 \right) C + \left(-\frac{\omega^2}{8s^2 x} S^2 + \frac{\omega^2}{4s^2} SS' - \frac{1}{8s} S'^2 \right) C',$$

$$C_{02}^1 = -\tfrac{2}{9} s\omega^2 x C + \tfrac{1}{9} \left(7s^2 x + s\omega^2 x^2 \right) C',$$

$$C_{04}^1 = \left(-\tfrac{479}{1350} s^3 \omega^2 x - \tfrac{29}{225} s^2 \omega^4 x^2 - \tfrac{1}{162} s\omega^6 x^3 \right) C$$
$$- \left(\tfrac{479}{1350} s^4 x + \tfrac{136}{675} s^3 \omega^2 x^2 + \tfrac{7}{450} s^2 \omega^4 x^3 \right) C',$$

$$P_{00}^1 = \omega C,$$

$$P_{20}^1 = \frac{a}{x^2} S' + \frac{a\omega^2}{2sx^2} S - \frac{\omega^5}{8s^3 x} C^3 + \frac{\omega^3}{8s^2 x} C^2 C' + \left(\frac{\omega}{4sx} - \frac{\omega^3}{8s^2} \right) CC'^2$$
$$- \frac{3\omega}{8s} C'^3 - \frac{\omega^5}{8s^3 x} CS^2 - \frac{3\omega^3}{8s^2 x} C'S^2 + \left(\frac{\omega}{4sx} - \frac{3\omega^3}{8s^2} \right) CS'^2$$
$$- \frac{3\omega}{8s} C'S'^2 + \frac{\omega^3}{2s^2 x} CSS' + \frac{\omega^3}{4s^2} C'SS',$$

$$P_{02}^1 = \tfrac{5}{18} s\omega^3 x C + \left(\tfrac{5}{18} s^2 \omega x + \tfrac{1}{9} s\omega^3 x^2 \right) C',$$

$$P_{04}^1 = \left(-\tfrac{283}{2700} s^3 \omega^3 x + \tfrac{43}{1800} s^2 \omega^5 x^2 - \tfrac{1}{162} s\omega^7 x^3 \right) C$$
$$+ \left(-\tfrac{283}{2700} s^4 \omega x + \tfrac{103}{1350} s^3 \omega^3 x^2 + \tfrac{1}{25} s^2 \omega^5 x^3 \right) C',$$

$$Q_{00}^1 = -\omega S,$$

$$Q_{20}^1 = \frac{a}{x^2} C' + \frac{a\omega^2}{2sx^2} C + \frac{\omega^5}{8s^3 x} S^3 - \frac{\omega^3}{8s^2 x} S^2 S' + \left(\frac{\omega^3}{8s^2} - \frac{\omega}{4sx} \right) SS'^2$$
$$+ \frac{3\omega}{8s} S'^3 + \frac{3\omega^3}{8s^2 x} C^2 S' - \frac{\omega^3}{4s^2} CC'S' + \frac{3\omega}{8s} S'C'^2 + \frac{\omega^5}{8s^3 x} C^2 S$$
$$- \frac{\omega^3}{2s^2 x} SCC' + \left(-\frac{\omega}{4sx} + \frac{3\omega^3}{8s^2} \right) SC'^2,$$

$$Q_{02}^1 = -\tfrac{5}{18} s\omega^3 x S + \left(-\tfrac{5}{18} s^2 \omega x - \tfrac{1}{9} s\omega^3 x^2 \right) S',$$

$$Q_{04}^1 = \left(\tfrac{283}{2700} s^3 \omega^3 x - \tfrac{43}{1800} s^2 \omega^5 x^2 + \tfrac{1}{162} s\omega^7 x^3 \right) S$$
$$+ \left(\tfrac{283}{2700} s^4 \omega x - \tfrac{103}{1350} s^3 \omega^3 x^2 - \tfrac{1}{25} s^2 \omega^5 x^3 \right) S'.$$

Second Harmonic:

$$S_{10}^2 = -\frac{\omega}{2s}SS' + \frac{\omega}{2s}CC',$$

$$S_{12}^2 = \left(\frac{4\omega^3}{3} + \frac{7\omega^5 x}{18s}\right)S^2 + \frac{7\omega^3 x}{3}SS' + \left(-\frac{2s\omega x}{9} - \frac{7\omega^3 x^2}{18}\right)S'^2$$

$$+ \left(-\frac{4\omega^3}{3} - \frac{7\omega^5 x}{18s}\right)C^2 - \frac{7\omega^3 x}{3}CC' + \left(\frac{2s\omega x}{9} + \frac{7\omega^3 x^2}{18}\right)C'^2,$$

$$C_{10}^2 = -\frac{\omega}{2s}SC' - \frac{\omega}{2s}S'C,$$

$$C_{12}^2 = \left(\frac{8\omega^3}{3} + \frac{7\omega^5 x}{9s}\right)SC + \frac{7\omega^3 x}{3}SC'$$

$$+ \frac{7\omega^3 x}{3}S'C + \left(-\frac{4s\omega x}{9} - \frac{7\omega^3 x^2}{9}\right)S'C',$$

$$P_{10}^2 = -\frac{\omega^2}{s}SC' - \frac{\omega^2}{s}S'C - \frac{1}{2}S'C',$$

$$P_{12}^2 = \left(\frac{23\omega^4}{6} + \frac{14\omega^6 x}{9s}\right)CS + \frac{49\omega^4 x}{18}SC'$$

$$+ \frac{49\omega^4 x}{18}S'C + \left(-\frac{41s\omega^2 x}{18} - \frac{14\omega^4 x^2}{9}\right)S'C',$$

$$Q_{10}^2 = -\frac{\omega^2}{s}CC' + \frac{\omega^2}{s}SS' - \frac{1}{4}C'^2 + \frac{1}{4}S'^2,$$

$$Q_{12}^2 = \left(\frac{23\omega^4}{12} + \frac{7\omega^6 x}{9s}\right)C^2 + \frac{49\omega^4 x}{18}CC' - \left(\frac{41s\omega^2 x}{36} + \frac{7\omega^4 x^2}{9}\right)C'^2$$

$$- \left(\frac{23\omega^4}{12} + \frac{7\omega^6 x}{9s}\right)S^2 - \frac{49\omega^4 x}{18}SS' + \left(\frac{41s\omega^2 x}{36} + \frac{7\omega^4 x^2}{9}\right)S'^2.$$

Third Harmonic:

$$S_{20}^3 = -\frac{\omega^4}{8s^3 x}S^3 - \frac{\omega^2}{8s^2 x}S^2 S' + \frac{3\omega^2}{8s^2}SS'^2 + \frac{1}{24s}S'^3$$

$$+ \left(\frac{3\omega^4}{8s^3 x}C^2 + \frac{\omega^2}{4s^2 x}CC' - \frac{3\omega^2}{8s^2}C'^2\right)S$$

$$+ \left(\frac{\omega^2}{8s^2 x}C^2 - \frac{3\omega^2}{4s^2}CC' - \frac{1}{8s}C'^2\right)S',$$

$$C_{20}^3 = \frac{\omega^4}{8s^3 x}C^3 + \frac{\omega^2}{8s^2 x}C^2 C' - \frac{3\omega^2}{8s^2}CC'^2 - \frac{3}{72s}C'^3$$

$$+ \left(-\frac{3\omega^4}{8s^3 x}S^2 - \frac{\omega^2}{4s^2 x}SS' + \frac{3\omega^2}{8s^2}S'^2\right)C$$

$$+ \left(-\frac{\omega^2}{8s^2 x} S^2 + \frac{3\omega^2}{4s^2} SS' + \frac{1}{8s} S'^2 \right) C',$$

$$\begin{aligned}
P_{20}^3 = {} & \frac{3\omega^5}{8s^3 x} C^3 + \frac{5\omega^3}{8s^2 x} C^2 C' + \left(\frac{\omega}{4sx} - \frac{9\omega^3}{8s^2} \right) CC'^2 - \frac{3\omega}{8s} C'^3 \\
& - \frac{9\omega^5}{8s^3 x} CS^2 - \frac{5\omega^3}{8s^2 x} C'S^2 - \frac{5\omega^3}{4s^2 x} CSS' \\
& + \left(-\frac{\omega}{2sx} + \frac{9\omega^3}{4s^2} \right) SS'C' + \left(-\frac{\omega}{4sx} + \frac{9\omega^3}{8s^2} \right) CS'^2 + \frac{9\omega}{8s} S'^2 C',
\end{aligned}$$

$$\begin{aligned}
Q_{20}^3 = {} & \frac{3\omega^5}{8s^3 x} S^3 + \frac{5\omega^3}{8s^2 x} S^2 S' + \left(\frac{\omega}{4sx} - \frac{9\omega^3}{8s^2} \right) SS'^2 - \frac{3\omega}{8s} S'^3 \\
& - \frac{5\omega^3}{8s^2 x} C^2 S' + \left(-\frac{\omega}{2sx} + \frac{9\omega^3}{4s^2} \right) CC'S' + \frac{9\omega}{8s} C'^2 S' \\
& - \frac{9\omega^5}{8s^3 x} C^2 S - \frac{5\omega^3}{4s^2 x} CC'S + \left(-\frac{\omega}{4sx} + \frac{9\omega^3}{8s^2} \right) C'^2 S.
\end{aligned}$$

4. SPECIAL CASES

Setting $a_{12} = -1/\omega$, $a_{21} = 1/\omega$, $a_{11} = a_{22} = 0$, we have $S(x) = -\omega^{-1} Y_0(\xi)$ and $C(x) = \omega^{-1} J_0(\xi)$. Then the main approximation for $\eta(x,t)$ is

$$J_0(\xi) \sin{(\omega t)} + Y_0(\xi) \cos{(\omega t)}$$

which is equivalent to $\sqrt{2/(\pi x)} \sin\left(2\omega \sqrt{x/s} - \frac{1}{4}\pi + \omega t \right)$ for $x \to +\infty$. So this case can be considered as a progressive wave, whereas the case $a_{11} = 1$, $a_{22} = a_{12} = a_{21} = 0$ is a standing wave considered in [4].

5. CONCLUSIONS

Periodic solution with the accuracy of $(\varepsilon^2, \varepsilon\mu^2, \mu^4)$ to equations (6.1)–(6.4) is presented. Expressions for $S_{\alpha\beta}^i$, $C_{\alpha\beta}^i$ $P_{\alpha\beta}^i$, and $Q_{\alpha\beta}^i$ can be proved by substitution into system (6.1)–(6.4) (using expression (6.8) for the potential). (Some system of computer algebra may be recommended.) We conjecture that these expressions are only the lowest terms of a certain expanded exact solution to system (6.1)–(6.4).

References

[1] Mei, C C and Le Méhauté B (1966) *Note on equation of long waves over an uneven bottom*, J. Geophys. Res. **7**, 393–400.

[2] Madsen, P A and Schäffer, H A (1998) *Higher-order Boussinesq-type equations for surface gravity waves: derivation and analysis*, Phil. Trans. Roy. Soc. Lond. A **8**, 441–455.

[3] Mei, C C (1983) *The Applied Dynamics of Ocean Surface Waves*, New York: Wiley.

[4] Shermenev, A and Shermeneva, M (2000) *Long periodic waves on an even beach*, Phys. Rev. E **61**, 6000–6002.

[5] Carrier, G F and Greenspan, H P (1957) *Water waves of finite amplitude on a sloping beach*, J. Fluid Mech. **4**, 97–109.

WAVES TRAPPED UNDER A MOVING PRESSURE DISTRIBUTION

J.-M. Vanden-Broeck

School of Mathematics

University of East Anglia

Norwich NR4 7TJ, UK

J.Vanden-Broeck@uea.ac.uk

Abstract   The steady nonlinear free surface flow due to a moving distribution of pressure is considered. Both gravity and surface tension are included in the free surface condition. Numerical solutions are obtained by a boundary integral equation method. Previous investigations have shown that there are solutions with trains of waves in the far field and others with decaying oscillatory tails. In this paper we show that, in the absence of surface tension, there are particular solutions for which waves are trapped under the support of the distribution of pressure. These solutions are related to the "waveless solutions" considered by previous authors. It is shown how these solutions can be used to construct numerically accurate free surface flows which satisfy the radiation condition.

1.   INTRODUCTION

A classical problem in fluid mechanics is the prediction of the waves generated by an object moving at a constant velocity U on or below a free surface. The object can be a ship, a submarine or an insect. Here we model an object on the free surface by a distribution of pressure with a compact support or decaying fast in the far field. This can be viewed as an inverse approach since the shape of the object is defined at the end of the calculations by the shape of the streamline under the support of the distribution of pressure. For simplicity, we assume in this paper that the fluid is incompressible, inviscid and of infinite depth and that the flow is irrotational and two–dimensional. We take into account the effects of gravity g and surface tension T. Surface tension is usually negligible in ship hydrodynamics. However it can be important for small objects such as insects and probes.

We take a frame of reference moving with the pressure distribution. The flow at infinite depth is then characterized by a uniform stream

61

I.D. Abrahams et al. (eds.),
IUTAM Symposium on Diffraction and Scattering in Fluid Mechanics and Elasticity, 61–68.

with a constant velocity U. This problem has been considered by many previous investigators. There are both steady and unsteady solutions.

Steady solutions were first calculated by Lamb [1]. Lamb assumed that the magnitude of the pressure distribution is small. Then the equations can be linearized by assuming a small perturbation around a uniform stream. The resulting linear equations are solved by using Fourier transforms. A radiation condition is imposed to render the solution unique. Lamb's results can be described in terms of the minimum phase velocity c_{min} of linear gravity–capillary waves, which is defined by

$$c_{min}^2 = 2\sqrt{\frac{Tg}{\rho}}. \tag{7.1}$$

Here ρ is the density of the fluid. For $U > c_{min}$, there are two trains of waves in the far field: one train of long waves (dominated by gravity) behind the distribution of pressure and a train of shorter waves (dominated by surface tension) at the front. For $U < c_{min}$, the free surface is characterized by oscillations of decaying amplitude in the far field. Lamb's solutions predict an infinite displacement of the free surface as $U \to c_{min}$. Therefore a nonlinear theory is needed to study the flow when U is close to c_{min}. This was done by Vanden-Broeck and Dias [2] who solved the fully nonlinear problem numerically for $U < c_{min}$. They found that there are multiple solutions for U close to c_{min}. One of these solutions is similar to Lamb's solution in the sense that it is a perturbation of a uniform stream. Others are perturbations of gravity capillary solitary waves in water of infinite depth. These solitary waves are characterized by decaying oscillatory tails and have attracted a lot of attention in recent years (see [3] for a review). Interestingly, the branches of solutions have turning points and for U very close to c_{min}, there are no steady solutions of the type considered here. Corresponding unsteady solutions were calculated by Milewski and Vanden-Broeck [4].

In this paper we reexamine the steady solutions for $U > c_{min}$. We ask the question whether or not there are solutions for which the waves are trapped below the support of the distribution of pressure. We show that such solutions exist when the surface tension is neglected. One application of these solutions is the design of "drag free" objects. This application was considered by Lamb [1] in the linear case and by Schwartz [5], Forbes [6] and Vanden-Broeck and Tuck [7] in the nonlinear regime. Tuck and Scullen [8] investigated a related problem for submerged objects. In all these solutions, only one or two wavelengths are trapped. We show that an arbitrary number of wavelengths can be trapped. We also indicates how such solutions can be used to construct numerically accurate "wavy flows" which satisfy the radiation condition.

2.   FORMULATION

We consider the steady two–dimensional free–surface flow due to a pressure distribution acting on the surface of a fluid of infinite depth. At large depth, the flow approaches a uniform stream with a constant velocity U. We choose cartesian coordinates with the x-axis parallel to the velocity U and the y-axis directed vertically upwards. The acceleration of gravity g is acting in the negative y-direction. The effect of the surface tension is neglected in the calculations.

We introduce the potential function $\phi(x, y)$ and the streamfunction $\psi(x, y)$. We choose $\psi = 0$ on the free surface. We assume that the pressure distribution is symmetric about $x = 0$ and we choose $\phi = 0$ at $x = 0$.

We define dimensionless variables by choosing U as the unit velocity and U^2/g as the unit length. We use ϕ and ψ as independent variables and denote by $x(\phi)$ and $y(\phi)$ the values of x and y on the free surface $\psi = 0$. On the free surface Bernoulli's equation yields

$$\frac{1}{2}\frac{1}{x_\phi^2 + y_\phi^2} + y + P(\phi) = \frac{1}{2}. \tag{7.2}$$

Here $P(\phi)$ is the prescribed distribution of pressure. We give P as a function of ϕ. This is an inverse approach and the value of P as a function of x can be found at the end of the calculations after $x(\phi)$ has been calculated. We assume that $P(\phi) \to 0$ as $|\phi| \to \infty$. Then the choice of the Bernoulli's constant on the right hand side of (7.2) fixes the level $y = 0$ as the level of the free surface at infinity.

Following Vanden-Broeck and Dias [2], we use Cauchy's integral formula to derive the equation

$$x_\phi = 1 - \frac{1}{\pi}\int_{-\infty}^{\infty}\frac{y_\zeta}{\zeta - \phi}d\zeta. \tag{7.3}$$

The integral in (7.3) is a Cauchy principal value.

Equations (7.2) and (7.3) define a nonlinear integral equation for the unknown $y(\phi)$. This equation is solved numerically in the next section.

3.   NUMERICAL RESULTS

If the magnitude of the pressure $P(\phi)$ is small, the flow is a small perturbation around a uniform stream with constant velocity 1 and (7.2) can be linearized as

$$1 - x_\phi + y + P = 0. \tag{7.4}$$

This linear problem was solved by Lamb [1], using Fourier transforms. In particular, he obtained solutions for

$$P = p_0 \quad -L < x < L \tag{7.5}$$

and $P = 0$ otherwise. Here p_0 is a constant. We note that in the framework of the linear theory, x can be replaced by ϕ in (7.5). The corresponding free surface profile, satisfying the radiation condition, is flat as $x \to -\infty$ and has in general a train of waves as $x \to \infty$. However for $L = n\pi$, the amplitude of the wave vanishes and the free surface is flat as $x = \pm\infty$. Here n is a positive integer. These particular solutions have received a special attention because the corresponding wave resistance is zero (see Schwartz [5], Forbes [6] and Vanden-Broeck and Tuck [7]).

Here we consider further these solutions for n large. In order to calculate nonlinear solutions, we need to smooth the distribution of pressure (7.5). We first note that the derivative of $P(\phi)$ with respect to ϕ is

$$P_\phi = p_0\delta(\phi + L) - p_0\delta(\phi - L). \tag{7.6}$$

Here δ denotes the Dirac generalized function. We smooth (7.6) by writing

$$P_\phi = p_0\delta_\nu(\phi + L) - p_0\delta_\nu(\phi - L) \tag{7.7}$$

where

$$\delta_\nu(\phi) = \frac{\nu}{\sqrt{\pi}} \exp(-\nu^2\phi^2) \tag{7.8}$$

and ν is a large parameter. We note that (7.7) is smooth and approaches (7.6) as $\nu \to \infty$.

Next we differentiate (7.2) with respect to ϕ. This yields

$$-\frac{x_\phi x_{\phi\phi} + y_\phi y_{\phi\phi}}{(x_\phi^2 + y_\phi^2)^2} + y_\phi + P_\phi = 0 \tag{7.9}$$

where P_ϕ is defined by (7.7).

We solve the system defined by (7.3) and (7.9) and seek solutions which are flat as $|\phi| \to \infty$. The procedure follows closely the work in [6] and [2].

We introduce the mesh points

$$\phi_I = -M + \frac{2(I-1)M}{N-1}, \quad I = 1,\ldots,N. \tag{7.10}$$

Here we truncated the infinite domain of ϕ at $\pm M$. Next we introduce the unknowns

$$Y_I' = y_\phi(\phi_I), \quad I = 1,\ldots,N. \tag{7.11}$$

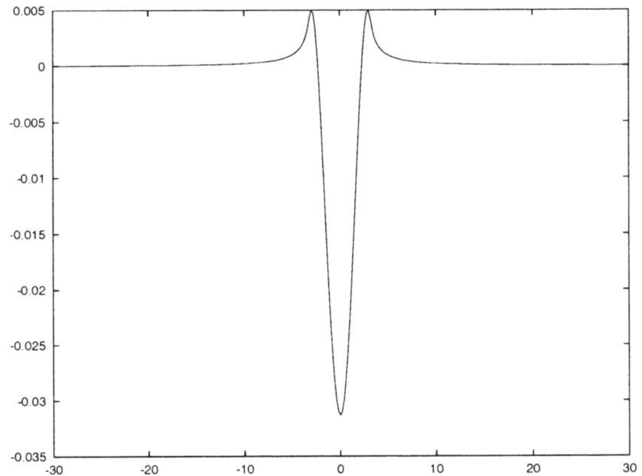

Figure 1   Free surface profile for $p_0 = 0.01$, $\nu = 3$, $M = 40$ and $L = \pi$ in the initial guess.

We use (7.3) to evaluate x_ϕ at the midpoints

$$\phi_I^m = \frac{(\phi_I + \phi_{I+1})}{2}, \quad I = 1, \ldots, N - 1. \tag{7.12}$$

The integral in (7.3) is evaluated by the trapezoidal rule with a summation over the points (7.10).

We then satisfy (7.9) at the midpoints (7.12). This leads to $N - 1$ equations. The second order derivatives in (7.9) are evaluated by central difference formulae. Two more equations are obtained by imposing

$$Y_1' = Y_N' = 0. \tag{7.13}$$

We now have a system of $N + 1$ nonlinear algebraic equations for the $N + 1$ unknowns Y_I' and L. This system is solved by Newton's method. In most of the calculations, we chose $\nu = 3$, $p_0 = 0.01$, $N = 1200$. We repeated the calculations for various values of M and N and checked that all the results presented are independent of M and N within graphical accuracy.

For the initial guess, we use $Y_I' = 0$ and $L = n\pi$.

Three typical profiles corresponding to different values of n in the initial guess are shown in Figs. 1–3. The value of n in the initial guess selects the solution which the Newton iterations converge to. The profiles in Figs. 1 and 2 are similar to those previously computed by Schwartz [5], Forbes [6] and Vanden-Broeck and Tuck [7]. The profile in Fig. 3 is new in the sense that it exhibits a long train of waves trapped under the support of the distribution of pressure. The number of oscillations under the pressure distribution increases as the value of n increases in

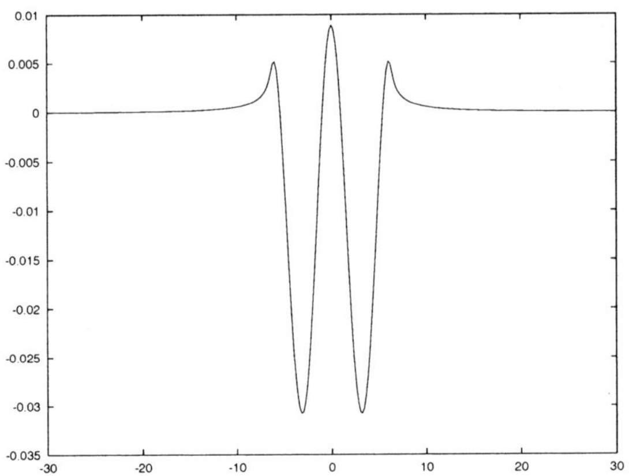

Figure 2   Same as Fig. 1 with $L = 2\pi$ in the initial guess.

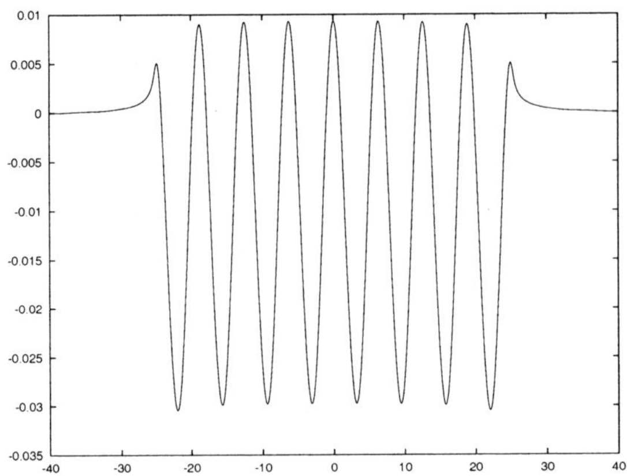

Figure 3   Same as Fig. 1 with $L = 8\pi$ in the initial guess.

the initial guess. The train of waves can be made arbitrarily long by choosing n sufficiently large.

We note that the solutions shown in Figs. 1–3 are special solutions for which the free surface is flat as $x \to -\infty$ and as $x \to \infty$. The solutions have in general a flat free surface as $x \to -\infty$ and a train of periodic waves as $x \to \infty$. We will refer to solutions with waves as $x \to \infty$ as "wavy solutions" and to those with flat free surfaces as $|x| \to \infty$ as "trapped solutions".

Wavy solutions have been calculated by Schwartz [5], Forbes [6], Asavanant and Vanden-Broeck [9] and others. One difficulty in calculating these solutions is the imposition of the radiation condition. If it is not done properly, spurious waves appear as $x \to -\infty$.

There is a relation between the wavy solutions and the trapped solutions as $n \to \infty$. To understand the limit as $n \to \infty$, we define a new variable

$$\bar{\phi} = \phi + L \tag{7.14}$$

and rewrite (7.7) as

$$P_\phi = p_0 \delta_\nu(\bar{\phi}) - p_0 \delta_\nu(\bar{\phi} - 2L). \tag{7.15}$$

As $n \to \infty$, $L \to \infty$ and for a fixed value of $\bar{\phi}$, the flow reduces to the one due to the pressure disturbance

$$P_\phi = p_0 \delta_\nu(\bar{\phi}). \tag{7.16}$$

This flow has a train of waves for large positive values of $\bar{\phi}$ and is therefore a wavy flow. We note that no spurious waves are present on the free surfaces of Figs. 1–3 for $|x|$ large. This suggests that the limit as $n \to \infty$ in trapped flows can be used to generate wavy flows which satisfy accurately the radiation condition (i.e. without spurious waves).

4.   CONCLUSIONS

We have considered nonlinear gravity-capillary free surface flows due to a moving obstacle. After reviewing the various types of solutions which can occur, we have computed particular solutions in the absence of surface tension. These solutions, which we called trapped flows, are characterised by a flat free surface as $|x| \to \infty$. Trapped flows can be labeled by an index n. We have shown that as $n \to \infty$, the trapped flows approach the usual wavy solutions.

Acknowledgments

The author would like to thank Professor Grimshaw for suggesting the problem of trapped waves.

References

[1] Lamb, H (1945) *Hydrodynamics*, New York: Dover.

[2] Vanden-Broeck, J-M and Dias, F (1992) *Solitary waves in water of infinite depth and related free surface flows*, J. Fluid Mech. **240**, 549–557.

[3] Dias, F and Kharif, C (1999) *Nonlinear gravity and capillary–gravity waves*, Ann. Rev. Fluid Mech. **31**, 301–346.

[4] Milewski, P and Vanden-Broeck, J-M (1999) *Time dependent gravity–capillary flows past an obstacle*, Wave Motion **29**, 63–79.

[5] Schwartz, L W (1981) *Nonlinear solution for an applied overpressure on a moving stream*, J. Eng. Math. **15**, 147–156.

[6] Forbes, L K (1982) *Nonlinear drag–free flow over a submerged semi-elliptical body*, J. Eng. Math. **16**, 171–180.

[7] Vanden-Broeck, J-M and Tuck, E O (1985) *Waveless free surface pressure distributions*, J. Ship Res. **29**, 151–158.

[8] Tuck, E O and Scullen D C (1998) *Tandem submerged cylinders each subject to zero drag*, J. Fluid Mech. **364**, 211–220.

[9] Asavanant, J and Vanden-Broeck, J-M (1994) *Free–surface flows past a surface–piercing object of finite length*, J. Fluid Mech. **273**, 109–124.

II

ENERGY TRAPPING AND PROPAGATION AROUND SUBMERGED BODIES AND IN PERIODIC ARRAYS

LOCALISED OSCILLATIONS NEAR SUBMERGED OBSTACLES

M. McIver

Department of Mathematical Sciences, Loughborough University
Loughborough, Leicestershire LE11 3TU, UK
m.mciver@lboro.ac.uk

1.   INTRODUCTION

In previous work McIver [1] showed that a trapped mode could be
supported by two specially shaped obstacles which are held a prescribed
distance apart on the surface of water of infinite depth. Physically a
trapped mode is a localised oscillation of the fluid which does not radiate any waves to infinity and which has finite energy. Mathematically it
is an eigenfunction associated with a certain operator and the trapped
mode found in [1] has the property that its corresponding eigenvalue
is embedded in the continuous spectrum of that operator. In general
embedded eigenvalues are unstable to small perturbations in the body
geometry and it is difficult to prove their existence or absence for a
given body using standard variational techniques. Despite this, further
examples of embedded trapped modes were found in both two and three
dimensions [2, 3, 4] and in each case the mode was found to be supported by one or more bodies which isolate a portion of the free surface.
However Evans (private communication) gave an argument based on a
wide spacing approximation, which suggests that trapped modes at a
certain frequency are possible for two suitably spaced, identical bodies
provided that the transmission coefficient for the single body is zero at
that frequency. There are submerged bodies which have zeros of transmission and so this argument suggests that trapped modes will exist for
pairs of such bodies. The purpose of this work is to explicitly construct
a pair of submerged bodies which support trapped modes, with the use
of the approach described by McIver [1]. Further details of this method
and its use for constructing submerged bodies is available in [5].

I.D. Abrahams et al. (eds.),
IUTAM Symposium on Diffraction and Scattering in Fluid Mechanics and Elasticity, 71–78.
© 2002 *Kluwer Academic Publishers. Printed in the Netherlands.*

2.   FORMULATION

Two-dimensional Cartesian coordinates are chosen so that the origin is in the mean free surface and the y-axis points vertically downwards. A trapped mode potential $\mathrm{Re}[\phi\, e^{-i\omega t}]$ satisfies

$$\nabla^2 \phi = 0 \tag{8.1}$$

and the linearised free surface condition

$$K\phi + \frac{\partial \phi}{\partial y} = 0 \quad \text{on} \quad y = 0, \tag{8.2}$$

where $K = \omega^2/g$, ω is the angular frequency of oscillation and g is the acceleration due to gravity. In addition

$$\partial \phi / \partial n = 0 \tag{8.3}$$

on any body surface, and the mode must have finite energy, i.e.

$$\int_\Omega |\nabla \phi|^2 \, dV + K \int_F |\phi|^2 \, dx < \infty. \tag{8.4}$$

where Ω is the fluid domain and F is the mean free surface. In particular, this last condition means that the disturbance does not radiate any waves to infinity and so

$$\phi \to 0 \quad \text{as} \quad x^2 + y^2 \to \infty, \quad y \geq 0. \tag{8.5}$$

A potential is constructed from a symmetric combination of submerged vertical and horizontal dipoles and, it is shown numerically for a particular set of parameters, that this combination represents a trapped mode supported by a pair of submerged bodies.

The potential due to a submerged vertical dipole with singular point at (ξ, η), $\eta > 0$ is given by

$$\phi_s(x, y; \xi, \eta) = \frac{y - \eta}{K[(x - \xi)^2 + (y - \eta)^2]} - \frac{y + \eta}{K[(x - \xi)^2 + (y + \eta)^2]}$$
$$- 2\mathrm{Re} \int_0^\infty \frac{e^{-k(y+\eta)+ik(x-\xi)}}{k - K} \, dk. \tag{8.6}$$

This potential is real, symmetric about $x = \xi$, and satisfies (8.1) and (8.2). The far-field behaviour of ϕ_s represents a standing wave and is given by

$$\phi_s \sim 2\pi \, \mathrm{sgn}\, K(x - \xi) e^{-K(y+\eta)} \sin K(x - \xi) \quad \text{as} \quad |x - \xi| \to \infty. \tag{8.7}$$

(If ϕ_s is allowed to be complex it is possible to apply a radiation condition and force ϕ_s to represent outgoing waves at infinity. For the purpose of this work however, it is more convenient to define ϕ_s so that it is real.) The harmonic conjugate to ϕ_s is denoted by ψ_s and satisfies the Cauchy–Riemann equations $\partial\psi_s/\partial y = \partial\phi_s/\partial x$, $\partial\psi_s/\partial x = -\partial\phi_s/\partial y$, and is given by

$$\psi_s(x,y;\xi,\eta) = \frac{x-\xi}{K[(x-\xi)^2+(y-\eta)^2]} - \frac{x-\xi}{K[(x-\xi)^2+(y+\eta)^2]}$$
$$- 2\,\mathrm{Im}\int_0^\infty \frac{e^{-k(y+\eta)+ik(x-\xi)}}{k-K}\,dk. \tag{8.8}$$

The potential due to a submerged horizontal dipole with singular point at (ξ,η), $\eta > 0$ is given by

$$\phi_a(x,y;\xi,\eta) = \frac{x-\xi}{K[(x-\xi)^2+(y-\eta)^2]} + \frac{x-\xi}{K[(x-\xi)^2+(y+\eta)^2]}$$
$$+ 2\,\mathrm{Im}\int_0^\infty \frac{e^{-k(y+\eta)+ik(x-\xi)}}{k-K}\,dk. \tag{8.9}$$

This potential is real, antisymmetric about $x = \xi$, satisfies (8.1) and (8.2) and has the far-field behaviour

$$\phi_a \sim 2\pi\,\mathrm{sgn}\,K(x-\xi)e^{-K(y+\eta)}\cos K(x-\xi) \quad \text{as} \quad |x-\xi| \to \infty. \tag{8.10}$$

Its harmonic conjugate is given by

$$\psi_a(x,y;\xi,\eta) = -\frac{y-\eta}{K[(x-\xi)^2+(y-\eta)^2]} - \frac{y+\eta}{K[(x-\xi)^2+(y+\eta)^2]}$$
$$- 2\,\mathrm{Re}\int_0^\infty \frac{e^{-k(y+\eta)+ik(x-\xi)}}{k-K}\,dk. \tag{8.11}$$

The combination of vertical and horizontal dipoles

$$\phi(x,y;Ka,Kh) = \sin Ka[\phi_s(x,y;a,h) + \phi_s(x,y;-a,h)]$$
$$- \cos Ka[\phi_a(x,y;a,h) - \phi_a(x,y;-a,h)]. \tag{8.12}$$

is formed and its harmonic conjugate, the stream function is given by

$$\psi(x,y;Ka,Kh) = \sin Ka[\psi_s(x,y;a,h) + \psi_s(x,y;-a,h)]$$
$$- \cos Ka[\psi_a(x,y;a,h) - \psi_a(x,y;-a,h)]. \tag{8.13}$$

Lines on which ψ is constant are streamlines of the flow. This choice for ϕ represents a symmetric combination of two dipoles which are a horizontal distance $2a$ apart and which are submerged at a depth h. The orientation of the dipoles depends on their horizontal spacing. Because of the symmetry properties of the individual dipoles, the combination of potentials in (8.12) is symmetric and so it is only necessary to investigate its properties in the region $x \geq 0$. The potential has a singular point at (a, h) and from (8.7) and (8.10)

$$\phi(x, y; Ka, Kh) \sim 2\pi e^{-K(y+\eta)}[\sin Ka(\sin K(x - a) + \sin K(x + a))$$
$$- \cos Ka(\cos K(x - a) - \cos K(x + a))] = 0 \text{ as } x \to \infty. \tag{8.14}$$

Thus ϕ satisfies (8.1), (8.2) and (8.5). If there is a streamline of the flow which surrounds the singular point then this streamline may be interpreted as a submerged body. In this case, the potential represents a symmetric trapped mode which is supported by the submerged body and its reflection in the line $x = 0$. In the next section such a mode will be shown to exist for a particular value of Ka and Kh.

3.   CONSTRUCTION OF THE SUBMERGED BODIES

Because of numerical inaccuracies it is impossible to demonstrate that a particular streamline is closed and instead the following approach is taken. The value of Ka is fixed at $Ka = \pi/4$ and the argument principle is used to show that the flow has two stagnation points in $x > 0$, $y > 0$ for a range of values of Kh. Numerical experimentation shows that at $Kh = 0.04$ the value of the streamfunction ψ at the higher stagnation point S_1 is less than the value of ψ at the other stagnation point S_2, and at $Kh = 0.07$ the value of ψ at S_1 is greater than its value at S_2. There is also a qualitative difference in the streamline patterns for $Kh = 0.04$ and $Kh = 0.07$ as illustrated in Figs. 1 and 2, respectively.

Continuity requires that there is a value of Kh, $0.04 < Kh < 0.07$ such that the values of ψ at both stagnation points are the same. Numerically this value of Kh is given by $Kh = 0.0547846$. Fig. 3 illustrates the streamline pattern for this value of Kh and Fig. 4 displays the variation of the stream function on the free surface in $x \geq 0$. As $\psi_1 = \psi_2$ two parts of the streamline which emanate from S_1 go into S_2 and form a closed streamline which surrounds the singularity. The argument principle may be used to show that both stagnation points lie below the free surface. Thus the closed portion of the streamline which passes through the stagnation points forms a submerged body because from Fig. 4, there are only two places at which $\psi(x, 0)$ takes the value at the stagnation points, and from Fig. 3 these must correspond to the points at which the parts of the streamlines which emanate from the stagnation points

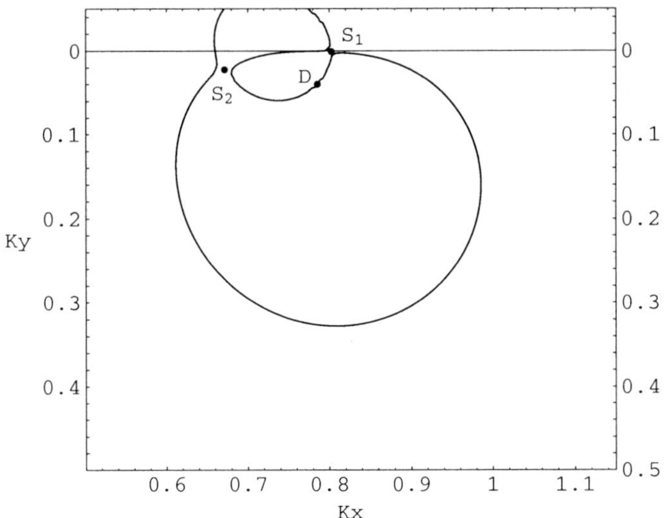

Figure 1   Streamline pattern, $Kh = 0.04$. S_1 and S_2 are stagnation points, D is the singular point.

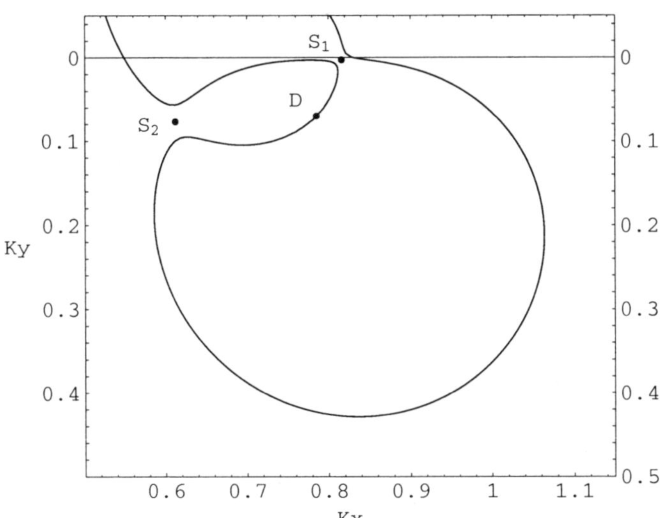

Figure 2   Streamline pattern, $Kh = 07$. S_1 and S_2 are stagnation points, D is the singular point.

cross the free surface, rather than points at which other parts of the body intersect the free surface. By symmetry, there is a similar body in the region $x < 0$. Together these streamlines represent two submerged bodies which support trapped modes.

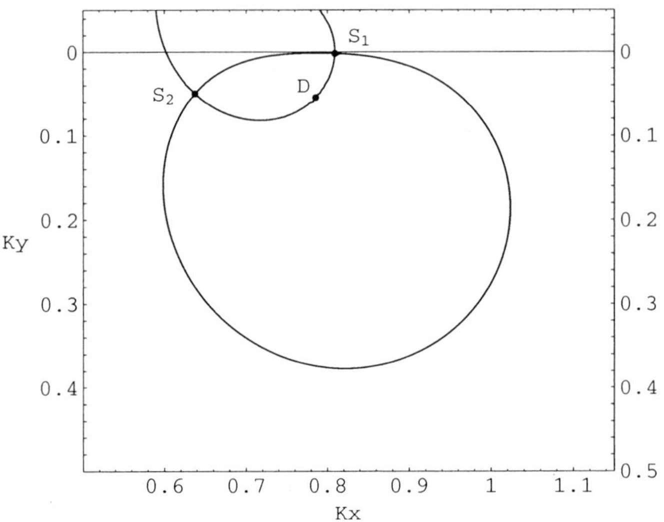

Figure 3   Streamline pattern, $Kh = 0.0547846$. S_1 and S_2 are stagnation points, D is the singular point.

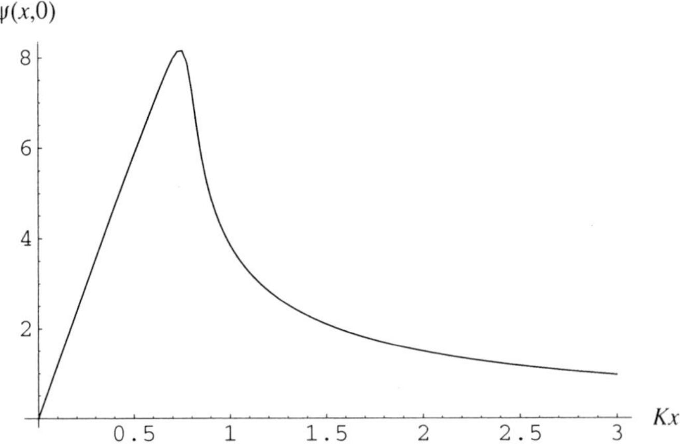

Figure 4   Variation of the streamfunction along the free surface, $Kh = 0.0547846$.

4.　DISCUSSION

The analysis of the previous section shows that a pair of submerged bodies which support a trapped mode exist, but from Fig. 3 it is clear that the upper edges of the bodies are extremely close to the free surface. Numerical experimentation suggests that trapped modes may be constructed for other values of Ka in the range $0 < Ka < \pi/2$, but that the depth of submergence of the highest point of the bodies tends to zero as $Ka \to 0$ and $Ka \to \pi/2$. If this criterion to describe depth of submergence is used, it is found that the most deeply submerged bodies correspond, at least approximately, to $Ka = \pi/4$, although it is not clear why this should be so.

From the work of Simon and Ursell [6] it is known that bodies submerged in deep water cannot support trapped modes if they are sufficiently deeply submerged, more precisely if they are contained within lines which make an angle of $\pm\pi/4$ to the free surface at a point. One question which resulted from this work was whether it is possible to reduce the angle which the lines make with the free surface and still preserve uniqueness. The results presented here show that there is a non-zero lower limit on this angle, albeit possibly very small. (In order to calculate a lower bound for this angle the position of the highest point of the critical streamline would have to be calculated numerically. This point is not necessarily the same as the highest stagnation point although from Fig. 3 it is clearly close.)

The existence of the trapped mode found here for a pair of submerged bodies is more vulnerable to body perturbations than that found for surface-piercing bodies in [1]. In the latter case for specific values of Ka it was possible to interpret a range of different streamlines as the boundaries of surface-piercing bodies which support trapped modes. In the work discussed here, there is only one streamline for specific values of Ka and Kh which may be interpreted as the boundary of a submerged body. In addition it has been shown that if one of the parameters is fixed and an arbitrarily small perturbation given to the other then the trapped mode associated with submerged bodies is lost. However other streamlines exist which may be interpreted as the boundaries of surface-piercing bodies and from Fig. 3, these are streamlines which go from one point on the free surface to another and enclose the stagnation points as well as the singularity. Such surface-piercing bodies are also possible for a range values of Kh, as may be seen from Figs. 1 and 2. In these cases although they are not illustrated on the figures, further streamlines exist which join two points on the free surface and surround the stagnation points and the dipole.

References

[1] McIver, M (1996) *An example of non-uniqueness in the two-dimensional linear water wave problem*, J. Fluid Mech. **315**, 257–266.

[2] Linton, C M and Kuznetsov, N G (1997) *Non-uniqueness in two-dimensional water wave problems: numerical evidence and geometrical restrictions*, Proc. Roy. Soc. London A **453**, 2437–2460.

[3] McIver, P and McIver, M (1997) *Trapped modes in an axisymmetric water-wave problem*, Quart. J. Mech. Appl. Math. **50**, 165–178.

[4] Evans, D V and Porter, R (1998) *An example of non-uniqueness in the two-dimensional linear water wave problem involving a submerged body*, Proc. Roy. Soc. London A **454**, 3145–3165.

[5] McIver, M (1997) *Trapped modes supported by submerged obstacles*, Proc. Roy. Soc. London A **456**, 1851–1860.

[6] Simon, M J and Ursell, F (1984) *Uniqueness in linearized two-dimensional water-wave problems*, J. Fluid Mech. **148**, 137–154.

APPROXIMATIONS TO EMBEDDED TRAPPED MODES IN WAVE GUIDES

P. McIver*

Department of Mathematical Sciences, Loughborough University

Loughborough, Leicestershire LE11 3TU, UK

p.mciver@lboro.ac.uk

1.   INTRODUCTION

This work is concerned with fluid motions in an infinitely-long, two-dimensional acoustic waveguide containing an obstacle. Cartesian coordinates (x, y) are chosen so that the x axis lies along the centre of the guide and the walls of the guide are at $y = \pm d$. Under the usual assumptions of linear acoustics, time-harmonic motions of radian frequency ω may be described by a velocity potential $\phi(x, y)e^{-i\omega t}$. Non-trivial solutions for ϕ are sought within the guide (excluding the obstacle) that, on the guide walls, satisfy either a homogeneous Neumann condition (a 'Neumann guide') or a homogeneous Dirichlet condition (a 'Dirichlet guide'). In addition, the solutions are required to satisfy a homogeneous Neumann condition on the obstacle and to decay to zero as $|x| \to \infty$. Such a solution corresponds to a 'trapped mode', that is a free oscillation of the fluid with finite energy. The physical significance of the existence of a trapped mode at a particular frequency is that it corresponds to an 'acoustic resonance' in a forced problem.

A natural frequency parameter for this problem is kd, where $k = \omega/c$ and c is the sound speed. In a Neumann guide, solutions of the governing equations that do not decay as $|x| \to \infty$ exist for $kd \in [0, \infty)$ and so any trapped-mode frequencies will be embedded in this continuous spectrum. However, if the obstacle is symmetric about $y = 0$, and the fluid motion is assumed to be antisymmetric about $y = 0$, then the lowest point of the continuous spectrum for this restricted problem corresponds to $kd = \pi/2$ and the search for trapped modes for $kd \in (0, \pi/2)$ is made easier as the trapped-mode frequencies are no longer embedded in the

---

*This work is part of a collaboration with C. M. Linton, M. McIver and J. Zhang that is funded by EPSRC grant GR/M30937.

I.D. Abrahams et al. (eds.),
IUTAM Symposium on Diffraction and Scattering in Fluid Mechanics and Elasticity, 79–86.
© 2002 *Kluwer Academic Publishers. Printed in the Netherlands.*

continuous spectrum. In fact, it is now well established [1] that, with the symmetries as stated above, trapped modes exist for a wide class of obstacles with kd in this range. In a Dirichlet guide the continuous spectrum corresponds to $kd \in [\pi/2, \infty)$ so there is the possibility of obtaining non-embedded trapped-mode solutions without the need to impose particular symmetries.

This work is concerned with trapped modes that are genuinely embedded in the continuous spectrum, in the sense that for the circumstances considered there is no known way to restrict the problem so as to remove the part of the continuous spectrum for the appropriate range of kd. Here an approximate theory for 'slender' structures [2] is adapted and used to obtain explicit results. These results have been very useful in guiding other work (to be reported on elsewhere) that uses more general methods. Trapped modes are sought for the following situations.

(a) Neumann guide: No obstacle symmetry about $y = 0$; $kd \in (0, \pi/2)$.

(b) Neumann guide: Obstacles symmetric about $y = 0$ and motions antisymmetric about $y = 0$; $kd \in (\pi/2, 3\pi/2)$.

(c) Dirichlet guide: No obstacle symmetry about $y = 0$; $kd \in (\pi/2, \pi)$.

2.   FORMULATION

An obstacle is placed in a waveguide with walls $y = \pm d$. Either Neumann or Dirichlet boundary conditions are imposed on the guide walls, but a Neumann boundary condition is used for the obstacle. In addition, restriction is made to geometries and fluid motions that are symmetric about $x = 0$ so that attention may be confined to $x \geq 0$. The surface of the obstacle in $x \geq 0$ is described by $x = \epsilon f(y) \geq 0$, $y \in (-d, d)$, where f is single valued. Here $\epsilon = a/b$, where a measures the deviation of the obstacle from $x = 0$ and b the length in the y direction; by assumption $\epsilon \ll 1$. A normal vector to the surface is

$$(n_x, n_y) = (1, -\epsilon f'(y)). \tag{9.1}$$

A slender obstacle has $n_y = O(\epsilon)$, that is $f'(y) = O(1)$, as $\epsilon \to 0$.

Trapped modes are non-trivial solutions of the field equation

$$(\nabla^2 + k^2)\phi = 0, \quad x > \epsilon f(y), \quad |y| < d, \tag{9.2}$$

subject to the wall boundary condition

$$\phi_y = 0 \quad \text{or} \quad \phi = 0 \quad \text{on} \quad y = \pm d, \quad x > \epsilon f(\pm d), \tag{9.3}$$

the obstacle boundary condition

$$\phi_n = n_x \phi_x + n_y \phi_y = 0 \quad \text{on} \quad x = \epsilon f(y), \quad |y| < d, \tag{9.4}$$

and

$$\phi \to 0 \quad \text{as} \quad x \to \infty. \tag{9.5}$$

In the above, subscripts are used to denote differentiation and n is a coordinate measured in a direction normal to the surface of the obstacle. In the limit $\epsilon \to 0$ equations (9.2)–(9.4) are approximated by

$$(\nabla^2 + k^2)\phi = 0, \quad \text{for} \quad x > 0, \quad |y| < d, \tag{9.6}$$

$$\phi_y = 0 \quad \text{or} \quad \phi = 0 \quad \text{on} \quad y = \pm d, \quad x > 0, \tag{9.7}$$

$$\phi_x = \epsilon \left(f'(y)\phi_y - f(y)\phi_{xx} \right) + O(\epsilon^2) \quad \text{on} \quad x = 0, \quad |y| < d, \tag{9.8}$$

where the last equation is obtained by Taylor expansion about $x = 0$.

A solution of (9.5)–(9.8) is sought using separation of variables. A set of eigenfunctions $\{\Psi_n(y), n = 0, 1, \ldots\}$ is introduced that satisfy the wall boundary conditions (9.7) and also the orthogonality relations

$$\frac{1}{d} \int_{-d}^{d} \Psi_m(y)\Psi_n(y)\, dy = \delta_{mn}, \quad m, n = 0, 1, \ldots. \tag{9.9}$$

The forms of these eigenfunctions will be specified later when individual cases are considered. In every case, either $\Psi_0(y) \equiv 0$ or the part of the solution corresponding to $n = 0$ is a propagating mode which violates (9.5). Hence, for the trapped-mode problem, there is no term corresponding to $n = 0$ and separation of variables gives a solution

$$\phi(x, y) = e^{-\kappa_1 x}\Psi_1(y) + \chi(x, y), \quad x > 0, \quad |y| < d, \tag{9.10}$$

where

$$\chi(x, y) = \sum_{n=2}^{\infty} A_n e^{-\kappa_n x}\Psi_n(y), \tag{9.11}$$

and each κ_n, $n = 1, 2, \ldots$, is a positive real number that depends on the frequency parameter k. By construction, for all $x \geq 0$

$$\int_{-d}^{d} \chi(x, y)\Psi_n(y)\, dy = 0, \quad n = 0, 1, \tag{9.12}$$

which, in particular, implies that

$$\lim_{x \to 0} \int_{-d}^{d} \frac{\partial \chi}{\partial x}(x, y)\Psi_n(y)\, dy = 0, \quad n = 0, 1. \tag{9.13}$$

It remains to satisfy the obstacle boundary condition (9.8). A solution for χ is sought in the form

$$\chi(x, y) = \chi_0(x, y) + \epsilon\chi_1(x, y) + O(\epsilon^2) \quad \text{as} \quad \epsilon \to 0, \tag{9.14}$$

where each χ_n is strictly of order unity, so that (9.8) becomes

$$- \kappa_1 \Psi_1(y) + \frac{\partial \chi_0}{\partial x}(0, y) + \epsilon \frac{\partial \chi_1}{\partial x}(0, y) = \epsilon f'(y) \left(\Psi_1'(y) + \frac{\partial \chi_0}{\partial y}(0, y) \right)$$
$$- \epsilon f(y) \left(\kappa_1^2 \Psi_1(y) + \frac{\partial^2 \chi_0}{\partial x^2}(0, y) \right) + O(\epsilon^2) \quad \text{as} \quad \epsilon \to 0, \quad |y| < d.$$
$$(9.15)$$

If $\lim_{\epsilon \to 0} \kappa_1 \neq 0$, then to leading order (9.15) gives

$$\frac{\partial \chi_0}{\partial x}(0, y) = \kappa_1 \Psi_1(y), \quad |y| < d, \tag{9.16}$$

which contradicts (9.13) with $n = 1$ and hence $\lim_{\epsilon \to 0} \kappa_2 = 0$. If the latter holds then the leading-order terms in (9.15) give

$$\frac{\partial \chi_0}{\partial x}(0, y) = 0, \quad |y| < d, \tag{9.17}$$

and so, after taking account of the other boundary conditions, it follows that $\chi_0(x, y) \equiv 0$. If $\kappa_1 = o(\epsilon)$ as $\epsilon \to 0$, then to leading order

$$\frac{\partial \chi_1}{\partial x}(0, y) = f'(y)\Psi_1'(y), \quad |y| < d, \tag{9.18}$$

which contradicts (9.13) except possibly for special $f(y)$ that satisfy

$$\int_{-d}^{d} f'(y)\Psi_1'(y)\Psi_n(y) \, dy = 0, \quad n = 0, 1. \tag{9.19}$$

Under the assumption that both of conditions (9.19) are not satisfied, it follows that solutions under the stated assumptions are possible only if κ_1 is strictly of order ϵ and the leading-order approximation to (9.15) is

$$\epsilon \frac{\partial \chi_1}{\partial x}(0, y) = \kappa_1 \Psi_1(y) + \epsilon f'(y)\Psi_1'(y). \tag{9.20}$$

The conditions (9.13) then yield

$$\int_{-d}^{d} f'(y)\Psi_1'(y)\Psi_0(y) \, dy = 0 \tag{9.21}$$

and

$$\kappa_1 d = -\epsilon \int_{-d}^{d} f'(y)\Psi_1'(y)\Psi_1(y) \, dy. \tag{9.22}$$

Trapped modes exist provided (9.21) is satisfied and κ_1 is positive (solutions with non-positive κ_1 violate equation (9.5)). The frequency of a trapped mode follows through the dependence of κ_1 on k.

When $\Psi_0(y) \equiv 0$ equation (9.21) is satisfied for any shape function $f(y)$. Thus for a specified shape (e.g. an ellipse of given eccentricity) there will be a range of different sizes for which trapped modes exist and whose frequency follows from (9.22). On the other hand when $\Psi_0(y) \neq 0$ a geometrical parameter must be adjusted to satisfy (9.21). For example, with a specific obstacle a trapped mode may exist only for a given position in the guide, or for a given shape and position of obstacle a trapped mode may exist only for a given size.

3.   NEUMANN GUIDE: ASYMMETRIC OBSTACLE

For Neumann guides where the potential has zero normal derivative on $y = \pm d$, the eigenfunctions are

$$\Psi_0(y) = 2^{-1/2}, \quad \Psi_n(y) = \cos\left[\frac{n\pi}{2d}(y+d)\right], \quad n = 1, 2, \ldots, \qquad (9.23)$$

and

$$\kappa_n = [(n\pi/2d)^2 - k^2]^{1/2}. \qquad (9.24)$$

For $kd < \pi/2$, only the $n = 0$ mode propagates and solutions have the structure described in section 2. Hence, from equation (9.21) trapped modes exist only if

$$\int_{-d}^{d} f'(y) \cos\frac{\pi y}{2d}\, dy = 0 \qquad (9.25)$$

and from (9.22) the corresponding trapped-mode frequency follows from

$$\kappa_1 d = -\frac{\epsilon\pi}{4d}\int_{-d}^{d} f'(y) \sin\frac{\pi y}{d}\, dy. \qquad (9.26)$$

For the limit $\epsilon \to 0$ and in the vicinity of the obstacle, trapped-mode solutions are perturbations of the standing wave that can exist in the guide for $kd = \pi/2$. If the obstacle is symmetric about $y = 0$ then (9.25) is satisfied identically and (9.26) reduces to equation (4.5) in [2].

APPLICATIONS

1 Consider an obstacle with a line of symmetry at $y = y_o$ (for example, an off-centre ellipse) that is contained completely within the guide so that $f(y) \neq 0$ only for $|y - y_o| < b$, and such that $f'(y)$ is one signed for $y \in (y_o, y_o + b)$. It is easy to show that the existence condition (9.25) can be satisfied only if $y_o = 0$ so that the problem reduces to the symmetric (about $y = 0$) case investigated in [1].

Hence, for this class of symmetric obstacles there are no genuinely embedded trapped modes for $kd \in (0, \pi/2)$.

2 If the obstacle does not have a line of symmetry parallel to the x axis the existence condition can be satisfied for many shape functions. For example, a triangular obstacle is described by

$$x = \epsilon f(y) \equiv \frac{a}{b}(y - y_\mathrm{o} - b)[H(y - y_\mathrm{o} - b) - H(y - y_\mathrm{o})], \quad y \in (-d, d),$$
(9.27)

where H denotes the Heaviside step function and the triangle has height b and the base is at $y = y_\mathrm{o}$. This shape does not satisfy the slender-body assumption but is used here as the results are readily calculated. In this case, the existence condition reduces to

$$\tan \frac{\pi y_\mathrm{o}}{2d} = \frac{\sin(\pi b/2d) - \pi b/2d}{2\sin^2(\pi b/4d)}$$
(9.28)

and for a sufficiently small $b/d > 0$ there is one solution for $y_\mathrm{o} \in (-d, 0)$ so that the obstacle straddles $y = 0$ (if b/d is too large the obstacle extends outside the guide). As $b/d \to 0$, $y_\mathrm{o} \sim -b/3$ and $\kappa_1 d \sim \pi^2 ab/8d^2 > 0$ as required for trapped modes. Numerical computations give qualitative support for these conclusions.

4.   NEUMANN GUIDE: SYMMETRIC OBSTACLE

As in section 3, Neumann conditions are imposed on the guide walls but the obstacle now has symmetry about the centre line. If symmetry of the obstacle is not imposed then for $kd > \pi/2$ there are at least two propagating modes and the formulation of section 2 does not apply. When $kd < \pi/2$ this is a special case of that considered in section 3. Here trapped-mode solutions are sought for $kd \in (\pi/2, 3\pi/2)$ and obstacles that are symmetric, and motions that are antisymmetric, about $y = 0$ so that

$$\phi = 0 \quad \text{on} \quad y = 0, \quad x > 0.$$
(9.29)

The eigenfunctions for this problem are

$$\Psi_n(y) = \sin\left[\frac{(2n + 1)\pi y}{2d}\right], \quad n = 0, 1, \ldots,$$
(9.30)

and

$$\kappa_n = [((2n + 1)\pi/2d)^2 - k^2]^{1/2}.$$
(9.31)

For $kd \in (\pi/2, 3\pi/2)$ there is one propagating mode corresponding to $n = 0$ and so again the formulation of section 2 applies. The condition for the existence of trapped modes is

$$\int_0^d f'(y) \cos \frac{3\pi y}{2d} \sin \frac{\pi y}{2d} \, dy = 0 \qquad (9.32)$$

and the trapped-mode frequency follows from

$$\kappa_1 d = -\frac{3\pi\epsilon}{2d} \int_0^d f'(y) \sin \frac{3\pi y}{d} \, dy. \qquad (9.33)$$

In both of the last equations the symmetry has been used to reduce the range of integration to $(0, d)$. For the limit $\epsilon \to 0$ and in the vicinity of the obstacle, trapped-mode solutions are perturbations of the standing wave that can exist in the guide for $kd = 3\pi/2$.

APPLICATIONS

1 The shape-function

$$x = \epsilon f(y) \equiv \frac{a}{b} \left(b - |y|\right) H\left(b - |y|\right), \quad y \in (-d, d), \quad b \le d, \quad (9.34)$$

describes a 'diamond' with total width $2a$ in the x direction and $2b$ in the y direction. From (9.21) trapped modes exist whenever $b/d = 1/2$ and then from (9.22) $\kappa_1 d = a/2b$. The value $b/d = 1/2$ corresponds to the limit as $a/d \to 0$ of a family of trapped modes for the shapes (9.34). In general, for a given a/d trapped modes exist only for discrete b/d when $kd \in (\pi/2, 3\pi/2)$.

2 The shape-function

$$x = \epsilon f(y) \equiv \frac{a}{b} \left(b^2 - y^2\right)^{1/2} H\left(b - |y|\right), \quad y \in (-d, d), \quad b \le d$$

describes an ellipse of axis length $2a$ in the x direction and $2b$ in the y direction. This ellipse violates the assumption that $n_y = O(\epsilon)$ as $\epsilon \to 0$ when $1 - y/b = O(\epsilon^2)$ which would require the inclusion of additional terms in (9.15). However, the corresponding additional contributions to the integrals in (9.21)–(9.22) are negligible in the limit $\epsilon \to 0$ and so the slender-obstacle theory may still be applied.

The existence condition (9.21) reduces to

$$J_1(2\pi b/d) - J_1(\pi b/d) = 0 \qquad (9.35)$$

where J_1 is a Bessel function. There is one root $b/d = B \approx 0.392$ for $b/d \in (0, 1]$ and then from (9.22) $\kappa_1 d = 3\pi^2 a J_1(3\pi B)/4d > 0$. Numerical computations for the full problem support the hypothesis that the value $b/d = B$ is the limit as $a/d \to 0$ of a family of trapped modes for ellipses of varying aspect ratio b/a.

5.   DIRICHLET GUIDE

In a Dirichlet guide the potential is zero on $y = \pm d$ and so the eigenfunctions are

$$\Psi_n(y) = \sin\left[\frac{(n+1)\pi}{2d}(y+d)\right], \quad n = 0, 1, \ldots, \tag{9.36}$$

and

$$\kappa_n = \left[((n+1)\pi/2d)^2 - k^2\right]^{1/2}. \tag{9.37}$$

For $kd \in (\pi/2, \pi)$ only the $n = 0$ mode propagates and the formulation of section 2 applies. Geometrical symmetry need not be imposed because, unlike the Neumann guide, there is no propagating mode for $kd < \pi/2$. The condition for the existence of trapped modes is

$$\int_{-d}^{d} f'(y) \cos\frac{\pi y}{d} \cos\frac{\pi y}{2d}\, dy = 0 \tag{9.38}$$

and the trapped-mode frequency follows from

$$\kappa_1 d = -\frac{\pi\epsilon}{2d}\int_{-d}^{d} f'(y)\sin\frac{2\pi y}{d}\, dy. \tag{9.39}$$

As $\epsilon \to 0$ and in the vicinity of the obstacle, trapped-mode solutions are perturbations of the standing wave that may exist for $kd = \pi$.

APPLICATION

- For the triangle given by (9.27) the existence condition (9.38) has real solutions such that as $b/d \to 0$ then $y_o \sim -b/3$. It then follows from (9.39) that $\kappa_1 d \sim \pi^2 ab/2d^2 > 0$ as required for trapped modes. This type of mode seems to be new and no numerical computations have yet been performed to support its existence.

References

[1] Evans, D V, Levitan, M and Vassiliev, D (1994) *Existence theorems for trapped modes*, J. Fluid Mech. **261**, 21–31.

[2] Evans, D V and McIver, P (1991) *Trapped modes over symmetric thin bodies*, J. Fluid Mech. **223**, 509–519.

TRAPPED MODES ABOUT TUBE BUNDLES IN WAVEGUIDES

R. Porter, D. V. Evans

School of Mathematics

University of Bristol, Bristol BS8 1TW, UK

richard.porter@bris.ac.uk, D.V.Evans@bristol.ac.uk

1.   INTRODUCTION

A possible design for a heat exchanger involves packing a rectangular array of rigid hollow cylinders in a parallel-walled duct; the transfer of heat conducted from a fluid flowing in the cylinders to a fluid flowing through the duct. It is important to identify under what conditions the configuration may be subject to resonant acoustic modes which may be generated by various means so that design parameters can be chosen to avoid this undesirable effect. These resonant modes of oscillation are associated with the occurrence of so-called trapped modes in waveguides. Trapped modes are characterised by a persistent oscillation in the vicinity of a geometric structure which does not radiate energy away to infinity. They therefore also represent a non-uniqueness in a corresponding forcing problem.

The existence of acoustic trapped modes in parallel-walled waveguides is well-established. For example, Evans *et al.* [1], proved that trapped modes exist for a large class of cylinder cross-sections which are symmetric about the centre of the guide. The modes, which are antisymmetric about the centreline of the guide, are shown to exist for frequencies below the first cut-off frequency for the guide.

The purpose of the present paper is to investigate the trapped modes occurring in a rectangular array of $N \times M$ circular cylinders. An approximate method, based on large numbers of cylinders and applying the homogenisation theory described by Blevins [2] is compared with results computed using an exact formulation of the problem. This latter approach uses multipole expansions for the potential about each cylinder as its basis and combines the methods of Evans and Porter [3] who investigated the trapped modes about an array of N cylinders all placed on the centreline of a waveguide and Porter and Evans [4] using Rayleigh-

I.D. Abrahams et al. (eds.),

IUTAM Symposium on Diffraction and Scattering in Fluid Mechanics and Elasticity, 87–94.

© 2002 *Kluwer Academic Publishers. Printed in the Netherlands.*

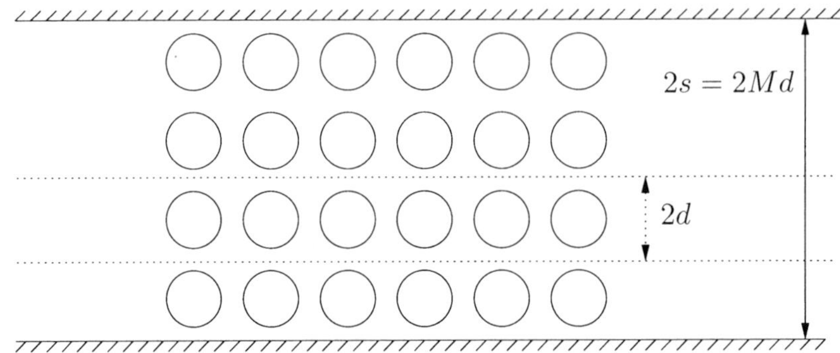

Figure 1   Rectangular array of $N \times M$ cylinders in a waveguide, width $2s = 2Md$.

Bloch theory to obtain results for a single row of M cylinders spanning a waveguide.

2.   FORMULATION

A cross-section through the geometry is shown in Fig. 1. Situated in a parallel hard-walled waveguide of width $2s$ is a rectangular periodic array of $N \times M$ identical cylinders of radius a separated from neighbouring cylinders by $2b(2d)$ in the direction parallel (perpendicular) to the guide walls whilst the walls are a distance d from the centre of the nearest cylinders. By taking images in the walls, the configuration of $N \times M$ cylinders may be regarded as being part of an array of N infinite rows of cylinders. Thus, rather than seeking the solution to the problem of $N \times M$ array of cylinders in the guide directly, the approach we adopt here involves seeking the most general trapped mode solution for N rows of cylinders – namely Rayleigh-Bloch waves – and through particular choices of a parameter, β (to be introduced), we reconstruct the solution for the $N \times M$ array in the guide. This was the approach adopted for a single row of cylinders by Porter and Evans [4].

Moreover, the problem of wave trapping by N rows of infinite periodic cylinders can be reduced to a problem in a single strip of the domain of width $2d$ containing one period of the array by placing appropriate boundary conditions on the sides of the strip. Thus, we consider the strip $(x, y) \in S_0 = \{-\infty < x < \infty, -d \le y \le d\}$ consisting of N cylinders placed on the centreline at $(x_j, 0)$, $x_j = 2jb$, $j = 1, \ldots, N$. Local polar coordinates, (r_j, θ_j), are used about the jth cylinder, such that $x = x_j + r_j \cos \theta_j$, $y = r_j \sin \theta_j$.

Potential theory is used to describe the flow. Suppressing a time harmonic factor of $e^{i\omega t}$, where $\omega/2\pi$ is the frequency, from the potential, then a reduced complex potential $\phi(x, y)$ describing Rayleigh-Bloch waves satisfies:

$$(\nabla^2 + k^2)\phi(x, y) = 0, \qquad \text{in the fluid}, \tag{10.1}$$

where $k = \omega/c$, c is the speed of sound;

$$(\partial/\partial r_j)\phi(r_j, \theta_j) = 0, \qquad \text{on } r_j = a, \; j = 1, \ldots, N, \tag{10.2}$$

being the boundary condition on acoustically-hard cylinders;

$$\phi(x, y) \to 0, \qquad \text{as } |x| \to \infty, \tag{10.3}$$

the condition that trapped waves do not radiate energy to infinity; and the periodicity conditions on the sides of the strip $y = \pm d$,

$$\phi(x, d) = e^{2i\beta d}\phi(x, -d), \qquad \phi_y(x, d) = e^{2i\beta d}\phi_y(x, -d). \tag{10.4}$$

The potential in the field outside the strip S_0 can be determined from

$$\phi(x, y + 2md) = e^{2im\beta d}\phi(x, y), \qquad m \in \mathbb{Z}. \tag{10.5}$$

Here, β is known as the dominant Rayleigh-Bloch wavenumber. In seeking a Rayleigh-Bloch surface wave it is usual to confine attention to values of wavenumber k given by

$$0 < k < \beta \le \pi/2d. \tag{10.6}$$

This condition on the wavenumber k ensures that no wave radiation to infinity is possible and (10.3) is satisfied. Also, $k(\pi/d - \beta) = k(\beta) = k(n\pi/d + \beta)$ for $n \in \mathbb{Z}$ (see [4]).

The formulation has much in common with the one described in [3] in which multipole potentials are constructed satisfying Neumann conditions on $y = \pm d$ and a Dirichlet condition on the centreline. Here, multipole potentials are designed instead to satisfy (10.4), a process which is described in detail in [5]. However, unlike in [5], by taking k in the range defined by (10.6) each of the multipoles satisfies (10.3). Thus, the general form of the solution can be written

$$\phi(x, y) = \sum_{k=1}^{N} \sum_{n=0}^{\infty} \left\{ a_n^k \phi_n^k(r_k, \theta_k) + i b_n^k \psi_n^k(r_k, \theta_k) \right\} \tag{10.7}$$

where a_n^k and b_n^k are coefficients to be determined and we tacitly assume that $b_0^k = 0$ throughout. See (10.A.1) and (10.A.2) in the Appendix for the definition of the multipole potentials ϕ_n^k and ψ_n^k in terms of the local polar coordinate system which can then be used to write

$$\phi(r_j, \theta_j) = \sum_{n=0}^{\infty} \left\{ Y_n(kr_j)\{ia_n^j \cos n\theta_j - b_n^j \sin n\theta_j\} \right.$$

$$+ \sum_{m=0}^{\infty} J_m(kr_j)\{a_n^j(iC_{mn}^{jj} \cos m\theta_j - D_{mn}^{jj} \sin m\theta_j)$$

$$\left. + b_n^j(iS_{mn}^{jj} \cos m\theta_j - T_{mn}^{jj} \sin m\theta_j)\} \right\}$$

$$+ \sum_{\substack{k=1 \\ k \neq j}}^{N} \sum_{n=0}^{\infty} \sum_{m=0}^{\infty} J_m(kr_j)\{a_n^k(i(C_{mn}^{jk} + E_{mn}^{jk}) \cos m\theta_j - D_{mn}^{jk} \sin m\theta_j)$$

$$+ b_n^k(iS_{mn}^{jk} \cos m\theta_j - (T_{mn}^{jk} + U_{mn}^{jk}) \sin m\theta_j)\}. \qquad (10.8)$$

in terms of (r_j, θ_j) only and where C_{mn}^{jk}, T_{mn}^{jk}, D_{mn}^{jk}, E_{mn}^{jk}, S_{mn}^{jk}, U_{mn}^{jk} are all given in the Appendix. Applying the no-flow condition, (10.2), and equating coefficients of $\cos m\theta_j$ and $\sin m\theta_j$, $m = 0, 1, \ldots$ for $j = 1, \ldots, N$ results in the *real* coupled infinite systems of linear equations:

$$a_m^j + \frac{J_m'(ka)}{Y_m'(ka)} \sum_{n=0}^{\infty} \left\{ (a_n^j C_{mn}^{jj} + b_n^j S_{mn}^{jj}) + \sum_{\substack{k=1 \\ k \neq j}}^{N} (a_n^k(C_{mn}^{jk} + E_{mn}^{jk}) + b_n^k S_{mn}^{jk}) \right\} = 0,$$

$$(10.9)$$

and

$$b_m^j + \frac{J_m'(ka)}{Y_m'(ka)} \sum_{n=0}^{\infty} \left\{ (a_n^j D_{mn}^{jj} + b_n^j T_{mn}^{jj}) + \sum_{\substack{k=1 \\ k \neq j}}^{N} (a_n^k D_{mn}^{jk} + b_n^k(T_{mn}^{jk} + U_{mn}^{jk})) \right\} = 0,$$

$$(10.10)$$

for $m = 0, 1, \ldots$, and $j = 1, \ldots, N$. Rayleigh-Bloch waves correspond to values of $k(\beta)$ that correspond to the vanishing of the real infinite determinant of the coupled systems above.

3.   HOMOGENISATION THEORY

Here we give a brief account of the homogenisation theory developed by Blevins [2] as an approximation to the localised modes of oscillation for a large number of cylinders. See [2] for further details.

The averaged effect of a large array of cylinders occupying a rectangular region of the waveguide is to reduce the speed of sound in that region to a value, c_1, say. More formally, each cylinder of cross-section area v occupies its own cell, of area V, defining a solidity factor, $\sigma = v/V$.

Then each cylinder may be replaced by a 'virtual cylinder' of fluid of the same density as in the surrounding cell. Certain forces and fluxes can then be chosen to make this virtual cylinder fixed and rigid. This process introduces added inertia coefficients a_x, a_y associated with motions longitudinal and transverse to the guide respectively. The reduced velocity potential, ϕ, is shown to satisfy

$$\frac{\phi_{xx}}{1 + \sigma a_x} + \frac{\phi_{yy}}{1 + \sigma a_y} + k^2 \phi = 0, \qquad \text{in the region occupied by cylinders.}$$

Experimental work of Blevins [2] suggests that values of $a_x = a_y = 1$ are good approximations in this continuum model, whence

$$(\nabla^2 + k_1^2)\phi = 0, \qquad \text{in } |x| \le Nb,$$

where $k_1 = (1 + \sigma)^{\frac{1}{2}} k = \omega/c_1$. Exterior to the region occupied by the cylinders, $|x| \ge Nb$, (10.1) is satisfied by ϕ, whilst $\phi_y = 0$ on $y = \pm s = \pm Md$ are the wall conditions and, finally, ϕ must satisfy (10.3) for trapped modes. It is well-known that trapped modes are present in a guide having a rectangular region of increased wavenumber since it is equivalent to the case of trapping over a long elevated rectangular shelf in water waves. The method of solution for this model is simple involving eigenfunction expansions in the two domains. For example, modes symmetric about the centreline of the guide ($y = 0$) and through the line of geometric symmetry ($x = 0$) are described by

$$\phi_n^s = \begin{cases} A_n^s \cos(l_n y) \cos \alpha_n x, & |x| \le Nb, \\ B_n^s \cos(l_n y) e^{-\gamma_n (|x| - Nb)}, & |x| \ge Nb, \end{cases}$$

where $l_n = n\pi/Md$, $\alpha_n = (k_1^2 - l_n^2)^{\frac{1}{2}}$, $\gamma = (l_n^2 - k^2)^{\frac{1}{2}}$, for $n = 1, 2, \ldots$. For each value of n, trapped modes can be found for k such that $k < l_n < k_1 = k(1 + \sigma)^{\frac{1}{2}}$. Matching ϕ and ϕ_x across $x = Nb$, gives the condition for trapped modes in this case as

$$\alpha_n \tan(\alpha_n Nb) = \gamma_n,$$

whilst for solutions antisymmetric about $x = 0$ the condition is

$$-\alpha_n \cot(\alpha_n Nb) = \gamma_n.$$

A further set of modes must also be considered, corresponding to antisymmetric motions about $y = 0$. It is a routine matter to determine these trapped modes numerically. For circular cylinders, the solidity factor is $\pi a^2/4bd$.

4.   RESULTS

In section 2, the problem of trapped modes was formulated for N infinite rows of circular cylinders of radius a and period $2d$ in terms of a Rayleigh-Bloch wavenumber, β. Trapped modes for an $N \times M$ array of cylinders in a guide of width $2Md$ as sketched in Fig. 1 with Neumann conditions on the walls can be reconstructed by choosing particular values of β, a process which is described in detail in [4]. Thus, trapped modes in a waveguide are given by real values of k determined by seeking non-trivial solution to the coupled system (10.9), (10.10) for values of $\beta_n = n\pi/2Md$, $n = 1, \ldots, M$. Results for a 12×12 array of equally-spaced cylinders ($b = d$) are shown in Fig. 2. Curves showing the variation of non-dimensional wavenumber, kd, are plotted against varying cylinder size. Thus, for sufficiently small cylinders ($a/d \ll 1$) there are only M modes present, approximated by $k \approx \beta_n$, $n = 1, \ldots, M$. As the cylinder size increased, further modes appear through the cut-off wavenumber $k = \beta_n$, $n = 1, \ldots, M$.

Results computed for the same configuration of cylinders from the continuum model of section 3 are depicted in Fig. 3. They show similar behaviour, although closer inspection and comparison with Fig. 2 reveals that the results are only really accurate in the region $ka \lesssim 1$. These observations for a 12×12 array are also confirmed for other sizes of array. One of the assumptions used in the homogenisation theory was that the cylinders are widely-spaced compared to the wavelength, which justifies the use of $a_x = a_y = 1$ as added inertia coefficients. It is believed that improvements to the continuum model can be made by incorporating the effects of closely-spaced cylinders upon the added inertia coefficients.

Appendix: Multipole expansions

The details involved in deriving multipole potentials may be found in [5]. Here we simply give expressions for factors that eventually arise from the use of multipole potentials. In terms of local polar coordinates, the multipoles can be expressed as

$$\phi_n^k(r_k, \theta_k) = iY_n(kr_k)\cos n\theta_k + \sum_{m=0}^{\infty}\{iC_{mn}^{jk}\cos m\theta_j - D_{mn}^{jk}\sin m\theta_j\}J_m(kr_j)$$

$$(10.\mathrm{A}.1)$$

$$\psi_n^k(r_k, \theta_k) = iY_n(kr_k)\sin n\theta_k + \sum_{m=0}^{\infty}\{S_{mn}^{jk}\cos m\theta_j + iT_{mn}^{jk}\sin m\theta_j\}J_m(kr_j)$$

$$(10.\mathrm{A}.2)$$

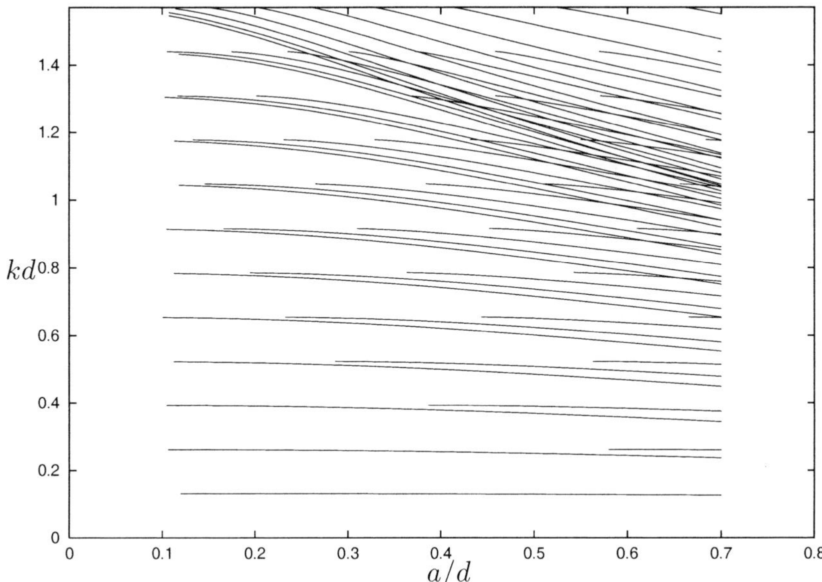

Figure 2   Trapped mode wavenumber ka variation with cylinder size a/d for a 12×12 array as given by the exact theory.

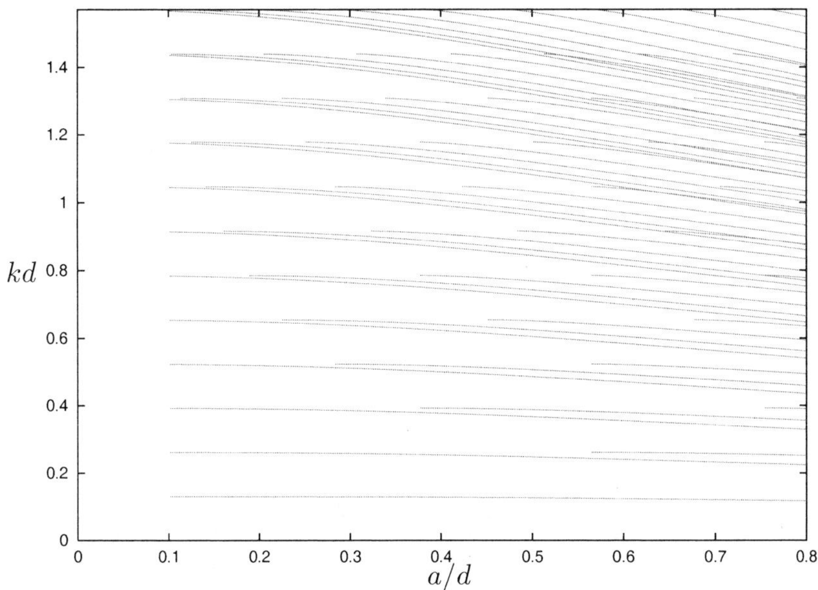

Figure 3   Trapped mode wavenumber ka variation with cylinder size a/d for a 12×12 array as predicted by the continuum model of section 3.

for $n = 1, \dots$ and $j \neq k$, provided $r_j < 2d$. Here,

$$C_{mn}^{jk} = -\frac{2\varepsilon_m}{\pi}\Re\int_0^\infty [\cos 2\beta d - e^{-2k\gamma d}]\frac{\cos\Phi}{\Delta}c_n(t)\,c_m(t)\,dt,$$

$$D_{mn}^{jk} = -\frac{4}{\pi}\int_0^\infty \sin 2\beta d\,\frac{\sin\Phi}{\Delta}c_n(t)\,s_m(t)\,dt,$$

$$S_{mn}^{jk} = \frac{2\varepsilon_m}{\pi}\int_0^\infty \sin 2\beta d\,\frac{\sin\Phi}{\Delta}s_n(t)\,c_m(t)\,dt,$$

$$T_{mn}^{jk} = \frac{4}{\pi}\Re\int_0^\infty [\cos 2\beta d - e^{-2k\gamma d}]\frac{\cos\Phi}{\Delta}s_n(t)\,s_m(t)\,dt,$$

$\Phi = k(x_j - x_k)t + (m-n)\pi/2$, $\Delta = \gamma(t)\{\cosh 2k\gamma(t)d - \cos 2\beta d\}$
$\gamma(t) = -i(1-t^2)^{1/2}$ for $t \leq 1$, $\gamma(t) = (t^2-1)^{1/2}$ for $t > 1$,

$$c_n(t) = \cos[n(\sin^{-1}t - \tfrac{1}{2}\pi)], \quad s_n(t) = i\sin[n(\sin^{-1}t - \tfrac{1}{2}\pi)] \quad \text{for } t \leq 1,$$

$$c_n(t) = \cosh[n\cosh^{-1}t], \quad s_n(t) = \sinh[n\cosh^{-1}t] \quad \text{for } t > 1.$$

Also, $\varepsilon_m = 2$, $m \geq 1$ and $\varepsilon_0 = 1$. Finally, terms emerge that are derived from applying Graf's addition theorem to express the first terms in (10.A.1), (10.A.2) in terms of (r_j, θ_j) for $j \neq k$:

$$E_{mn}^{jk} = \tfrac{1}{2}\varepsilon_m\Big[Y_{n-m}(k|x_j-x_k|)+(-1)^mY_{n+m}(k|x_j-x_k|)\Big]\mathrm{sgn}\,(x_j-x_k)^{n+m},$$

$$U_{mn}^{jk} = \Big[Y_{n-m}(k|x_j-x_k|) - (-1)^mY_{n+m}(k|x_j-x_k|)\Big]\mathrm{sgn}\,(x_j-x_k)^{n+m},$$

for $j \neq k$. Notice that C_{mn}^{jk}, T_{mn}^{jk}, D_{mn}^{jk}, E_{mn}^{jk}, S_{mn}^{jk}, U_{mn}^{jk} are all real factors. They also possess various symmetry relations between the elements that can be used to increase numerical efficiency.

References

[1] Evans, D V, Levitin, M and Vassiliev, D (1994) *Existence theorems for trapped modes*, J. Fluid Mech. **261**, 21–31.

[2] Blevins, R D (1986) *Acoustic modes of heat exchanger tube bundles*, J. Sound & Vib. **109**, 19–31.

[3] Evans, D V and Porter, R (1997) *Trapped modes about multiple cylinders in a channel*, J. Fluid Mech. **339**, 331–356.

[4] Porter, R and Evans, D V (1999) *Rayleigh-Bloch surface waves along periodic gratings and their connection with trapped modes in channels*, J. Fluid Mech. **386**, 233–258.

[5] Linton, C M and Evans, D V (1993) *The interaction of waves with a row of circular cylinders*, J. Fluid Mech. **251**, 687–708.

PHONONIC BAND STRUCTURES FOR ARRAYS OF CIRCULAR CAVITIES IN AN ELASTIC MEDIUM

V. V. Zalipaev, A. B. Movchan, C. G. Poulton
Department of Mathematical Sciences
University of Liverpool, Liverpool L69 3BX, UK
abm@liv.ac.uk

R. C. McPhedran
School of Physics, University of Sydney
Sydney 2006, Australia
r.mcphedran@physics.usyd.edu.au

Abstract   We study the propagation of elastic waves in a two-dimensional solid containing a doubly periodic array of circular holes. The method of multipoles expansions is employed here, so that it takes into account a coupling between shear and dilatational waves (this coupling occurs via the traction boundary conditions). As a result, we obtain an infinite system of linear algebraic equations, which can be truncated and solved numerically in order to determine the frequencies of the propagating modes. The algorithm has been implemented as a computer code used to construct the dispersion diagrams and analyse the filtering properties of the composite structure.

1.   INTRODUCTION

We begin with the reference of the classical background ideas established in 1892 when Rayleigh published his paper [1] on the multipoles expansion method applied to a scalar two-dimensional problem for a doubly periodic array of circular holes. For two-dimensional static problems of linear elasticity the multipoles expansion method was further developed by McPhedran and Movchan [2]. In this case, the mathematical formulation involved a vector field of displacement whose components are linked via the boundary conditions. In a recent paper [3], a generalisation of the original Rayleigh idea was developed for models of elastic

95

I.D. Abrahams et al. (eds.),
IUTAM Symposium on Diffraction and Scattering in Fluid Mechanics and Elasticity, 95–104.
© 2002 *Kluwer Academic Publishers. Printed in the Netherlands.*

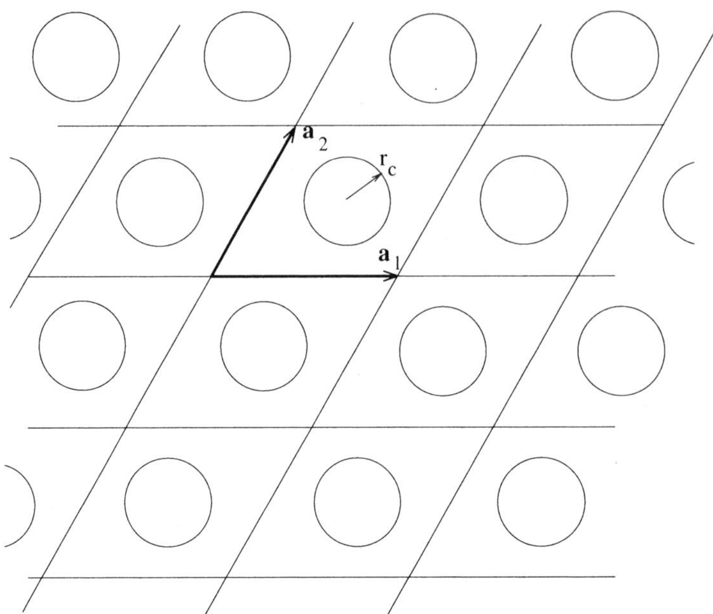

Figure 1   A parallelogram array of circular holes.

wave propagation through a two-dimensional square array of circular holes.

The basic ideas of our algorithm are similar to those employed in the method of Korringa, Kohn and Rostoker traditionally used for scalar equations of quantum mechanics in periodic structures. The advance into the vector problems of elasticity is important. In this short paper we give an outline of the model describing propagation of waves in an elastic medium containing a hexagonal array of circular holes. For this type of periodic structure, we show the presence of phononic band gap for the anti-plane problem and evaluate the intervals of vibration frequency for which all propagating modes in the material are suppressed. In addition we give examples of phononic band diagrams for the in-plane problem. Finally, we derive explicit analytical formulae for the asymptotics of eigen-frequencies associated with arrays of slightly non-circular cavities.

2.   GOVERNING EQUATIONS

We study the problem of in-plane propagation of time-harmonic waves within an isotropic elastic medium, which contains a set of infinitely long circular cylindrical cavities of radius r_c directed along the z-axis (Fig. 1). The propagation of waves in the elastic medium outside the cavities is described by the Navier vector equation

$$\mu \triangle \mathbf{u}(\mathbf{r}) + (\lambda + \mu)\nabla(\nabla \cdot \mathbf{u}(\mathbf{r})) + \rho\omega^2\mathbf{u}(\mathbf{r}) = \mathbf{0}, \qquad (11.1)$$

where the material density is ρ, and the Lamé coefficients are λ and μ. The displacement $\mathbf{u}$ does not depend on z, and it satisfies the traction boundary condition on the surface of the cavities:

$$\sigma_{ij}(\mathbf{u})n_j = 0 \quad \text{on } r = r_c.$$

As the problem under consideration is two-dimensional, that is, the displacement vector $\mathbf{u}$ depends only on (x, y), (11.1) can be decoupled into an equation for anti-plane shear (Helmholtz equation) and the equations for the plane strain problem. For the u_z component we have

$$(\triangle + k_b^2)u_z = 0, \quad (\partial u_z/\partial r)|_{r=r_c} = 0, \qquad (11.2)$$

where $k_b = \omega/v_b$ and $v_b^2 = \mu/\rho$. The components u_r and u_θ satisfy a coupled in-plane vector problem.

We introduce the well-known scalar and vector potentials for dilatational and shear waves

$$\mathbf{u}(\mathbf{r}) = \nabla\varphi + \nabla \times \boldsymbol{\psi}, \qquad (11.3)$$

where $\boldsymbol{\psi} = (0, 0, \psi)$,

$$(\triangle + k_a^2)\varphi(x, y) = 0, \quad (\triangle + k_b^2)\psi(x, y) = 0,$$

$k_a = \omega/v_a$ and $v_a^2 = (\lambda + 2\mu)/\rho$. Then the boundary conditions for the elastic potentials can be written in the form:

$$\left[-\frac{2}{r_c^2}\frac{\partial\varphi}{\partial\theta} + \frac{2}{r_c}\frac{\partial^2\varphi}{\partial\theta\partial r} + k_b^2\psi + \frac{2}{r_c}\frac{\partial\psi}{\partial r} + \frac{2}{r_c}\frac{\partial^2\psi}{\partial\theta^2} \right]\Bigg|_{r=r_c} = 0,$$

$$\left[2\mu\left(\frac{1}{r_c}\frac{\partial^2\psi}{\partial\theta\partial r} - \frac{1}{r_c^2}\frac{\partial\psi}{\partial\theta} - \frac{1}{r_c}\frac{\partial\varphi}{\partial r} - \frac{1}{r_c^2}\frac{\partial^2\varphi}{\partial\theta^2} \right) - k_a^2(2\mu + \lambda)\varphi \right]\Bigg|_{r=r_c} = 0.$$

$$(11.4)$$

Unlike the anti-plane problem, these two equations lead to a coupling between shear and dilatational waves.

Finally, we assume that $\mathbf{u}(\mathbf{r})$ satisfies the following quasi-periodicity condition

$$\mathbf{u}(\mathbf{r} + \mathbf{R}_p) = \mathbf{u}(\mathbf{r})\, e^{i\mathbf{k}_0 \cdot \mathbf{R}_p}, \qquad (11.5)$$

where $\mathbf{R}_p = p_1\mathbf{a}_1 + p_2\mathbf{a}_2$. Here $\mathbf{a}_1, \mathbf{a}_2$ are the fundamental translation vectors of the parallelogram array, $\mathbf{k}_0$ is the Bloch wave-vector, and the multi-index $p = (p_1, p_2)$ has integer components $p_{1,2}$. Our aim is to construct the dispersion equations (that relate ω and $\mathbf{k}_0$), thence to obtain dispersion diagrams and so to analyse phononic band gaps which may exist within the material.

3.   METHOD OF ANALYSIS

According to [3], inside every cell of the array the solution for the u_z component of the displacement and elastic potentials for the in-plane problem can be expanded in terms of multipoles

$$u_z(\mathbf{r}) \;=\; \sum_{l=-\infty}^{\infty} \left[A_l J_l(k_b r) + B_l Y_l(k_b r) \right] e^{il\theta},$$

$$\varphi(\mathbf{r}) \;=\; \sum_{l=-\infty}^{\infty} \left[A_l^{(a)} J_l(k_a r) + B_l^{(a)} Y_l(k_a r) \right] e^{il\theta},$$

$$\psi(\mathbf{r}) \;=\; \sum_{l=-\infty}^{\infty} \left[A_l^{(b)} J_l(k_b r) + B_l^{(b)} Y_l(k_b r) \right] e^{il\theta}.$$

Using the boundary conditions (11.2) and (11.4), we deduce

$$A_l = -M_l B_l, \quad M_l = \left[Y_l'(k_b r_c) \right] / J_l'(k_b r_c),$$

$$\begin{pmatrix} A_l^{(a)} \\ A_l^{(b)} \end{pmatrix} = \begin{pmatrix} M_l^{(aa)} & M_l^{(ab)} \\ M_l^{(ba)} & M_l^{(bb)} \end{pmatrix} \begin{pmatrix} B_l^{(a)} \\ B_l^{(b)} \end{pmatrix},$$

and the expressions for the coefficients $M_l^{(aa)}$, $M_l^{(ab)}$, $M_l^{(ba)}$ and $M_l^{(bb)}$ are given in [3]. Application of the generalised Rayleigh method leads to the following system of algebraic equations

$$z_l + \sum_{m=-\infty}^{\infty} (-1)^{l+m} \mathrm{sgn}(M_l) \frac{S_{m-l}^{Y}(k_b, \mathbf{k}_0)}{\sqrt{|M_l M_m|}} z_m = 0; \qquad (11.6)$$

$$x_l + \sum_{m=-\infty}^{\infty} \left(D_{lm}^{(aa)} x_m + D_{lm}^{(ab)} y_m \right) = 0,$$

$$y_l + \sum_{m=-\infty}^{\infty} \left(D_{lm}^{(ba)} x_m + D_{lm}^{(bb)} y_m \right) = 0, \qquad (11.7)$$

for anti-plane and in-plane problems, correspondingly [3]. Here $z_l = B_l \sqrt{|M_l|}$,

$$D_{lm}^{(aa)} \;=\; -\frac{\mathrm{sgn}(M_l^{(aa)})}{\Delta_m} \left| \frac{M_m^{(bb)}}{M_l^{(aa)}} \right|^{1/2} (-1)^{l+m} S_{m-l}^{Y}(k_a, \mathbf{k}_0),$$

$$D_{lm}^{(ab)} \;=\; \frac{\mathrm{sgn}(M_l^{(aa)}) \mathrm{sgn}(M_m^{(aa)}) M_m^{(ab)}}{|M_l^{(aa)} M_m^{(aa)}|^{1/2} \Delta_m} (-1)^{l+m} S_{m-l}^{Y}(k_a, \mathbf{k}_0),$$

$$D_{lm}^{(ba)} = \frac{\text{sgn}(M_l^{(bb)})\text{sgn}(M_m^{(bb)})M_m^{(ba)}}{|M_l^{(bb)}M_m^{(bb)}|^{1/2}\Delta_m}(-1)^{l+m}S_{m-l}^Y(k_b,\mathbf{k}_0),$$

$$D_{lm}^{(bb)} = -\frac{\text{sgn}(M_l^{(bb)})}{\Delta_m}\left|\frac{M_m^{(aa)}}{M_l^{(bb)}}\right|^{1/2}(-1)^{l+m}S_{m-l}^Y(k_b,\mathbf{k}_0),$$

$$\Delta_m = |M_m^{(aa)}M_m^{(bb)}|^{1/2}\left(1-\frac{M_m^{(ab)}M_m^{(ba)}}{M_m^{(aa)}M_m^{(bb)}}\right),$$

$$x_l = \sqrt{|M_l^{(aa)}|}\left(B_l^{(a)}+\frac{M_l^{(ab)}}{M_l^{(aa)}}B_l^{(b)}\right),$$

$$y_l = \sqrt{|M_l^{(bb)}|}\left(B_l^{(b)}+\frac{M_l^{(ba)}}{M_l^{(bb)}}B_l^{(a)}\right).$$

In the expressions of coefficients of the systems (11.6) and (11.7), only the array sums $S_l^Y(k,\mathbf{k}_0)$ are dependent on the geometry of the array structure. Their definition is given by

$$S_l^Y(k,\mathbf{k}_0) = \sum_p{}' Y_l(kR_p)\, e^{i\Phi_p l + i\mathbf{k}_0\cdot\mathbf{R}_p}. \tag{11.8}$$

Here $\Phi_p = \arg(\mathbf{R}_p)$, and the dash indicates that $p = (p_1,p_2) \neq (0,0)$. As this series is slowly convergent, we shall use the following formula, derived in [4], to calculate the array sums:

$$\begin{aligned}
S_l^Y(k,\mathbf{k}_0)J_{l+q}(kz) &= -\delta_{l0}\left[Y_q(kz)+\frac{1}{\pi}\sum_{n=1}^{q}\frac{(q-n)!}{(n-1)!}\left(\frac{2}{kz}\right)^{q-2n+2}\right] \\
&\quad -\frac{4i^l}{A}\sum_p\left(\frac{k}{Q_p}\right)^q\frac{J_{l+q}(Q_p z)e^{il\theta_p}}{Q_p^2-k^2},
\end{aligned} \tag{11.9}$$

where $A = |\mathbf{a}_1\times\mathbf{a}_2|$ denotes the area of the unit cell. The above formula is characterised by faster convergence achieved via integration with respect to z. The integer parameter q in (11.9) gives the 'convergence acceleration index'. The reciprocal unit cell is defined by the vectors

$$\mathbf{b}_1 = 2\pi(\mathbf{a}_2\times\mathbf{e})/A, \quad \mathbf{b}_2 = 2\pi(\mathbf{e}\times\mathbf{a}_1)/A, \quad \mathbf{e} = (\mathbf{a}_1\times\mathbf{a}_2)/A,$$

and so we have the reciprocal array vectors, translated by $\mathbf{k}_0$ to satisfy (11.5)

$$\mathbf{Q}_p = p_1\mathbf{b}_1 + p_2\mathbf{b}_2 + \mathbf{k}_0, \quad \theta_p = \arg(\mathbf{Q}_p).$$

The array sums satisfy the identity

$$S^Y_{-l}(k, \mathbf{k}_0) = S^{Y\,*}_{l}(k, \mathbf{k}_0), \tag{11.10}$$

and hence it is sufficient to calculate them only for $l \geq 0$.

Numerical tests consistently demonstrate that it is acceptable to keep $q = 20$ for $l \neq 0$, and $q = 3$ for $l = 0$. Higher values for q lead to numerical instability arising in the calculation of (11.9). Using the identity (11.9), one should choose appropriately the parameter z. It must satisfy the inequality

$$0 < z < \min\{|\mathbf{a}_1|, |\mathbf{a}_2|, |\mathbf{a}_1 \pm \mathbf{a}_2|\}.$$

In the systems (11.6) and (11.7) the off-diagonal elements decay exponentially fast away from the main diagonal. Thus, to obtain the phononic band diagrams for plane waves propagating through the array of circular holes, we truncate the above systems and evaluate the determinant of the truncated matrix. Zeros of the determinant of the truncated matrix give the values k_a, k_b, for a given $\mathbf{k}_0$.

4.   PHOTONIC AND PHONONIC BAND GAPS: NUMERICAL RESULTS

Here we discuss the results of numerical calculations describing the propagation of elastic waves through arrays of circular holes of hexagonal lattices. In Fig. 2 we show the primitive cells of corresponding reciprocal lattice. The phononic band diagram represents the dispersion dependence of frequency ω versus the magnitude of the Bloch vector calculated at points of the triangular contour GMK. The triangle GMK corresponds to the irreducible region of the first Brillouin zone.

The aim of the numerical computations is to evaluate the phononic band gap states corresponding to the range of frequencies for which no elastic waves can propagate within the composite. The corresponding numerical results have been obtained for the anti-plane shear and the in-plane problem. For all calculations we use $\lambda = 2.3$, $\mu = 1$ and $\rho = 1$. We included all multipoles up to the fifth order in our calculations so that the dimensions for truncated matrix (11.6) and (11.7) are 11 and 22, respectively.

The phononic band diagrams for the hexagonal array are shown in Figs. 3 and 4 for anti-plane and in-plane problems, correspondingly. These were calculated for different values of radius of circular cavity r_c in the unit cell. The horizontal axis in these diagrams corresponds to the arc length evaluated along the trajectory GMK.

For the anti-plane shear problem in the array, Fig. 3 shows the presence of the phononic band gap appearing as we increase the radius r_c of

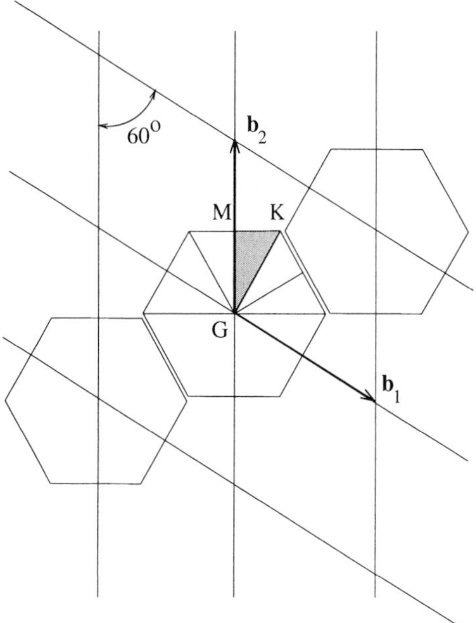

Figure 2   Reciprocal hexagonal array.

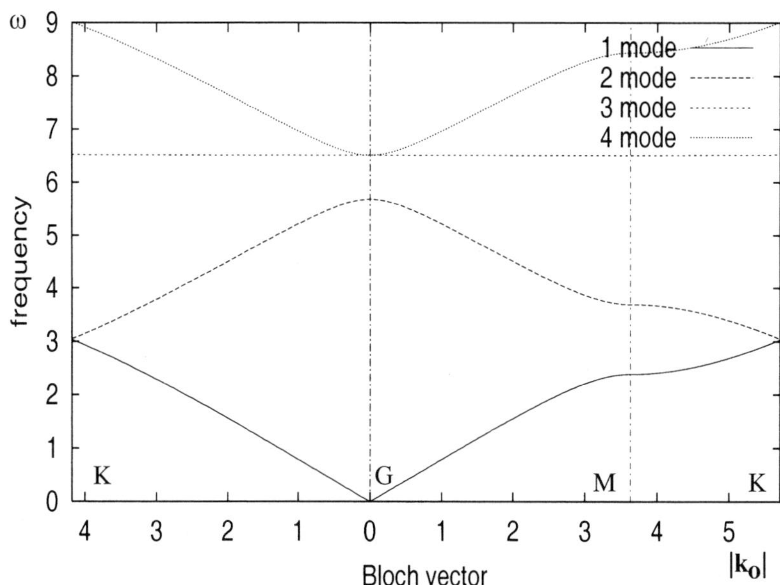

Figure 3 Dispersion diagram for anti-plane shear waves in the hexagonal array ($|\mathbf{a_1}| = |\mathbf{a_2}| = 1$ and $\chi = 60^{\circ}$) of circular holes of radius $r_c = 0.4$; the normalized shear modulus and density are $\mu = 1$ and $\rho = 1$ respectively.

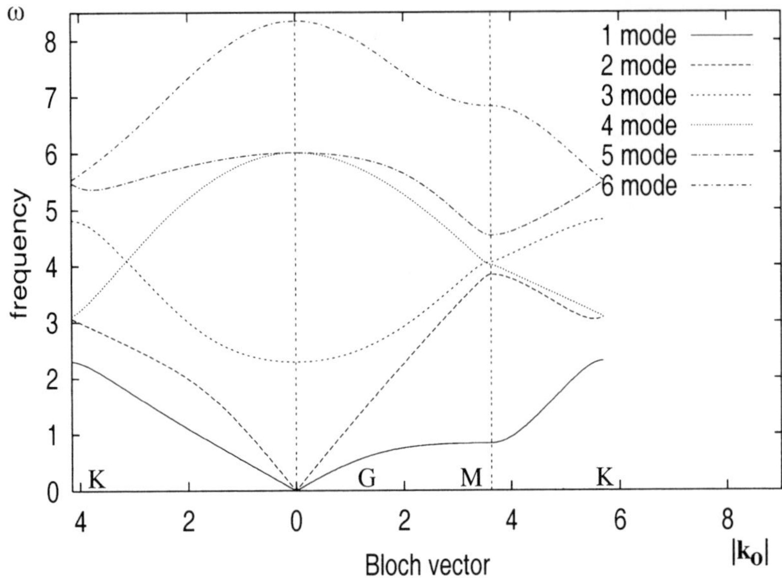

Figure 4   Dispersion diagram for in-plane dilatational and shear waves in the hexagonal array ($|\mathbf{a_1}| = |\mathbf{a_2}| = 1$ and $\chi = 60^\circ$) of circular holes of radius $r_c = 0.42$; the normalized elastic moduli and density are $\mu = 1$, $\lambda = 2.3$ and $\rho = 1$ respectively.

the holes. In contrast, the in-plane propagation of elastic waves in the hexagonal array does not exhibit band gaps for the range of frequencies calculated in the numerical tests (Fig. 4). It is noted that for the hexagonal array r_c is constrained such that $r_c < 0.5$, so that the circular holes do not overlap.

All anti-plane phononic band diagrams exhibit a single acoustic mode, represented by a curve in the neighbourhood of the point G. In a small neighbourhood of the origin, this curve is symmetric with respect to the vertical axis. This reflects the fact that the homogenised material is isotropic with respect to anti-plane shear waves. This effect has also been noted for a square array of circular cavities in [3].

In contrast with the scalar case, there are two acoustic modes in the vicinity of the point G for all phononic band diagrams associated with the in-plane elastic waves. However, for the hexagonal array the corresponding doubly-periodic elastic structure is still isotropic, with respect to the homogenization of the material.

5.   EIGENVALUES FOR AN ARRAY OF SLIGHTLY NON-CIRCULAR HOLES

It is well known that there is a limitation for the generalized Rayleigh method, in that it has been applied only to arrays of cavities or inclusions of circular shape. However, in the case of a small perturbation of the circular shape one can derive a formula for the first approximation of an eigenvalue of the relevant spectral problem. Below, we present the main points of derivation of this result for the case of the scalar anti-plane problem for the square array.

Let us consider the problem (11.2), (11.5) where for simplicity we change the spectral parameter notation for k. Let Γ be a small perturbation of the circular contour Γ_0, that is $\Gamma = \{(r, \theta) : r = r_c(1 + \epsilon f(\theta)), \theta \in [0, 2\pi]\}$ and $0 < \epsilon \ll 1$. Then, we write

$$u = u^{(0)} + \epsilon u^{(1)} + O(\epsilon^2), \quad k = k^{(0)} + \epsilon k^{(1)} + O(\epsilon^2), \qquad (11.11)$$

where u, $u^{(0)}$ and k, $k^{(0)}$ are perturbed and unperturbed eigenfunctions and eigenvalues, respectively. Both functions satisfy the quasi-periodicity condition (11.5) for the same value of the Bloch vector $\mathbf{k}_0$. Substituting the expressions (11.11) into the Helmholtz equation in (11.2) and equating like terms of ϵ^1, we obtain $(\triangle + (k^{(0)})^2)u^{(1)} = -2k^{(1)}k^{(0)}u^{(0)}$. Multiplying both sides of the last equality by $(u^{(0)})^*$ and integrating them over the domain of the primitive cell Ω_0 with unperturbed contour Γ_0 yields the formula for the first approximation of an eigenvalue,

$$k^{(1)} = -\frac{\int_{\Omega_0} (u^{(0)})^*(\triangle + (k^{(0)})^2)u^{(1)} \, dx}{2k^{(0)} \int_{\Omega_0} |u^{(0)}|^2 \, dx}. \qquad (11.12)$$

Now we simplify the numerator in (11.12) using Green's formula

$$\int_{\Omega_0} [(u^{(0)})^*\triangle u^{(1)} - \triangle(u^{(0)})^*u^{(1)}]dx = \int_{\Gamma_0} \left[(u^{(0)})^*\frac{\partial u^{(1)}}{\partial n} - u^{(1)}\frac{\partial(u^{(0)})^*}{\partial n} \right] ds,$$

where we take into account the fact that the contribution of the curvilinear integral over the outer boundary of the square is cancelled because of the quasi-periodicity condition. As $u^{(0)}$ satisfies the Hemholtz equation and a homogeneous Neumann boundary condition, we obtain

$$k^{(1)} = -\frac{\int_{\Gamma_0} (u^{(0)})^*\frac{\partial u^{(1)}}{\partial n} \, ds}{2k^{(0)} \int_{\Omega_0} |u^{(0)}|^2 \, dx}. \qquad (11.13)$$

Finally, evaluating asymptotically the normal derivative $\partial u^{(1)}/\partial n$ over the contour Γ_0, we come to the the desired formula for the first approximation of an eigenvalue

$$k^{(1)} = -\frac{\int_0^{2\pi}\left[(r_c k^{(0)}|u^{(0)}|)^2 + (u^{(0)})^*\frac{\partial}{\partial\theta}\left(f(\theta)\frac{\partial u^{(0)}}{\partial\theta}\right)\right]d\theta}{2k^{(0)}\int_{\Omega_0}|u^{(0)}|^2\,dx}. \qquad (11.14)$$

Now we can provide a perturbed eigenvalue using the formulae $(11.11)_2$ in which $k^{(0)}$ is evaluated by means of the generalised Rayleigh method and $k^{(1)}$ is to be calculated by (11.14). This result is expected to be quite accurate even for ϵ values close to 0.2.

6.   CONCLUDING REMARKS

We have studied the eigenvalue problem for a class of doubly-periodic elastic structures. This analysis is based on a new development of the Rayleigh method for elastic dynamic materials, and it is valid for a general range of wavelengths. This method provides analytical series representation for the fields, and it is characterised by high accuracy and fast convergence. The main achievement of the paper is the analysis of dispersion diagrams showing the presence of phononic band gaps. Finally, explicit asymptotic formulae have been presented for the eigenvalues associated with arrays of slightly non-circular holes.

Acknowledgments

This work was supported by EPSRC research grant (GR/M93994). R. C. McPhedran acknowledges the support of the visiting fellowship EPSRC grant (GR/N33065) during his visit to Liverpool University.

References

[1] Rayleigh, Lord (1892) *On the influence of obstacles arranged in rectangular order upon the properties of a medium*, Phil. Mag. **34**, 481–502.

[2] McPhedran, R C and Movchan, A B (1994) *The Rayleigh multipole method for linear elasticity*, J. Mech. Phys. Solids **42**, 711–727.

[3] Poulton, C G, Movchan, A B, McPhedran R C, Nicorovici, N A and Antipov, Y A (2000) *Eigenvalue problems for doubly periodic structures and phononic band gaps*, Proc. Roy. Soc. A **456**, 2543–2559.

[4] Chin, S K, Nicorovici, N A, and McPhedran, R C (1994) *Green's function and lattice sums for electromagnetic scattering by a square array of cylinders*, Phys. Rev. E **49**, 4590–4602.

III

WAVE PROPAGATION THROUGH INHOMOGENEOUS AND ACTIVE MEDIA

THE SPLIT-OPERATOR TECHNIQUE IN ACOUSTICAL PHYSICS

D. Bosquetti*, J. Sánchez-Dehesa

Departamento de Física Teórica de la Materia Condensada

Facultad de Ciencias (C-5), Universidad Autónoma de Madrid

Madrid E-28049, Spain

diogenes@uamca3.fmc.uam.es, jose.sanchezdehesa@uam.es

Abstract   We present a new finite-difference scheme to study the scattering of sound in the time-domain. The scheme is based on a clever application of the split-operator technique previously employed in quantum mechanics to solve the time-dependent Schrödinger equation. This scheme has advantages in comparison with the usual finite-difference time-domain algorithms. Its main characteristics are: it has a precision of the order $(\Delta t)^3$; it can work with non-homogeneous space-discretizations; and it verifies the energy conservation law.

1.   INTRODUCTION

The goal of this work is to look for an efficient numerical integration scheme to solve the second order differential equation which defines the sound propagation in inhomogeneous systems. This problem is of current interest since it has been shown that periodic structures consisting of rigid cylinders in air exhibit bands of frequencies at which the sound cannot propagate [1]. The method usually employed to treat such problems was to consider infinite periodic systems and to compare the acoustic band gaps with the frequency bands at which sound attenuation is found in zero-order transmission experiments. Nevertheless, that procedure fails in actual structures because of their finite dimension. In other words, the attenuations associated to the existence of uncoupled bands, or the ones due to a possible energy transfer to Bragg waves of higher order, do not appear as stop gaps in the calculated dispersion relation of the acoustic bands [1]. This drawback is overcame in the transfer matrix method [2], which is able to calculate transmission properties of finite systems. Nevertheless, this method has a lack of

---

*Partial funding provided by grant MAT97-0698-C04.

I.D. Abrahams et al. (eds.),
IUTAM Symposium on Diffraction and Scattering in Fluid Mechanics and Elasticity, 107–114.
© 2002 *Kluwer Academic Publishers. Printed in the Netherlands.*

precision at high frequencies. On the other hand, the finite-difference time domain (FDTD) method introduced by Yee [3] has been extensively used to simulate the wave-propagation of electromagnetic waves. In the same spirit the FDTD method has been employed to solve underwater acoustic problems [4] and in dealing with seismic wave propagation [5].

Here, a new discretization scheme is introduced to solve the differential equation for sound propagation in non-homogeneous media. In order to do that, a non-conventional set of canonical variables is used to define a phase-space where the Liouville theorem applies. Afterwards, the corresponding first order differential equation associated to the spectral density is solved by the split-operator technique [8]. Finally, the operator of time evolution is employed to obtain the final differential equation in space-domain which can be discretized by using a non-uniform mesh.

2.   PROBLEM FORMULATION

Consider a linear non-dissipative acoustic medium defined by some space-dependent mass density $\rho(\mathbf{r})$ and sound velocity $c(\mathbf{r})$, which are time independent. The pressure field $\Phi(\mathbf{r}, t)$ and particle velocity $\mathbf{v}(\mathbf{r}, t)$ satisfy Newton's law of motion and the continuity equation:

$$\rho(\mathbf{r})\partial\mathbf{v}(\mathbf{r}, t)/\partial t \ = \ -\nabla\Phi(\mathbf{r}, t), \tag{12.1}$$

$$\partial\Phi(\mathbf{r}, t)/\partial t \ = \ -\rho(\mathbf{r})c^2(\mathbf{r})\nabla \cdot \mathbf{v}(\mathbf{r}, t). \tag{12.2}$$

These equations can be combined to obtain the well known second order partial differential equation for the scalar field,

$$\partial^2\Phi(\mathbf{r}, t)/\partial t^2 = \Gamma(\mathbf{r})\Phi(\mathbf{r}, t), \tag{12.3}$$

where

$$\Gamma(\mathbf{r}) \equiv c^2(\mathbf{r})\left\{\nabla^2 - \nabla\ln[\rho(\mathbf{r})] \cdot \nabla\right\}. \tag{12.4}$$

Also, the energy is conserved since the system is non-dissipative.

For reasons which will become clear later, it is convenient to cast this equation as a pair of first order partial differential equations. For this, it is enough to define a quantity $\Lambda(\mathbf{r}, t) = \partial\Phi(\mathbf{r}, t)/\partial t$, and to rewrite (12.3) as

$$\partial\Phi(\mathbf{r}, t)/\partial t \ = \ \Lambda(\mathbf{r}, t), \tag{12.5a}$$

$$\partial\Lambda(\mathbf{r}, t)/\partial t \ = \ \Gamma(\mathbf{r})\Phi(\mathbf{r}, t). \tag{12.5b}$$

The variables (Φ, Λ) are canonical and define a multidimensional phase space with a spectral density $\sigma(\Phi, \Lambda; t)$. Besides, since the system is conservative, σ verifies the Liouville theorem: $d\sigma(\Lambda, \Phi, t)/dt = 0$. σ can have an explicit time dependence plus an implicit time dependence through Φ and Λ. Therefore, making use of (12.5a) and (12.5b), the Sturm–Liouville equation is obtained:

$$\frac{\partial \sigma(\Phi, \Lambda, t)}{\partial t} = - \left[\Lambda\left(\mathbf{r}, t\right) \frac{\partial}{\partial \Phi} + \Gamma(\mathbf{r}) \Phi\left(\mathbf{r}, t\right) \frac{\partial}{\partial \Lambda} \right] \sigma(\Phi, \Lambda, t). \qquad (12.6)$$

Now, the time evolution of the spectral density between some initial time t_i up to a final time t_f can be studied. A convenient way to study such evolution is to integrate (12.6) from t_i to t_f and to define a time propagator operator $U(t_f, t_i)$,

$$\sigma\left(\Phi, \Lambda, t_i\right) = U(t_f, t_i)\sigma(\Phi, \Lambda, t_f), \qquad (12.7)$$

where

$$U(t_f, t_i) \equiv \exp\left\{ -\int_{t_i}^{t_f} [A\left(\mathbf{r}, t\right) + B\left(\mathbf{r}, t\right)]\, dt \right\}, \qquad (12.8)$$

$A\left(\mathbf{r}, t\right) \equiv \Lambda\left(\mathbf{r}, t\right) \partial/\partial\Phi$ and $B\left(\mathbf{r}, t\right) \equiv \Gamma(\mathbf{r})\Phi(\mathbf{r}, t)\partial/\partial\Lambda$.

The total interval $[t_i, t_f]$ can be divided in a number M of equal subintervals, Δt, so that:

$$\exp\left\{ -\int_{t_i}^{t_f} [A + B]\, dt \right\} = \sum_{j=1}^{M} \exp\left\{ -\int_{t_i+(j-1)\Delta t}^{t_i+j\Delta t} [A + B]\, dt \right\}, \qquad (12.9)$$

where $t_f = t_i + M\Delta t$.

The time subinterval, Δt, can be chosen small enough so that the time dependence on the parameters $\Gamma(\mathbf{r})\Phi\left(\mathbf{r}, t\right)$ and $\Lambda\left(\mathbf{r}, t\right)$ can be neglected, i.e., $\Phi\left(\mathbf{r}, t + \Delta t\right) \approx \Phi\left(\mathbf{r}, t\right)$ and $\Lambda\left(\mathbf{r}, t + \Delta t\right) \approx \Lambda\left(\mathbf{r}, t\right)$ In this limit, the total time operator is

$$U(t_f, t_i) = \prod_{j=1}^{M} U(t_i + j\Delta t, t_i + (j - 1)\Delta t). \qquad (12.10)$$

Moreover, in case that both operators, A and B, were time independent, or they had a smooth variation over the total interval, (12.10) will consist of M applications of the same operator $U(\Delta t) \equiv \exp\left[-\Delta t\left(A + B\right)\right]$. Therefore, the total evolution operator $U(t_f, t_i) = [U(\Delta t)]^{M}$, and the exponential form of the Sturm–Liouville equation is

$$\sigma(\Phi, \Lambda, t_f) = \{\exp\left[-\Delta t\left(A + B\right)\right]\}^{M} \sigma(\Phi, \Lambda, t_i), \qquad (12.11)$$

which defines the evolution between t_i and t_f.

3.   TIME EVOLUTION OF Φ AND Λ: THE SPLIT-OPERATOR TECHNIQUE

Let us consider the time evolution of variables (Φ, Λ) from some initial time t_n up to a final time t_{n+1} separated by a step interval Δt, $t_{n+1} = t_n + \Delta t$, n being an integer with $0 \leq n \leq M - 1$. As discussed before, (12.11) can be applied to this case:

$$\sigma\left(\Lambda, \Phi, t_{n+1}\right) = \exp\left[-\Delta t\left(A + B\right)\right]\sigma\left(\Lambda, \Phi, t_n\right). \tag{12.12}$$

The differential operators A and B do not commute and, therefore, the exponential cannot be put like a product of two exponentials; i.e., $\exp(-\Delta tA - \Delta tB) \neq \exp(-\Delta tA)\exp(-\Delta tB)$. However, the Glauber formula [7] can be extended to three operators A, B, C and, afterwards, the condition $A = C$ can be applied. In other words, we can use the following approach:

$$\begin{aligned}
\exp\left[-\Delta t\left(A + B\right)\right] &= \exp\left[-\frac{\Delta t}{2}A\right]\exp[-\Delta tB]\exp\left[-\frac{\Delta t}{2}A\right] \\
&\quad + O\left[(\Delta t)^3\right]
\end{aligned} \tag{12.13}$$

This is the so called symmetrically split-operator (SSOP) technique, which has been successfully used in quantum mechanics to solve numerically the time dependent Schrödinger equation [8]. Also, it can be applied to physical problems in classical mechanics [6].

Notice in (12.13) how a clever splitting of an operator has produced a substantial reduction of the error $\left((\Delta t)^3\right)$ associated to its sequential application. According to which operator, A or B, is split, two diferent schemes can be developed with analogue numerical results. Here, we analyze the following case:

$$\sigma\left(\Phi, \Lambda, t_{n+1}\right) \approx \exp\left[-\frac{\Delta t}{2}A\right]\exp\left[-\Delta tB\right]\exp\left[-\frac{\Delta t}{2}A\right]\sigma(\Phi, \Lambda, t_n).$$

It is known that the action of an exponential operator $\exp\left[\vartheta(\partial/\partial x)\right]$, where ϑ is a parameter, produces a shift of the variable by ϑ, namely, $\exp\left[\vartheta(\partial/\partial x)\right]f(x) = f\left(x + \vartheta\right)$. By using this property one obtains the time evolution of the canonical variables (Φ, Λ). After the application of the three exponential operators, the relationship between $(\Phi^{n+1}, \Lambda^{n+1})$, the quantities at the t_{n+1} temporal step, and (Λ^n, Φ^n), the corresponding ones at t_n, is found. It is given by the following 2×2 matrix equation:

$$\begin{pmatrix} \Phi^{n+1}\left(\mathbf{r}\right) \\ \Lambda^{n+1}\left(\mathbf{r}\right) \end{pmatrix} = \begin{pmatrix} \alpha_{11}\left(\mathbf{r}, \Delta t\right) & \alpha_{12}\left(\mathbf{r}, \Delta t\right) \\ \alpha_{21}\left(\mathbf{r}, \Delta t\right) & \alpha_{22}\left(\mathbf{r}, \Delta t\right) \end{pmatrix} \begin{pmatrix} \Phi^n\left(\mathbf{r}\right) \\ \Lambda^n\left(\mathbf{r}\right) \end{pmatrix}, \tag{12.14}$$

Table 1   Density, sound velocity and acoustic impedance of the homogeneous media.

	$\rho(\text{g/cm})$	$c(\text{cm/s})$	$Z(\text{g/s})$
medium 1	0.25	2.0	0.5
medium 2	1.0	1.0	1.0
medium 3	4.0	0.5	2.0

where the coefficients α_{11}, α_{12}, α_{21} and α_{22} are spatial differential operators:

$$\alpha_{11}(\mathbf{r}, \Delta t) = \alpha_{22}(\mathbf{r}, \Delta t) = 1 + \Delta t^2 \Gamma(\mathbf{r})/2, \tag{12.15}$$

$$\alpha_{12}(\mathbf{r}, \Delta t) = -\Delta t - \Delta t^3 \Gamma(\mathbf{r})/4, \tag{12.16}$$

$$\alpha_{21}(\mathbf{r}, \Delta t) = -\Delta t \Gamma(\mathbf{r}). \tag{12.17}$$

In conclusion, if we know the variables at some initial time t_i, their corresponding values at some final time t_f, which is separated by M equal time steps Δt, are

$$\begin{pmatrix} \Phi(\mathbf{r}, t_f) \\ \Lambda(\mathbf{r}, t_f) \end{pmatrix} = \begin{pmatrix} \alpha_{11}(\mathbf{r}, \Delta t) & \alpha_{12}(\mathbf{r}, \Delta t) \\ \alpha_{21}(\mathbf{r}, \Delta t) & \alpha_{22}(\mathbf{r}, \Delta t) \end{pmatrix}^M \begin{pmatrix} \Phi(\mathbf{r}, t_i) \\ \Lambda(\mathbf{r}, t_i) \end{pmatrix} \tag{12.18}$$

Now, this equation can be solved using a convenient discretization of the **r**-space.

4.   TEST CASE: SOUND PROPAGATION IN A LAYERED SYSTEM

The numerical performance of the SSOP algorithm is analyzed in this section by studying a simple one-dimensional problem whose analytical solution is known. Let us consider a layered system composed of three different homogeneous media, whose mass densities ρ_i $(i = 1, 3)$, sound velocities c_i, and acoustic impedances, $Z_i = \rho_i c_i$, are listed in Table 1. Also, medium 2 is considered finite, with length $L = 800\text{cm}$, and is placed in between media 1 and 3, both considered semi-infinite (Fig. 1).

The reflection (R) and transmission (T) coefficients at a given interface only depend on the impedance ratio, $Z_{\text{incident}}/Z_{\text{transmitted}}$, and they are well known [9]:

$$R_{2\to1} = \frac{Z_2/Z_1 - 1}{Z_2/Z_1 + 1} = +\frac{1}{3}; \quad T_{2\to1} = \frac{2}{Z_2/Z_1 + 1} = \frac{2}{3} \tag{12.19a}$$

$$R_{2\to3} = \frac{Z_2/Z_3 - 1}{Z_2/Z_3 + 1} = -\frac{1}{3}; \quad T_{2\to3} = \frac{2}{Z_2/Z_3 + 1} = \frac{4}{3} \tag{12.19b}$$

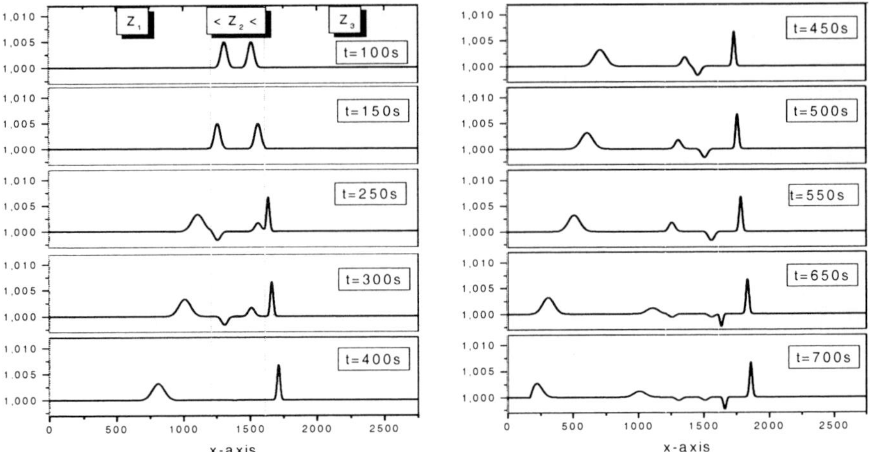

Figure 1   Time evolution of a pressure wave-packet in a layered acoustic structure. The vertical axis defines the pressure in arbitrary units. An initial gaussian wave-packet is put in the middle of medium 2 at $t = 0$ (not shown in the figure). Immediately after, it splits into two equal wave-packets which move in opposite directions (see $t = 100$s). At $t = 150$s, both packets reach the interfaces (the vertical dotted lines). Regarding their reflection and transmission, different phenomena occur at each interface because of the different impedance ratios (see text). Notice that at $t = 400$s, a total destructive interference occurs in medium 2. Finally, at $t = 600$s, a second transmission and reflection process appears in the system.

In both cases, the condition $T + R = 1$ is verified, but the negative sign of R indicates that $T > 1$.

In the system described we have studied the propagation in the time domain of a gaussian wave-packet which is put at the middle of medium 2 at $t = 0$ and it is subjected to the initial condition $\Lambda^0(x) = 0$. The corresponding equation (12.18) has been solved by using a non-uniform mesh in the x-axis; more points were defined in the regions close to the interfaces. The pressure pattern as a function of time is shown in Fig. 1. At $t = 100$s, it can be observed how the initial wave-packet is now divided in two equal packets travelling in opposite directions. After some time interval, $t = 150$s, both wave-packets arrive at the two interfaces at the same time. However, since the acoustic impedances are such that $Z_1 < Z_2 < Z_3$, two different phenomena occur. At the $2 \rightarrow 1$ interface ($Z_2/Z_1 = 2$), the left wave-packet passes to a medium with a lower acoustic impedance. In this case, a phase inversion takes place on the reflected part, while the transmitted part has a peak lower than the incident. On the other hand, at the $2 \rightarrow 3$ interface ($Z_2/Z_3 = 1/2$), the right wave-packet passes to a higher acoustic impedance media. Now,

Table 2   Transmission and reflection coefficients (absolute values)

Coefficient	SSOP technique	Exact (Eqs. (1.22)
$T_{2\to1}$	0.646	0.667
$T_{2\to3}$	1.354	1.333
$R_{2\to1}$	0.354	0.333
$R_{2\to3}$	0.354	0.333

Table 3   Average positions (in cm) of reflected (R) and transmitted (T) packets.

$t(s)$	$\langle x_R(t)\rangle_{2\to1}$	$\langle x_T(t)\rangle_{2\to1}$	$\langle x_R(t)\rangle_{2\to3}$	$\langle x_T(t)\rangle_{2\to3}$
250	1260	1100	1560	1610
350	1360	900	1460	1660
400	1410	800	1410	1685
500	1510	600	1310	1735

we have the opposite case: no phase inversion occurs on the reflected part while the transmitted part has a peak higher than the incident.

The theoretical predictions agree with the qualitative behaviour shown in Fig. 1. Moreover, regarding quantitative results, Table 2 presents the good agreement found between coefficients T and R calculated by using the SSOP technique with those obtained analytically.

Since the media are homogeneous, their group and phase velocities are equal inside a given medium. This fact is also accomplished by the SSOP algorithm. Let us define the average position of the reflected and transmitted wave-packets by $\langle x_R(t)\rangle$ and $\langle x_T(t)\rangle$, respectively. Table 3 shows their values at different times. From Table 3 the group velocity of the corresponding wave-packets at the $2\to1$ interface can be obtained:

$$v_{gR}\left(2\to1\right) \;=\; \frac{\langle x_R(t_2)\rangle_{2\to1} - \langle x_R(t_1)\rangle_{2\to1}}{t_2 - t_1} = 1.0 = c_2, \quad (12.20)$$

$$v_{gT}\left(2\to1\right) \;=\; \frac{\langle x_T(t_2)\rangle_{2\to1} - \langle x_T(t_1)\rangle_{2\to1}}{t_2 - t_1} = 0.5 = c_1. \quad (12.21)$$

Using an analogous procedure for the $2\to3$ interface, we also find that $v_{gR}\left(2\to3\right) = 1.0 = c_2$, and $v_{gT}\left(2\to3\right) = 0.5 = c_3$.

With regards to acoustic energy conservation, the method also verifies such property with maximum accuracy.

In summary, this work has introduced an algorithm based on the split-operator method which is suitable for treating the propagation of pressure waves through inhomogeneous structures having abrupt variations of mass density and velocity. Also, its high numerical accuracy has been shown in a simple one-dimensional layered system. Applications to physical systems of current interest will be presented elsewhere.

Acknowledgments

We acknowledge D. Caballero and P. Tarazona for useful discussions. D. Bosquetti acknowledges a grant provided by the Autonomous University of Madrid. The authors also thank Dr. F. Meseguer, Prof. G. E. Marques and Prof. E. Marega for their continuous support and interest in this work.

References

[1] Sánchez-Pérez, J V, Caballero, D, Martínez-Sala, R M, Rubio, C, Sánchez-Dehesa, J, Meseguer, F, Llinares, J, and Gálvez, F (1998) *Sound attenuation by a two-dimensional array of rigid cylinders*, Phys. Rev. Lett. **80**, 5325–5328.

[2] Sigalas, M M and Economou, E N (1996) *Attenuation of multiple scattered sound*, Europhys. Lett. **36**, 241–246.

[3] Yee, K S (1966) *Numerical solution of initial boundary value problems involving Maxwell's equations in isotropic media*, IEEE Trans. Antennas Propagat. **AP-14**, 302–307.

[4] Wang, S (1996) *Finite-difference time-domain approach to underwater acoustic scattering problems*, J. Acoust. Soc. Am. **99**, 1924–1928.

[5] Stephen, R A (1988) *A review of finite-difference methods for seismo-acoustic problems at seafloor*, Rev. Geophys. **26**, 445–458.

[6] Dattoli, G, Ottaviani, P L, Segreto, A and Torre, A (1996) *Symmetric-split-operator techniques and finite difference methods for the solution of classical and quantum evolution problems*, Nuovo Cimento B **111**, 825–839.

[7] Cohen-Tannoudji, C, Diu, B and Laloë, F (1977) *Mecanique Quantique*, Paris: Hermann.

[8] Fleit, M D, Fleck, J A and Steiger, A (1982) *Solution of the Schrödinger equation by a spectral method*, J. Comp. Phys. **47**, 412–415.

[9] Towne, D H (1988) *Wave Phenomena*, New York: Dover.

A UNIFIED MODEL FOR THE PROPERTIES OF COMPOSITE MATERIALS

P. R. Brazier-Smith

Thomson Marconi Sonar Ltd., Ashurst Drive

Cheadle Heath, Stockport SK3 0XB, UK

Peter.Brazier-Smith@tms-ltd.com

1.   INTRODUCTION

A model is described for calculating the properties of a specific kind of composite material consisting of an elastic substrate with embedded inclusions. The model requires that four scatter cross-sections be calculated from the scatter properties of the individual inclusions constituting the ensemble population. The composite properties then follow simply from these scatter cross-sections.

Central to the model is the concept of multiple scattering for which there is a large body of literature of which three references, namely Foldy [1], Sabina and Willis [2] and Kim et al [3], are relevant to the work here. The present author however believes that unresolved issues remain and attempts to address them here.

The basis of the analysis here is that, with inclusions randomly scattered throughout the substrate material, the problem may be considered a stochastic one where fields within the substrate are considered only through their averaged properties. Then two approaches are considered. In the first, inclusions are distributed uniformly in a substrate occupying all space. Then a solution is sought where incident and scattered fields are the same and therefore form a single self-consistent field. The analysis then proceeds by replacing the summations of scattered fields from distinct inclusions by integrals in an averaging process. The resulting homogenous spatial integral equations then reduce to wavenumber eigenvalue equations in phase space.

In the second approach, a substrate that occupies all space is again considered, but the region in which the inclusions are distributed is reduced to a semi-infinite half space bounded by a plane. A plane wave in

I.D. Abrahams et al. (eds.),
IUTAM Symposium on Diffraction and Scattering in Fluid Mechanics and Elasticity, 115–122.
© 2002 *Kluwer Academic Publishers. Printed in the Netherlands.*

the uniform substrate half space incident on the composite region is then considered. Allowing the plane wave to continue into the composite half space means that the inclusions are subject to the sum of that incident field and their mutually scattered fields. Again, replacing summations by integration as with the self-consistent field results in an inhomogenous integral equation. This reduces to two Wiener–Hopf equations in phase (wavenumber) space, one for incident shear wave and one for incident compression wave the kernels of which are the matrices already encountered in the first approach. As with ordinary Wiener–Hopf equations, solution must proceed by factorisation of the kernels which this time are matrices. For those familiar with this sort of problem, it is known that matrices do not readily factorise, but in these cases their factors do exist and have been found (courtesy of Prof. I. D. Abrahams). Although full final solution of the Wiener–Hopf equations is incomplete, this factorisation removes any formal obstacle to completion. Reduced solutions are relatively simple, where only one of the four scatter coefficients is non-zero, and are shown to recover the expected properties of the reflected and transmitted fields. Indeed, the transmitted field has a component *that exactly cancels the incident field in the composite half space*: such cancellation had to occur because the incident field cannot exist in the composite half space.

2.   SELF-CONSISTENT SCATTERING THEORY

The fundamental scattered field about an inclusion is the Green function for an isotropic elastic solid [2, 3]. Taking field variables to be time harmonic with convention $e^{-i\omega t}$, it is given by

$$G_{ij} = k_s^2 G_s \delta_{ij} - \frac{\partial^2}{\partial x_i \partial x_j}(G_p - G_s) \tag{13.1}$$

where

$$G_s = \frac{e^{ik_s R}}{4\pi \rho \omega^2 R}, \qquad G_p = \frac{e^{ik_p R}}{4\pi \rho \omega^2 R} \tag{13.2}$$

k_p and k_s are respectively the compression and shear wavenumbers at circular frequency ω, ρ is the material density and R is the radial distance from source to field point. It is convenient to define

$$G_m = \nabla G = k_p^2 \nabla(G_p) \tag{13.3}$$

as the displacement field about a monopole source. The fundamental displacement scattered field, u_s, about a single inclusion may then be written

$$u_s = a_d G \cdot u + a_m G_m S + a_c (\nabla G) : \Omega + a_s (\nabla G) : E. \qquad (13.4)$$

Here u is the incident displacement field at the inclusion site, S its divergence (condensation), Ω its anti-symmetrised covariant derivative (curl) and E the symmetrised covariant derivative (strain). Thus the scattered field depends not only upon the incident displacement field, but also on three further fields that are derived from the incident field.

Consider now that the substrate occupies all space and that the inclusions are identical and uniformly distributed. Self-consistency is then expressed through the equation

$$u^{(i)} = a_d \sum_{j \neq i} G \cdot u^{(j)} + a_m \sum_{j \neq i} G_m S^{(j)} + a_c \sum_{j \neq i} \nabla G : \Omega^{(j)} + a_s \sum_{j \neq i} \nabla G E^{(j)}.$$
$$(13.5)$$

This equation is to be complemented by three further equations obtained by taking its divergence, curl and symmetric covariant derivative:

$$S^{(i)} = a_d \sum_{j \neq i} G_m \cdot u^{(j)} + a_m \sum_{j \neq i} \nabla \cdot G_m S^{(j)} + a_s \sum_{j \neq i} \nabla^2 G E^{(j)} \qquad (13.6)$$

$$\Omega^{(i)} = a_d \sum_{j \neq i} \{G\}_a u^{(j)} + a_c \sum_{j \neq i} \{\nabla G\}_a : \Omega^{(j)} + a_s \sum_{j \neq i} \{\nabla G\}_a : E^{(j)} \qquad (13.7)$$

$$E^{(i)} = a_d \sum_{j \neq i} \{G\}_s \cdot u^{(j)} + a_m \sum_{j \neq i} \nabla G_m S^{(j)} + a_c \sum_{j \neq i} \{\nabla G\}_s : \Omega^{(j)}$$
$$+ a_s \sum_{j \neq i} \{\nabla G\}_s E^{(j)} \qquad (13.8)$$

where $\{\}_s$ and $\{\}_a$ indicate symmetrized and anti-symmetrized covariant derivatives. In the absence of further scattering modes than the four already considered, these equations are exact (as were those of Sabina and Willis at their equivalent point). However for an infinite number of randomly distributed scatterers, the solution of these equations does not represent a practical proposition and we must seek a means of replacing the summations by integration and select a suitable averaging process for the fields. For wavelengths that are long compared to the mean spacing of the scatterers, a suitable process is to carry out averaging integration of (13.5)–(13.8) where the weighting function is slowly varying on the interstitial spacing scale but compact on a wavelength scale. This results in the following set of equations in the averaged fields:

$$u = A_d \int G \cdot u \, dv + A_m \int G_m S \, dv + A_c \int \nabla G : \Omega \, dv + A_s \int \nabla G : E \, dv$$
$$(13.9)$$

$$S = A_d \int G_m u \, dv + A_m \int \nabla \cdot G_m S \, dv + A_s \int \nabla^2 G : E \, dv \qquad (13.10)$$

$$\Omega = A_d \int \{G\}_a \cdot u \, dv + A_c \int \{\nabla G\}_a : \Omega \, dv + A_s \int \{\nabla G\}_a : E \, dv \qquad (13.11)$$

$$E = A_d \int \{G\}_s \cdot u \, dv + A_m \int \nabla \cdot G_m S \, dv + A_c \int \{\nabla G\}_s : \Omega \, dv$$
$$+ A_s \int \{\nabla G\}_s : E \, dv \qquad (13.12)$$

where integration is over all space and the individual scatter coefficients have been replaced by scattering cross-sections, $A_m = a_m n$, $A_d = a_d n$, $A_c = a_c n$ and $A_s = a_s n$ where n is the scatterer concentration. We note that the fields, being ensemble averaged fields, should carry angled brackets, i.e. $\langle u \rangle$, $\langle S \rangle$ $\langle \Omega \rangle$ and $\langle E \rangle$, but for the sake of brevity they are dropped. However before continuing a note of explanation as to why we have four equations involving four variables is in order: it turns out that *the average of a gradient operator (e.g. divergence) on the displacement field is different from the result of the gradient operator on the averaged displacement field.* As an example, the proposition $\nabla \cdot \langle u \rangle \neq \langle \nabla \cdot u \rangle$ for a slowly varying convolution function is readily provable using Green's theorem: the act of moving the gradient operator from the field onto the Green function results in a commutator arising from surface integrals over the scatterers. For this reason the four variables of (13.9)–(13.12) are considered independent. We restrict ourselves to seeking plane wave solutions of the form

$$u_x = \hat{u}_x e^{ikx}; \; S = \hat{S} e^{ikx}; \; E_{xx} = \hat{E}_{xx} e^{ikx} \qquad (13.13)$$

for longitudinal waves and

$$u_z = \hat{u}_z e^{ikx}; \; \Omega_{xz} = \hat{\Omega}_{xz} e^{ikx}; \; E_{xz} = \hat{E}_{xz} e^{ikx} \qquad (13.14)$$

for shear waves polarised in the z-direction. With these restrictions, the integrals are calculable and for longitudinal waves result in the following matrix eigenvalue equation

$$A_L \left(\hat{u}_x, \hat{S}, \hat{E}_{xx} \right)^T = 0$$

where

$$A_L(k) = \begin{pmatrix} \dfrac{A_d k_p^2}{\rho \omega^2 (k^2 - k_p^2)} - 1 & \dfrac{-iA_m k_p^2 k}{\rho \omega^2 (k^2 - k_p^2)} & \dfrac{-iA_s k_p^2 k}{\rho \omega^2 (k^2 - k_p^2)} \\[2ex] \dfrac{-iA_d k_p^2 k}{\rho \omega^2 (k^2 - k_p^2)} & \dfrac{-A_m k_p^4}{\rho \omega^2 (k^2 - k_p^2)} - 1 & \dfrac{-A_s k_p^4}{\rho \omega^2 (k^2 - k_p^2)} \\[2ex] \dfrac{-iA_d k_p^2 k}{\rho \omega^2 (k^2 - k_p^2)} & \dfrac{-A_m k_p^4}{\rho \omega^2 (k^2 - k_p^2)} & \dfrac{-A_s k_p^4}{\rho \omega^2 (k^2 - k_p^2)} - 1 \end{pmatrix}. \qquad (13.15)$$

For transverse waves the matrix eigenvalue equation is

$$A_T \left(\hat{u}_z, \hat{\Omega}_{xz}, \hat{E}_{xz} \right)^T = 0$$

where

$$A_T(k) = \begin{pmatrix} \dfrac{A_d k_s^2}{\rho\omega^2 (k^2 - k_s^2)} - 1 & \dfrac{-iA_c k_s^2 k}{\rho\omega^2 (k^2 - k_s^2)} & \dfrac{iA_s k_s^2 k}{\rho\omega^2 (k^2 - k_s^2)} \\[2ex] \dfrac{-iA_d k_s^2 k}{2\rho\omega^2 (k^2 - k_s^2)} & \dfrac{-A_c k_s^4}{2\rho\omega^2 (k^2 - k_s^2)} - 1 & \dfrac{A_s k_p^4}{2\rho\omega^2 (k^2 - k_2^2)} \\[2ex] \dfrac{iA_d k_s^2 k}{\rho\omega^2 (k^2 - k_s^2)} & \dfrac{A_m k_s^4}{\rho\omega^2 (k^2 - k_s^2)} & \dfrac{-A_s k_s^4}{\rho\omega^2 (k^2 - k_s^2)} - 1 \end{pmatrix}.$$

$$(13.16)$$

The characteristic equations for the wavenumber, k, modified from k_p and k_s are then given by setting the respective determinants of the co-efficients of $\{\hat{u}, \hat{S}, \hat{\Omega}, \hat{E}\}$ in the matrix eigenvalue equations to zero. The resulting equations are

$$k^2 = k_p^2 \left(1 + \frac{A_d}{\rho\omega^2} \right) \left(1 - \frac{k_p^2}{\rho\omega^2} (A_m + A_s) \right) \qquad (13.17)$$

for compression waves and

$$k^2 = k_s^2 \left(1 + \frac{A_d}{\rho\omega^2} \right) \left(1 - \frac{k_s^2}{\rho\omega^2} (A_c + A_s) \right) \qquad (13.18)$$

for shear waves. The effective elastic constants and density of the composite material may then be adduced by noting that, for a homogenous material, the compression and shear wavenumbers are given by

$$k_p^2 = \frac{\rho\omega^2}{\lambda + 2\mu} \quad \text{and} \quad k_s^2 = \frac{\rho\omega^2}{\mu} \qquad (13.19)$$

Associating fractional density change with the term $(1 + A_d/(\rho\omega^2))$ common to (13.17) and (13.18), the elastic properties follow via these equations:

$$\rho' = \rho \left(1 + \frac{A_d}{\rho\omega^2} \right)$$

$$\lambda' + 2\mu' = (\lambda + 2\mu) \left(1 - \frac{k_p^2}{\rho\omega^2} (A_m + A_s) \right)^{-1} \qquad (13.20)$$

$$\mu' = \mu \left(1 + \frac{A_d}{\rho\omega^2} \right) \left(1 - \frac{k_s^2}{\rho\omega^2} (A_c + A_s) \right)^{-1}$$

3.   PLANE WAVE SCATTERING FROM A COMPOSITE HALF-SPACE

In the last section, a theory was developed for the properties of waves in a composite material occupying all space. From these properties the effective density and Lamé constants were adduced. However confirmation of these should be obtained from the effective reflection coefficients of plane waves in a uniform material occupying a semi-infinite half-space incident onto a plane interface of a composite with a substrate of the same material and occupying the other half-space. For two uniform isotropic different materials so joined, the reflection and transmission coefficients for normally incident shear or compression waves are

$$R = \frac{k_2\rho_1 - k_1\rho_2}{k_2\rho_1 + k_1\rho_2} \quad \text{and} \quad T = \frac{2k_2\rho_1}{k_2\rho_1 + k_1\rho_2} \tag{13.21}$$

where k_1 and k_2 are the respective wavenumbers in the half spaces bearing the incident and transmitted waves.

We shall examine compression waves here but note that parallel analysis exists for shear waves. Thus consider the inclusions to be distributed in the half space $x > 0$ in an otherwise uniform material and consider a plane compression wave at normal incidence of the form

$$\hat{u}e^{ik_px}, \quad \hat{S}e^{ik_px}, \quad \hat{E}_{xx}e^{ik_px}.$$

For the purposes of this study *the wave is considered to occupy all space.* This allows a set of coupled integral equations for the scattered wave fields $\{u, S, E_{xx}\}$ of similar form to (13.9)–(13.12) to be deduced, but with the following differences: (i) the field variables $\{u, S, E_{xx}\}$ are replaced by $u + \hat{u}e^{ik_px}$, $S + \hat{S}e^{ik_px}$, $E + \hat{E}e^{ik_px}$; (ii) the integrals are taken over the semi-infinite half space occupied by the composite; (iii) all terms with Ω are absent, as is the equivalent of (13.11). These equations readily Fourier transform with respect to x to give the following matrix equation in the three-component object $\xi \equiv \{\tilde{u}, \tilde{S}, \tilde{E}_{xx}\}$ (the tildes indicating transformed quantities):

$$F(s)\xi_+(s) + \xi_-(s) + H(s) = 0 \tag{13.22}$$

where the matrix $F(s) = -A_L(s)$ (see (13.15)) and the matrix H is given by

$$H(s) = \frac{-k_p^2}{\rho\omega^2(s^2 - k_p^2)(s + k_p)} \begin{pmatrix} iA_d\hat{u} & -sA_m\hat{S} & -sk_p^2A_s\hat{E}_{xx} \\ -sA_m\hat{u} & -ik_p^3A_m\hat{S} & -ik_p^3A_s\hat{E}_{xx} \\ -sk_p^2A_s\hat{u} & -ik_p^3A_s\hat{S} & -ik_p^3A_s\hat{E}_{xx} \end{pmatrix}.$$

The subscripts '+' and '−' indicate respectively the positive and negative half-range Fourier transforms of the subscripted quantity which correspondingly are analytic in the upper and lower half-planes. Equation (13.22) is therefore amenable to solution by the Wiener–Hopf method. The solution proceeds by defining a multiplicative factorisation of the kernel, $F = F_+/F_-$ where F_+ is analytic in the upper half plane, F_- analytic in the lower half plane and both in a strip containing the real axis. Equation (13.22) may then be written

$$\xi_+(s)F_+(s) + \xi_-(s)F_-(s) + H(s)F_-(s) = 0 \qquad (13.23)$$

whence *additive* split of $H(s)F_-(s)$ yields the Wiener–Hopf equation

$$\xi_+(s)F_+(s) + [H(s)F_-(s)]_+ + \xi_-(s)F_-(s) + [H(s)F_-(s)]_- = 0 \quad (13.24)$$

from which the first two and last two terms may separately be equated to zero by essentially an extension of Liouville's theorem.

In general, multiplicative splits of matrices do not exist. However in this case the matrix kernel (13.15) has been factorised as has the kernel (13.16) for shear waves (by courtesy I. D. Abrahams, private communication). Thus no formal obstacle exists for completion of the analysis, but its completion represents a non-trivial task and remains to be done. However, the analysis simplifies considerably if only one of the scatter cross-sections is non-zero leading to results in accord with (13.21). For example, if only A_d is non-zero, the reduced (13.22) yields

$$\tilde{u}_+(s)\frac{(s+k_t)(s-k_t)}{(s+k_p)(s-k_p)} + \tilde{u}_-(s) - \frac{i(k_t^2 - k_p^2)\hat{u}}{(s^2 - k_p^2)(s+k_p)} = 0 \qquad (13.25)$$

where $k_t = k_p\sqrt{1 + A_d/(\rho\omega^2)}$ is the wavenumber of the transmitted wave in accord with (13.17). The kernel factorises by inspection to give, after some algebra, the Fourier transforms of the transmitted and reflected fields as

$$\tilde{u}_+(s) = -\frac{i(k_t - k_p)\hat{u}}{(k_t + k_p)}\left[\frac{s + 2k_p + k_t}{(s+k_p)(s+k_t)}\right], \qquad (13.26)$$

$$\tilde{u}_-(s) = -\frac{i(k_t - k_p)\hat{u}}{(k_t + k_p)}\left[\frac{1}{s - k_t}\right]. \qquad (13.27)$$

Inverse transforming these equations and invoking the density change for the composite material of $\rho_2 = \rho_1\sqrt{1 + A_d/(\rho\omega^2)}$ leads to the transmission and reflection coefficients of (13.21) *but with an additional field of* $-\hat{u}e^{ik_p x}$ *on the transmitted side.* This field exactly cancels the incident field for $x > 0$. Recalling that the incident field was taken to occupy all space, this cancellation had to happen *because the incident field cannot exist in the region $x > 0$.* If one of the other scatter cross-sections is chosen to be non-zero similar analysis applies and gives results in accord with (13.21).

4. CLOSING REMARKS

The effective compression and shear wavenumbers of a composite material as given by equations (13.17) and (13.18) depend on the presence of scatterers through four parameters only, namely the monopole, dipole, shear and rotary scattering cross-sections. Thus this allows us to deduce composite material properties by solving the scattering response of an isolated inclusion as a separate physical problem followed by application of the equations referred to above. Of course solution of the single scattering problem may not be simple, but many solutions do exist, (e.g. the scattering theory for a shell in [4]) and the fact that the calculation of the composite properties is separated can only assist in building satisfactory models of composite materials.

Acknowledgments

The author is indebted to Professor I. D. Abrahams for much useful discussion and providing factorisations of the matrix kernels arising from the Wiener–Hopf analysis.

References

[1] Foldy, L L (1945) *Multiple scattering theory of waves*, Phys. Rev. **67**, 107–119.

[2] Sabina, F J and Willis, J R (1988) *A simple self-consistent analysis of wave propagation in particulate composites*, Wave Motion **10**, 127–142.

[3] Kim, J-Y, Ih, J-G and Lee, B-H (1995) *Dispersion of elastic waves in random particulate composites*, J. Acoust. Soc. Am. **97**, 1380–1388.

[4] Baird, A M, Kerr, F H and Townend, D J (1999) *Wave propagation in a viscoelastic medium containing fluid-filled microspheres*, J. Acoust. Soc. Am. **105**, 1527–1538.

ACOUSTIC MODELLING OF SIGNATURE REDUCTION MATERIALS FOR UNDERWATER APPLICATIONS

F. H. Kerr

QinetiQ, Winfrith Technology Centre
Dorchester, Dorset DT2 8XJ, UK
fhkerr@qinetiq.com

A. M. Baird

DSTL, Winfrith Technology Centre
Dorchester, Dorset DT2 8XJ, UK
ambaird@dstl.gov.uk

1.   INTRODUCTION

Air filled materials have a number of important applications within the underwater environment but are principally used for baffle treatments (to control sound transmission) and for signature reduction (to control reflection). Since sound propagates more readily between two media whose acoustic impedances are similar, the designer of multi-layered baffles and coatings is very much concerned with the need to control impedances at the boundaries of each adjacent type of medium and between the front face of the coating and the surrounding water. Typically, a good mismatch will occur when a low density rubber or polyurethane contains a high proportion of air-filled scatterers (thus lowering both sound speed and overall density) and a good match can be formed by using a high density matrix with a much lower air content.

In the absence of external pressure, each such layer can be modelled by dynamical effective medium theory using single or multiple scattering methods, given knowledge of the geometry and of the properties of the materials and inclusions. Then the sound propagation across a multilayered system can easily be predicted by iteration. At increasing pressures, however, the performance of the structure changes significantly. This is

I.D. Abrahams et al. (eds.),
IUTAM Symposium on Diffraction and Scattering in Fluid Mechanics and Elasticity, 123–132.

due mainly to the deformation of the cavities under hydrostatic loading and the consequent reduction in layer thickness.

DERA has developed an extensive capability for modelling the acoustic behaviour of air-filled materials as a function of both temperature and pressure. This paper describes some of these techniques, concentrating principally on two types of random materials: the first containing air-filled spherical cavities and the other containing small thin air-filled elastic-plastic shells (microspheres). We describe a two-step approach, firstly by adopting a phenomenological model to predict the static deformation of the material due to the pressure and then by modelling the sound propagation in the deformed medium using an effective medium method. Finally, theoretical predictions are compared with measurements and it is shown that, for these materials, the modelling of both frequency and depth dependence is very reliable.

2.   ACOUSTIC MODELLING

The problem of wave propagation in a medium containing randomly spaced inclusions has no exact analytical solution and is often approached by effective medium theories, i.e. by replacing the composite medium with a uniform 'effective' material which can be characterised by an appropriate choice of effective parameters. The composite layer can then be treated as homogeneous in the subsequent analysis.

Clearly, the crucial part of the acoustic modelling process lies in the estimation of the effective parameters; for reasons of space, we shall not describe those in detail but shall refer the reader to the published literature in which there are many examples. For the case of air-filled cavities, extensive work has been presented and reviewed [1, 2, 3, 4, 5]. In many cases, an analogous theory can be applied to shell-reinforced inclusions by replacing the scattering coefficients for scattering by a single air-filled cavity with the corresponding coefficients for scattering by a single spherical shell. Full details of this process are given in [3].

3.   PRESSURE DEPENDENCE

When the materials are subjected to hydrostatic loading, the properties of the air-filled layers are affected in a number of ways, causing a significant change in acoustic performance. This behaviour differs according to whether the layer contains cavities or microspheres. Therefore these two cases will be treated differently in the following analysis.

AIR-FILLED SPHERICAL CAVITIES

Intuitively, the most obvious result of applying hydrostatic pressure to an air-filled polymer is a decrease in the size of the cavities leading to a lower overall air content. Consequently, the air pressure within the individual inclusions is increased and the material becomes stiffer. The combined effect leads to an increase in both the density and wave speed and thus the impedance. Thus the mismatch is reduced as well as the effectiveness of the scattering mechanisms within the composite.

The method employed here for spherical air-filled cavities has been developed from models described in [6] and [7] for spherical voids. It is based upon the assumptions that the matrix substrate is incompressible, that the stresses in this matrix remain isotropic under hydrostatic pressure and that each cavity retains its spherical shape under compression. These assumptions are reasonable given the inherent symmetry of the problem. Finally, we shall limit our discussion to the case of the cavities being uniform in size, noting that the theory extends easily to a distribution of sizes by simple averaging.

Fig. 1 shows a unit cell consisting of a spherical cavity (before and after compression) surrounded by a substrate shell defined by half the nearest neighbour distance in the material. Initially, a cavity of radius a is contained within a unit cell of radius b where b is chosen to satisfy the condition $\phi = (a/b)^3$, ϕ denoting the volume fraction of cavities in the medium. After compression, the cavity radius has reduced to a', the volume fraction is ϕ' and the unit cell radius is now b' where $\phi' = (a'/b')^3$. By the assumption of matrix incompressibility, however, the volume shown shaded has remained constant.

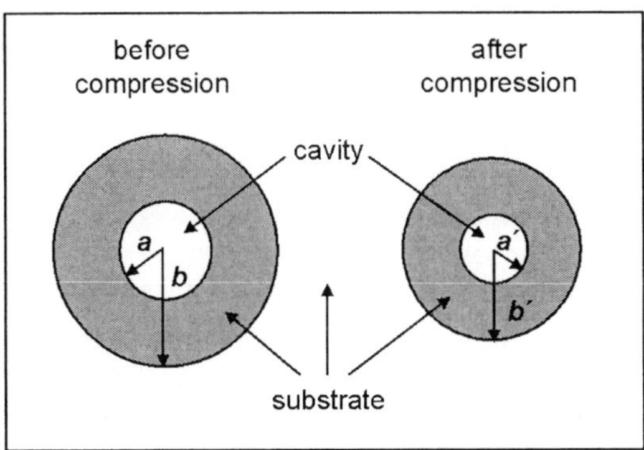

Figure 1   Unit cell geometry

If we choose the centre of the spherical co-ordinate system to be at the centre of the cavity and apply the assumptions of constant divergence and no vorticity, then the displacement field in the substrate s is radially symmetric with radial component

$$s_r = Ar + \tfrac{1}{3}B/r^2 \tag{14.1}$$

where A and B are constants. Note that, for incompressibility, we would normally require the divergence to be zero, in which case the constant A in (14.1) would vanish. However, it should be noted that $\nabla \cdot s$ would then be zero by virtue of the Lamé constant λ being considered as infinite. Indeed, it can be shown that $\nabla \cdot s \sim \lambda^{-1}$ and thus the term $\lambda \nabla \cdot s \equiv 3\lambda A$ which appears in the stress functions (below) is actually finite and non-zero. It is therefore convenient to retain the first term in (14.1) with the comment that the components involving A vanish anyway when the appropriate boundary conditions are applied.

From (14.1), the radial component of stress is easily obtained as

$$\tau_{rr} = \lambda \nabla \cdot s + 2\mu(\partial s_r/\partial r) = 3KA - \tfrac{4}{3}\mu B/r^3 \tag{14.2}$$

where K is the bulk modulus and λ and μ are the Lamé parameters. Applying boundary conditions to the compressed sphere, we require that the radial stress at the cavity wall be equal to the increase in air pressure within the cavity and that the radial stress at the unit cell boundary be equal to the isotropic external pressure in the substrate. For infinitesimal displacements, the constant B is shown to be approximately equal to $a'^3 - a^3$. Finally, Boyle's law is applied to relate the change in internal air pressure to that of the cavity volume. After some algebra, we obtain

$$(\phi' - \phi)\left\{1 - \phi' + \tfrac{3}{4}(\Gamma/\mu)\right\} + \tfrac{3}{4}(P/\mu)\phi'(1 - \phi) = 0 \tag{14.3}$$

where P represents the hydrostatic pressure and Γ denotes the air pressure within the cavities before compression. It is easily shown that this has exactly one solution for ϕ' in the interval $0 < \phi' < \phi$; this represents the reduced air fraction. The radius of the deformed cavity can then be found from the incompressibility assumption applied to the unit cell and the reduced layer thickness from the assumption that the volume of substrate material in the layer is constant.

Note that, although the theory of linear elasticity has been used (e.g. for the stress-displacement relations), the model has been found to be surprisingly accurate at all but the highest pressures. This is largely due to the requirement that the boundary conditions are satisfied on the deformed boundary.

MICROSPHERES

The treatment of materials containing elastic-plastic microspheres is much more intricate than that of those containing air-filled cavities. This is so, chiefly, for two reasons.

Firstly, the modelling of each individual microsphere is complicated by the extra layer represented by the shell wall. This is generally much stiffer than the surrounding substrate and, when subjected to hydrostatic loading, undergoes a deformation which is not merely elastic but elastic-plastic or perfectly plastic depending on the wall thickness and the amount of pressure applied. For a typical shell, the behaviour is initially elastic up to a point at which the deviatoric forces generated inside the shell exceed the yield stress of the shell material. At this point, a boundary between elastic and plastic deformation begins to migrate through the material from the inner to the outer shell boundary. We shall consider here only the case of thin walls for which the transition from elastic to fully plastic behaviour occurs very rapidly as a function of the external load and, indeed, may be assumed to happen instantaneously at a pressure which, in the following, will be known as the yield pressure. (Consequently, a separate treatment of the elastic-plastic regime is unnecessary.)

Secondly, there is the complication that both the radius and shell thickness vary in size; so for typical pressures, a shell-reinforced foam contains some inclusions which behave elastically and also some which behave plastically. Therefore its compressional behaviour is dependent on the properties of all of the microspheres rather than on the overall void fraction alone.

As for the air-filled cavities, we shall describe only the case of a distribution of identical microspheres, noting that the modelling extends easily to take account of the deformation behaviour of layers containng inclusions of different sizes and wall thicknesses.

The model used here is based largely on an extension of the models in [6] and [7] to account for the behaviour of a glassy shell containing air and embedded in an elastic or viscoelastic substrate. We shall discuss the modelling of the individual microspheres, splitting this process to deal with the cases of perfectly elastic and perfectly plastic shells and then defining a critical value ϵ^* for each pressure to serve as the boundary between the elastic region (shell fraction $\epsilon > \epsilon^*$) and the plastic region ($\epsilon < \epsilon^*$). In this context, the shell fraction is defined as the ratio of the shell wall volume to the total volume of the scatterer.

Elastic Region: The pressure dependency of a microsphere in its perfectly elastic region can be modelled by treating the inclusion as a fluid-filled spherical cavity as before but with an additional elastic layer to represent the shell wall. The assumptions of isotropy and incompressibility apply also to this interface layer. Further, we assume that the shape of the microsphere is not distorted under compression, i.e. that there is no vorticity or buckling. Boundary conditions are then similar to those from the previous case, i.e. continuity of radial stress at each elastic-elastic boundary and matching to the appropriate pressures at the cavity surface and at the unit cell wall. The solution process follows as before and it is found that

$$(\phi' - \phi)\left[1 - \phi' - \frac{\mu_x}{\mu} + \frac{1}{1 - \epsilon x}\left(\frac{\mu_x}{\mu} + \frac{3\Gamma}{4\mu}\right)\right] + \frac{3P}{4\mu}\phi'(1 - \phi) = 0 \quad (14.4)$$

where μ_x is the shear modulus in the shell and x is defined as

$$x = (\phi/\phi')(1 - \phi')/(1 - \phi). \quad (14.5)$$

Equation (14.4) can clearly be rearranged to form a cubic in ϕ'. It is easily shown that this cubic has exactly one root in the region $0 < \phi' < \phi$ and, having found the new void fraction, the modified shell fraction, radius, internal pressure and layer thickness follow by application of Boyle's law and the incompressibility conditions.

Plastic Region: In a plastic material, the radial and transverse stresses are related by the yield function. Various theoretical definitions for a yield point have been presented in the literature; for the present problem, the most satisfactory of these would appear to be the Tresca yield condition which in this case leads to a yield point χ defined by $2\chi = \sigma_{rr} - \sigma_{\theta\theta}$. Substituting that into the governing equation for equilibrium,

$$(\partial\sigma_{rr}/\partial r) + (2/r)(\sigma_{rr} - \sigma_{\theta\theta}) = 0 \quad (14.6)$$

gives a simple expression for the radial stress σ_{rr} in the shell. Once again, boundary conditions are imposed to ensure the continuity of radial stress. The algebra then leads to the following transcendental equation for ϕ'

$$\left(1 - \frac{\phi}{\phi'}\right)\left[(1 - \phi')(1 - \epsilon x) + \frac{3\Gamma}{4\mu}\right] + (1 - \phi)\left[\frac{3P}{4\mu} + \frac{\chi}{\mu}\ln(1 - \epsilon x)\right](1 - \epsilon x) = 0$$

$$(14.7)$$

where x is defined as in (14.5).

This time a more subtle argument is required to determine whether (14.7) has a unique solution. It turns out that this is the case provided that $3P + 4\chi\ln(1 - \epsilon) > 0$, i.e. at sufficiently large pressures. Moreover, we can show that this condition is satisfied for pressures above the critical pressure at which the shell yields (see next section) - therefore at the pressures which violate the condition, the shell is still operating in its elastic regime and deforms according to (14.4).

Yield Conditions: For thin shells, the transition from the onset of yield to full deformation occurs very rapidly as a function of the external load and thus may be assumed to happen instantaneously. We shall therefore associate the critical pressure as that at which the shell has fully yielded, i.e. at which yield occurs at the outer boundary (the deviatoric stresses being lowest there). So we must have $2\chi = \sigma_{rr} - \sigma_{\theta\theta}$ evaluated at the outer shell radius. This leads to

$$\chi = -\mu_x(1 - x) \tag{14.8}$$

and eventually to

$$P = \frac{4\mu}{3}(x - 1)\left[\frac{(1 - \phi)x}{\phi + (1 - \phi)x} + \frac{3\Gamma}{4\mu}\frac{1}{1 - \epsilon x}\right] - \frac{4\chi}{3}\ln(1 - \epsilon x). \tag{14.9}$$

Substituting (14.8) into (14.9) displays the pressure P^* as a function of the initial geometry and the material properties - this can be interpreted as the critical pressure at which the plastic deformation has occurred for a given shell thickness. It is easy to show that at this pressure, the uniqueness condition is satisfied so that (14.7) has a unique solution.

Finally, we note that (14.9) can be re-arranged in terms of a critical thickness ϵ^* at a specified pressure P. Since the materials typically contain different sizes of microspheres, this enables us to discriminate between the thinnest shells which are treated as elastic and those thicker shells which are assumed to be plastic at that pressure. It is then easily shown that $d\epsilon^*/dP > 0$, which indicates that as hydrostatic pressure increases, so does the critical shell thickness and hence also the region of plasticity, i.e. the material becomes more plastic at depth.

4.   COMBINED MODELLING

As shown in [8], the acoustic performance of materials containing air-filled cavities and voids can be modelled as a function of depth by substituting the depth-modified geometrical parameters into a suitable model for acoustic propagation in materials with inclusions. The same argument may be applied to materials containing microspheres using, for example, the model described in [3]. However, this pre-supposes that the plastic microspheres can be treated in exactly the same manner as those which are still operating in their elastic regime, i.e. that the acoustic properties of the shell (measured at atmospheric pressure) are unaffected by its yielding. This is unlikely to be true.

Consider, for example, a typical stress-strain curve associated with an idealised elastic/perfectly plastic material. While the material is operating in its elastic regime, the gradient of the curve represents the shear modulus. Above the transition pressure, further strain can be obtained with no additional stress, resulting in a zero slope. We can represent this by defining the material (in its plastic state) to have a zero shear modulus, i.e. by treating the yielded shells as fluids.

Clearly, the above non-rigorous argument takes no account of changes in the polymer chemistry and micromechanics - however, this very simple phenomenological model has been found to provide extremely accurate predictions of depth-dependency behaviour in a number of different polymer-microsphere materials.

To assess the effectiveness of the combined model, we constructed a two layer composite for acoustic testing in a pressurised pulse tube. A 17.7 mm thick base layer contained a high concentration (38% by volume) of microspheres to provide a Transmission Loss which would change noticeably with pressure within the dynamic range of the measurement system. This was covered by a 44.6 mm thick anechoic layer which contained a much lower concentration (around 4%), thus enabling the Echo Reduction predictions to be assessed during the same tests.

For each of the two layers, we have specified a constant inner radius of the microspheres to match a mean measured value of 50 μm and assumed that the shell thickness is distributed according to the Rayleigh distribution function

$$n(\epsilon) = \frac{n_0 \pi \epsilon}{2\bar{\epsilon}^2} \exp\left(-\tfrac{1}{4}\pi(\epsilon/\bar{\epsilon})^2\right) \tag{14.10}$$

where n_0 is a normalisation factor and $\bar{\epsilon}$ is the mean shell fraction (taken to be 0.025 here). Although chosen on heuristic grounds, initial measurements indicate that this distribution is not far from reality for the grade of microspheres used in the samples described here.

Figs. 2 and 3 display the measurements of Transmission Loss and Echo Reduction at five different pressures and figures 4 and 5 illustrate the corresponding model predictions. They clearly demonstrate the complex nature of these types of materials. For example, we see that the Transmission Loss initially increases in value as the shells become softer and the overall material less stiff with increasing pressure. Above a critical point, the assumed effect of increasing the pressure results in plastic deformation which leads to a reduction in acoustic performance. At that stage the material behaves similarly to an air-blown foam. Degradation is then attributable to a combination of factors (e.g. reduced volume fraction of air, reduced size of cavities with consequent increase in air pressure, reduced thickness of the composite and change in the overall acoustic impedance of the structure). We note that the absolute values predicted by the model are lower than the measured Transmission Loss

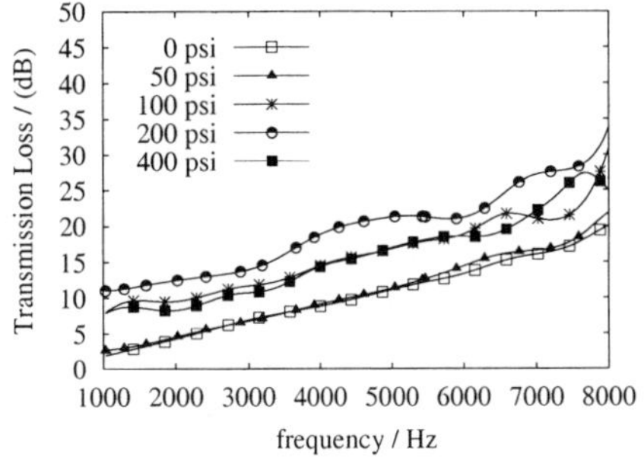

Figure 2   TL (measured)

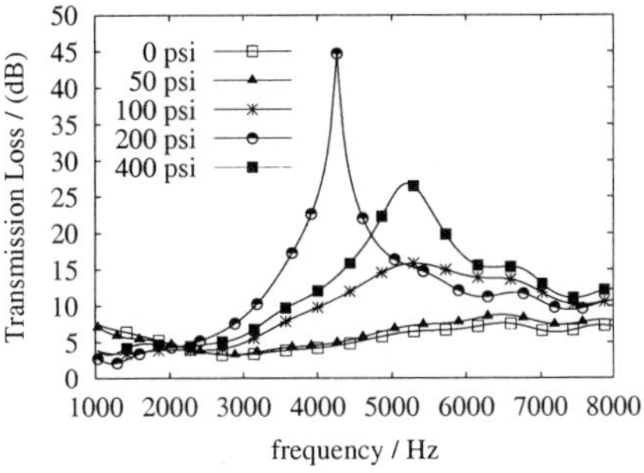

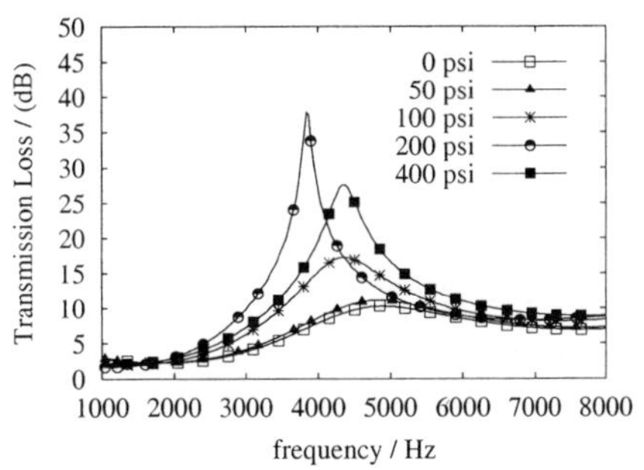

Figure 3   ER (measured)

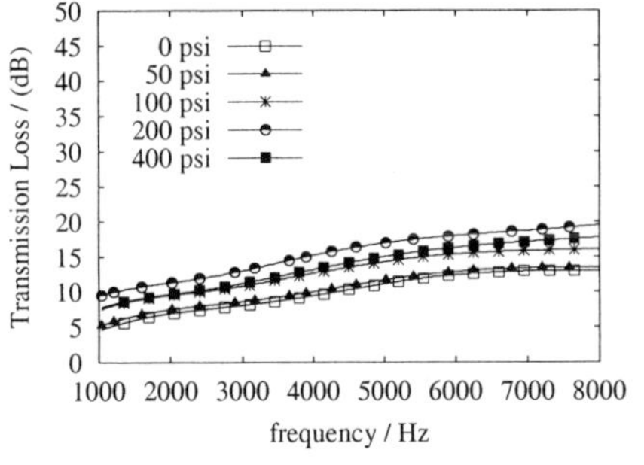

Figure 4   TL (modelled)

Figure 5   ER (modelled)

although the critical pressure is well predicted. The discrepancy is partly attributable to our choice of materials properties (particularly the size distribution function and our estimates for static shear and yield terms) and to the limiting of the measurements to one representative sample. It would be possible to improve the agreement by measuring and averaging over a large number of samples (thus randomising the microsphere distribution more).

Echo Reduction is similarly well predicted with the peaks moving first downwards in frequency as the material softens and then upwards again as the stiffening effect of the pressure collapse takes place. As before, the absolute values are slightly different from those found by experiment (as are the positions of the peaks). However, in summary, we would conclude that the model correctly predicts the nature of the pressure dependency and thus can be very suitable as a design tool.

5.   SUMMARY

We have described some of our analytical techniques for modelling the acoustic performance of materials which have been subjected to hydrostatic pressure. Our model takes a two-step approach, firstly by applying a simple phenomenological technique to predict the static deformation of the material due to the pressure and then by modelling the sound propagation in the deformed medium using an effective medium technique. This combined model has been validated against measurements taken in deep water in Norway and against smaller samples in pressurised conditions in a pulse tube and has been found to provide reliable results for a variety of materials.

Acknowledgments

The authors acknowledge the contribution of David Townend of DERA Farnborough and his team for constructing the test materials, for providing information on materials properties and for performing the Echo Reduction and Transmission Loss measurements at different pressures.

References

[1] Gaunaurd, G C (1989) *Resonance theory of the effective properties of perforated solids*, Appl. Mech. Rev. **42**, 143–192.

[2] Hashin, Z (1983) *Analysis of composite materials – a survey*, J. Appl. Mech. **50**, 481–505.

[3] Baird, A M, Kerr, F H and Townend, D J (1999) *Wave propagation in a viscoelastic medium containing fluid-filled microspheres*, J. Acoust. Soc. Am. **105**, 1527–1538.

[4] Gaunaurd, G C and Überall, H (1982) *Resonance theory of the effective properties of perforated solids*, J. Acoust. Soc. Am. **71**, 282–295.

[5] Waterman, P C and Truell, R (1961) *Multiple scattering of waves*, J. Math. Phys. **2**, 512–537.

[6] Christiansen, R M (1979) *Mechanics of Composite Materials*, New York: Wiley.

[7] Carroll, M M and Holt, A C (1972) *Pore-collapse relations for ductile porous materials*, J. Appl. Phys. **43**, 1626–1636.

[8] Gaunaurd, G, Callen, E and Barlow, J (1984) *Pressure effects on the dynamic effective properties of resonating perforated elastomers*, J. Acoust. Soc. Am. **76**, 173–177.

ACOUSTIC WAVE PHASE CONJUGATION IN ACTIVE MEDIA

Numerical simulations

A. Merlen, S. Ben-Khelil

Laboratoire de Mécanique de Lille, ura CNRS 1441

Bat M3 Cité scientifique, 59655 Villeneuve d'Ascq Cedex, France

alain.merlen@univ-lille1.fr, khelil@onera.fr

V. Preobrazhensky

Waves Research Center, Russian Academy of Science

38 Vavilov Street, Moscow 117942, Russia

preobr@newmail.ru

P. Pernod

Institut d'Electronique et de Micro Electronique du Nord

DOAE, umr CNRS 8520, 59655 Villeneuve d'Ascq Cedex, France

philippe.pernod@iemn.univ-lille1.fr

Abstract   The WAF computation scheme is applied to numerical simulations of wave propagations in time-dependent heterogeneous media (liquids and active solids). The parametric phase conjugation of a wide band ultrasound pulse is considered. The supercritical dynamics of the acoustic field is described, in linear and non linear pumping conditions, for one-dimensional systems containing a parametrically active solid.

1.   INTRODUCTION

The problem of wave propagation in a nonstationary medium when parameters depend on time is of fundamental interest because of its various applications in optics, acoustics and solid state physics. In acoustics, parametric wave phase conjugation (WPC) has been studied for liquids and solids and piezo-semiconductor systems. The modulation of the elastic parameters of solids is usually carried out by means of rf-, microwave- or optical pumping, distributed almost homogeneously in the

133

I.D. Abrahams et al. (eds.),
IUTAM Symposium on Diffraction and Scattering in Fluid Mechanics and Elasticity, 133–140.
© 2002 *Kluwer Academic Publishers. Printed in the Netherlands.*

active zone of the medium. There are no exact analytical solutions to the general problem of parametric WPC. The perturbation theory is applicable for relatively weak parametric interactions under the threshold of absolute parametric instability. Above the threshold (in a supercritical mode) multiscale asymptotic expansion methods (MSAE) can be used to describe narrow band resonance parametric interactions [1, 2]. Recently the problem of WPC has been discussed extensively in the context of ultrasound time reversal transformation for applications in nondestructive testing and medicine [3, 4]. The applicability of MSAE methods in practical conditions becomes problematical. For this reason, the development of numerical methods adapted to the problem seems to be a productive research orientation. On the basis of the propagation properties involved in the phenomenon it is possible to show that the mathematical problem falls within the scope of hyperbolic partial differential systems. Therefore the numerical background developed in the last decade within the frame of unsteady aerodynamics can be applied to this problem and, particularly, all the Godunov family schemes [5]. This paper presents an example of supercritical parametric WPC of a wide band acoustic pulse in linear and non linear pumping conditions.

2.   MATHEMATICAL FORMULATION

The basic idea of the present numerical approach comes from the natural formulation of acoustics in fluids. Nevertheless, here, the sound speed in the active medium is defined by a given function $C(t)$ such as $c^2 = C(t)^2 = c_0^2(1 + m\cos(\Omega t + \Psi))$ where m is a small parameter ($m \ll 1$) referred to as the 'modulation depth'. By linearisation for small m, the problem can be written in one dimension (1D) as

$$\left.\begin{array}{l} \dfrac{\partial w_1}{\partial t} + c_0 \dfrac{\partial w_1}{\partial x} = m\dfrac{\Omega}{4}(w_1 - w_2)\sin(\Omega t + \Psi) \\[4mm] \dfrac{\partial w_2}{\partial t} - c_0 \dfrac{\partial w_2}{\partial x} = -m\dfrac{\Omega}{4}(w_1 - w_2)\sin(\Omega t + \Psi) \end{array}\right\} \qquad (15.1)$$

where $w_1 = v + \theta$, $w_2 = v - \theta$, with $\theta = (p - p_0)/(\rho c)$ and p_0 being the uniform steady pressure of the medium at rest and p the instantaneous pressure. The problem finally comes down to two advection equations in opposite directions coupled by linear source terms. Apparently, the numerical treatment of such problems is well known but the need to manage high frequencies makes it less trivial than it seems at a first glance. The scheme has to be robust and very weakly dissipative.

For linear elasticity the formulation in previous form is less natural, but since the elastic properties do not depend on the space coordinates, the classical derivation of the wave equations in elastic bodies still holds. Briefly, in 1D the problem reduces to the wave equation for the compression waves. Letting ψ be the compression potential $\mathbf{u}_L = \operatorname{grad} \psi$ where $\mathbf{u}_L$ is the longitudinal displacement, the Navier equation gives $\triangle\psi - c_L^{-2}\partial^2\psi/\partial t^2 = 0$ with c_L the compression wave velocity. The following change of variables is introduced: $\theta_L = -c_L^{-1}\partial^2\psi/\partial t^2 = -\sigma_{xx}/(\rho c_L)$, $\mathbf{v}_L = \partial\mathbf{u}_L/\partial t = \operatorname{grad}(\partial\psi/\partial t)$ where σ_{xx} is the normal stress in the propagation direction. Consequently, $\operatorname{grad}\theta_L = -c_L^{-1}\partial\mathbf{v}_L/\partial t$ and the wave equation becomes:

$$\frac{\partial\theta_L}{\partial t} + c_L \operatorname{div} \mathbf{v}_L = \frac{\theta_L}{c_L}\frac{\partial c_L}{\partial t}. \tag{15.2}$$

Finally, after linearisation, and introducing again w_1 and w_2, these equations once again give system (15.1) but with a source term of the opposite sign. In one dimension no mode changes are expected between compression and shear waves, even at the interfaces, consequently no shear waves appear if no shear stress exists initially in the medium. For clarity of the present analysis, focused on the source term, we propose now that this condition is fulfilled. This can be done without any loss of generality, since the physical phenomenon behind the source terms is the magneto acoustic interaction, which is independent of the actual geometry of the sample. In the above presentation, the wave speed is slave of the magnetic field oscillation; this characterizes the linear pumping hypothesis. In real situations, the elasticity reacts on the electric circuits through its feedback on the magnetic field. The latter is slightly modified by the variation of the molecular spin that it contributes to produce inside the sample. This spin reorientation is the origin of the variation of stiffness in the solid. As a result, an interaction exists between the electric circuit providing the energy of the magnetic field and the variation of the elastic properties inside the sample. The numerical simulations have been an ideal means for testing different physical modelling of this interaction. Among these models, only one allows the simulations to reproduce non linear behaviours found in experiments. In this model the electric part is assimilated to a RLC circuit. According to this approach the wave speed in the solid has to be rewritten as

$$c^2 = c_0^2\left[1 + m\left(\cos(\Omega t + \Psi) + \frac{mMc_0^2}{4\pi P_e L}\int_0^L \frac{\partial}{\partial t}\left(\frac{\partial U}{\partial x}\right)^2 dx\right)\right]$$

where U is the displacement, M and L the mass and length of the sample and P_e the electric power. It can be seen that the non linear term is proportionnal to the ratio between the acoustic and the electric power.

This term becomes of the order of the sinusoidal oscillation as soon as the amplification by parametric resonance is sufficiently high. The time derivative in the integral can easily be changed into a spatial one by means of the advection operator and the problem can be written as a hyperbolic conservation system:

$$\frac{\partial U}{\partial t} + \frac{\partial F(U)}{\partial x} = S, \quad \forall x \in \mathbb{R}, \quad t \geq 0$$

$$U(x,0) = U_0(x) \quad \forall x \in \mathbb{R}$$

(15.3)

with

$$F(U) = \begin{pmatrix} c_0\, w_1 \\ -c_0\, w_2 \end{pmatrix}, \quad U = \begin{pmatrix} w_1 \\ w_2 \end{pmatrix},$$

$$S = -\begin{pmatrix} m(w_1 - w_2)\left(\frac{\Omega}{4}\sin(\Omega t + \Psi) - s\right) \\ -m(w_1 - w_2)\left(\frac{\Omega}{4}\sin(\Omega t + \Psi) - s\right) \end{pmatrix}$$

and

$$s = \frac{mMc_0^2}{8\pi P_e L}\left(\int_0^L \left(\frac{\partial w_1}{\partial x}\right)\left(\frac{\partial w_2}{\partial x}\right) dx + \frac{1}{4}\left[(w_1 - w_2)\frac{\partial(w_1 - w_2)}{\partial x}\right]_0^L \right).$$

System (15.3) is solved by an explicit finite volume method. The spatial domain is shared in N cells of length Δx and the time step is Δt. The numerical solution $U_i^n = U(i\Delta x, n\Delta t)$ is obtained at time $(n+1)\Delta t$ by

$$U_i^{n+1} = U_i^n - (\Delta t/\Delta x)[f_{i+\frac{1}{2}}(U^n) - f_{i-\frac{1}{2}}(U^n)] + \Delta t S(U_i^n). \qquad (15.4)$$

The choice of the numerical fluxes f at the cell interface characterises the scheme. Tests have been performed with the basic first order Godunov scheme [5] and its extension to second order with MUSCL (Monotone Upstream Centred Scheme for Conservation Laws) approach with limiter and finally with the second order WAF (Weighted Average Flux) and superbee limiter [6].

Numerical simulations help to explain phenomena which could not be accounted for by experiments or analytical results. For instance, the instantaneous stress field inside a sample of active medium is not available in experiments, or the effect of wave reflections on the sample boundary requires very complicated theoretical developments in the frame of purely analytical analysis. This issue is easily treated in the present approach by solving the classical problem of 'resolution of a discontinuity', which provides the numerical fluxes in the Godunov familly schemes. The interface separates two non-active media: R (Right) and

L (Left). For the 1D case, solid and fluid are treated in the same way. In medium R, of sound velocity c_R, the initial state is (U_R, θ_R) and (U_L, θ_L) in medium L with c_L as sound velocity. The continuity of stresses and normal velocity at the interface, of velocity v_i, gives $\theta_2 = \tau\theta_1$ and $v_2 = v_1 = v_i$ where $\tau = \rho_L c_L/(\rho_R c_R)$ is the transmission coefficient. Subscript 1 corresponds to the solution in medium R and subscript 2 in medium L. This solution is given by:

$$v_1 = v_2 = v_i = [\tau v_L + v_R + \tau\theta_L - \theta_R]/(1 + \tau),$$
$$\theta_1 = [v_L - v_R + \theta_L + \theta_R]/(1 + \tau).$$

When $\tau = 1$ the solution corresponds to the Riemann problem of the linear advection system in an homogeneous medium as given by Godunov. For $\tau \to 0$, medium R is infinitely rigid ($c_R \to \infty$), the solution is $\theta_1 = v_L + \theta_L$, $v_1 = v_i = v_2 = 0$, which corresponds to the 'half Riemann' problem for the linear advection problem often used for non transmittive boundary conditions. When $\tau \to \infty$, medium R is a vacuum and $\theta_1 = \theta_2 = 0$, $v_1 = v_2 = v_L + \theta_L$.

The present solution provides all the data needed for computing the fluxes at interfaces between non-active zones in any 1D situation. The boundary between active and passive zones is simulated in a very straightforward way by switching off the source terms in the passive zone. The same is done in an active zone as soon as the pumping has stopped.

3.   RESULTS

The numerical method and the interface problem have been mainly tested on the configuration presented below. From $x = 0$ to 0.5 cm the medium is water and the sample of magnetoacoustic ferrite is situated between $x = 0.5$ cm to $x = 4$ cm. The active zone lies between $x = 1$ cm and $x = 3.5$ cm. The mesh contains 1000 points in water and 7000 points in the sample. The boundary condition at $x = 0$ is non reflective as if the domain $x < 0$ were filled by water too. At $x = 4$ cm we assume no stress as if a vacuum would exist for $x > 4$ cm. The initial condition is given in the following form:

$$\begin{cases} w_1 = 2\sin(\frac{2\pi}{\lambda}(x - x_L)), & w_2 = 0 \quad \text{for } x_L < x < 0.5 \\ w_1 = w_2 = 0 & \text{elsewhere.} \end{cases} \tag{15.5}$$

Abscissa x_L is chosen such as $(0.5 - x_L) = 3\lambda$ where λ is the wave length: $\lambda = 2\pi c_0/\omega$ with $\omega = \pi 10^7 \text{s}^{-1}$ and $c_0 = 1500\text{m/s}$. The pumping begins at $t = 0$. In order to produce parametric resonance, the pumping frequency is $\Omega = 2\omega$. A few tests have also been performed in an infinite medium with a finite active zone of 3 cm in order to compare the numerical results with an analytical solution. Fig. 1 shows total agreement

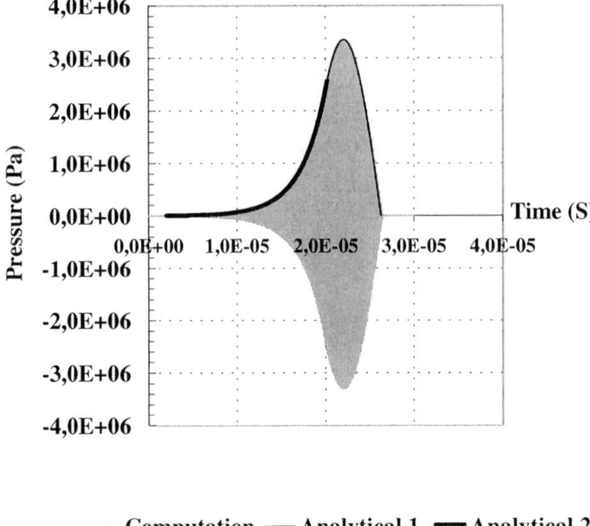

Figure 1   Comparison of the analytical and numerical solution in finite medium

between theory and numerical simulation in such an infinite medium for the stress versus time at a give point outside the active zone. In that case the pumping duration is $T = 19\mu s$ and $\Psi = \pi$, $m = 0.032$. Curve analytical 1 is the exponential growth, curve 2 is the sinusoidal amplitude of the travelling wave after the end of the pumping.

Fig. 2 shows the normal stress field (in Pa) inside the sample and at different times for $m = 0.041$. On the first figure ($t = 0.75\mu s$) the pumping has just been initiated and the direct amplified wave begins from the edge of the active zone. In the second one ($t = 7.75\mu s$) the conjugate wave can be clearly observed in the fluid. The impedance conversion of amplitude between stress and pressure is clear. In the solid, the incident wave has been just reflected at the end of the sample. The third ($t = 14.25\mu s$) illustrates the amplification process and the emission of the conjugate wave in the fluid. The incident wave (or its reflection on the edges of the sample) is no longer visible due to the high level of amplification of the conjugate and direct waves. In the last one ($t = 21.75\mu s$), the pumping is finished and all the waves are to be evacuated through the interface towards the fluid after many reflections at the end of the sample.

Fig. 3 shows the effect of the non linear term on the stress evolution at one point outside the active zone. In this case the pumping duration is $35\mu s$. This exhibits the saturation of the signal as in experiments.

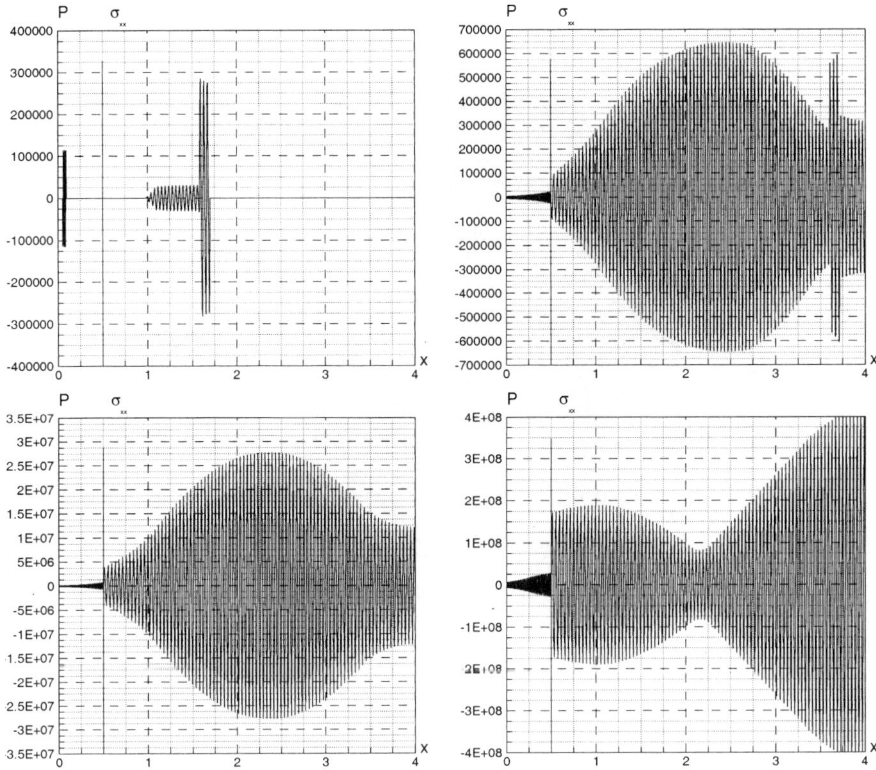

Figure 2 Numerical solution at different times: Stress in a finite sample (water-sample-vacuum)

4. CONCLUSION

In many respects, solving numerically the 1D problem, gives more information on the efficiency of the methods and on the physical behaviour of solutions than a direct resolution of a 3D case with all its inessential features due to purely geometrical effects or mesh problems. The present work validates the numerical approach, particularly for the interface problem. From the physical point of view, the simulations allowed the identification of the right model for the electro-acoustic coupling. The extension to 2D is now possible with inclusion of shear waves. This is the next step in process now. For 3D, a theoretical improvement is necessary in order to reduce the number of mesh cells by a hybrid asymptotic-numerical method.

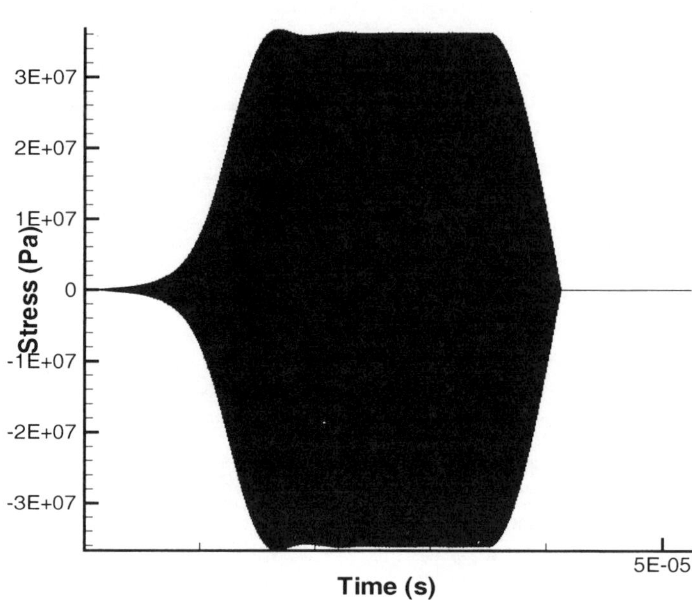

Figure 3   Non linear pumping: stress saturation

References

[1] Bobroff, D L and Hans, H A (1967) *Impulse response of active coupled wave systems*, J. Appl. Phys. **38**, 390–403.

[2] Brysev, A P, Bunkin, F V, Hamilton, M F, Krutyianskii, L M, Cunningham, K B, Preobrazhensky, V L, Pyl'nov, Yu V, Stakhovskii, A D and Younghouse, S J (1998) *Nonlinear propagation of a quasi-plane conjugate ultrasonic beam*, Acoust. Phys. **44**(6), 641–650.

[3] Fink, M (1996) *Time reversal of ultrasonic fields*, IEEE Trans. **UFFC-39**, 555–592.

[4] Brysev, A P, Krutyianskii, L M and Preobrazhenskii, V L (1998) *Wave phase conjugation of ultrasonic beams*, Physics-Uspekhi **41**(8), 793–805.

[5] Godunov, S K (1959) *A finite difference method for numerical computation of discontinuous solutions of the systems of the equations of fluid dynamics*, Mat. Sb. **47**, 357–393.

[6] Toro, E F (1992) *The weighted average flux method applied to the Euler equations*, Phil. Trans. Roy. Soc. Lond A **341**, 499–530.

WAVE PASSAGE THROUGH A STRING HAVING MULTIELEMENT INCLUSIONS WITH PARTIAL INTERIOR DYNAMICS

A. D. Sergeyev

Institute for Problems in Mechanical Engineering, RAS
Bolshoy pr. V.O. 61, St. Petersburg 199178, Russia
sergeyev@cards.lanck.net

D. A. Sergeyev

Railway Transport Institute, Technical University of Riga
Indrika Street, N8 LV-1004 Riga, Latvia
dijs@dzti.edu.lv

Abstract   Some theoretical problems connected with interpretation of experimental results are considered. The phenomenon of suppressing longitudinal oscillations by means of transverse ones and vice versa takes place in a very reliable mechanical model, which is often used not only in Mechanics, but in simulating various real physical processes. There exists a definite qualitative indeterminacy for the spatial multi-element system's observable responses onto the typical harmonic perturbation.

1.   INTRODUCTION

Suppose our physical object of inquiry is a structure, which may be simulated by a continuous system with compound discrete inertia inclusions placed at a finite distance from a disturbing source. Particular properties of the suggested object are substantial extension in one direction (conventionally called the *longitudinal* direction), and the presence of distributed interaction between the object and an environment along the longitudinal direction; the object has a concentrated nonhomogeneity with interior dynamics, and a periodic disturbance on the end of the distributed parameter system at a finite distance from the nonhomogeneity.

Theory predicts that in a system of such a type so-called trapped oscillations may exist. They are attributes of the system and take place just in the local vicinity of various inclusions. The localized vibrations

I.D. Abrahams et al. (eds.),
IUTAM Symposium on Diffraction and Scattering in Fluid Mechanics and Elasticity, 141–148.
© 2002 *Kluwer Academic Publishers. Printed in the Netherlands.*

actually do not depend on details of a stimulation, if before an original perturbation reaches a discrete inclusion it passes through some intermediate inhomogeneous media. In this case, it is mainly the inclusion's interior dynamical properties which determine the behaviour of the rigid box of the inclusion.

2.   SIMULATING THE OBJECT

Let a string of semi-infinite extent, laid on Winkler's basis, be the continuum to model distributed elastic-inertia properties of the object. Suppose the nonhomogeneity is an N-particle-oscillator placed inside a rigid box (platform), pictured in Fig. 2. In order to prevent infinite longitudinal displacements in the system formally, the rigid box is also supposed to be connected to some immovable body by an inertialess spring. The spring is not presented in the drawing.

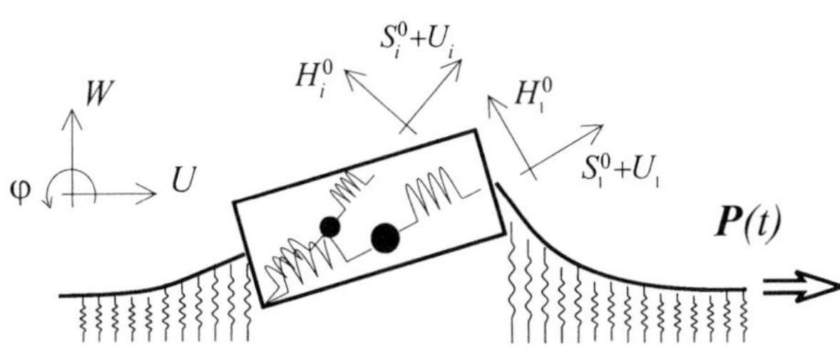

Figure 1   The physical model of the system under investigation.

The basis stiffness at the transverse direction is denoted by k_w. Suppose T_0 is a static tension of the string, A is its elastic modulus, s is a lagrangian coordinate, ρ is a string mass density. Introduce a longitudinal force $P(t)$ as an exterior perturbation in the system. This force acts on the end of the string with the lagrangian coordinate $s = s_\star$. Let the string be in contact with the platform at the points $s = -L_0/2$ and $s = s_\star - L_0/2$.

An individual oscillator is a particle of mass m_i. This particle, being attached to the inertialess spring, can move only along a rectilinear directrix, fixed in the platform; μ_i is the stiffness of the spring. The directrix with number i is placed at an angle α_i to the nondeformed string at the distance H_i^0 from the centre of mass of the platform, $U_i(t)$ is a small dynamic displacement of the particle along the directrix with respect to an equilibrium configuration. This configuration is characterized by the parameter S_i^0.

3.   HARMONIC LONGITUDINAL PERTURBATION

The following equations describe the elastic-inertia part of our system

$$\rho\,\ddot{u} = A\,u'', \qquad \rho\,\ddot{w} = T_0\,w'' - k_w w,$$

where, for example, $\ddot{u} = \partial^2 u/\partial t^2$ and $u'' = \partial^2 u/\partial x^2$. We have boundary conditions at the points of a contact between the string and the platform,

$$u(-L/2, t) = U, \qquad w(-L/2, t) = W - L\varphi/2$$
$$u(L/2, t) = U, \qquad w(L/2, t) = W + L\varphi/2, \qquad Au'(s_\star, t) = P(t).$$

Linearized equations of motion for the multielement inclusion are

$$m_\Sigma \ddot{U} + \mu_0\,U - m_{1,N}\,H_{1,N}\,\ddot{\varphi} + \sum_{i=1}^{N} m_i \ddot{U}_i \cos\alpha_i$$
$$= Au'\,(L/2) - Au'(-L/2), \quad (16.1)$$

$$m_\Sigma \ddot{W} + m_{1,N}\,S_{1,N}\,\ddot{\varphi} + \sum_{i=1}^{N} m_i \ddot{U}_i \sin\alpha_i$$
$$= T_0(w'(L/2) - w'(-L/2)), \quad (16.2)$$

$$-m_{1,N}\,H_{1,N}\,\ddot{U} + m_{1,N}\,S_{1,N}\,\ddot{W} + J_{0\Sigma}\ddot{\varphi} + LT_0\varphi - \sum_{i=1}^{N} m_i H_i^0 \ddot{U}_i$$
$$= \tfrac{1}{2}LT_0\left(w'(-L/2) + w'(L/2)\right), \quad (16.3)$$

$$m_i\left(\ddot{U}\cos\alpha_i + \ddot{W}\sin\alpha_1 - H_i^0\ddot{\varphi} + \ddot{U}_i\right) = -\mu_i U_i, \quad i = 1,\dots,N, \quad (16.4)$$

where $m_\Sigma = \sum_{i=0}^{N} m_i$,

$$m_{1,N}\,S_{1,N} = \sum_{i=1}^{N} m_i\left(S_i^0 \cos\alpha_i - H_i^0 \sin\alpha_i\right),$$

$$m_{1,N}\,H_{1,N} = \sum_{i=1}^{N} m_i\left(S_i^0 \sin\alpha_i + H_i^0 \cos\alpha_i\right), \quad (16.5)$$

$$J_{0\Sigma} = J_0 + \sum_{i=1}^{N} m_i\left(S_i^{0^2} + H_i^{0^2}\right). \quad (16.6)$$

If there is no perturbation source 'on the left' of the platform in the semi-infinite domain, we have there

$$u(s, t) = u(s + L/2 + c_u t), \qquad c_u = \sqrt{A/\rho},$$

and longitudinal displacement u is determined by the boundary condition $u(-L/2, t) = U(t)$, because $u'(-L/2, t)$ is connected with the longitudinal velocity of the platform one-to-one. Then

$$u'(s, t)\big|_{s=-L/2} = \dot{u}(s, t)\big|_{s=-L/2}\Big/ c_u = \dot{U}/c_u. \qquad (16.7)$$

Consider our system under the harmonic longitudinal force $P(t) = P_0 \exp(i\omega t)$. The necessary condition for a perturbation to be a propagating wave in the semi-infinite string, laid on the Winkler basis, is a frequency of perturbation source exceeds so-called cut-off frequency. Suppose the magnitude of the frequency ω is situated below the 'transverse' cut-off frequency $\omega_{bw} = \sqrt{k_w/\rho}$. According to the Winkler basis properties the longitudinal cut-off frequency is zero. Hence, let us find a stationary solution in the form $f(t) = f_0 \exp(i\omega t)$. In that case the equations of the string are

$$\rho\omega^2 u = -A\,u'', \qquad \rho\omega^2 w = -T_0 w'' + k_w w.$$

The contact condition $w(-L/2) = W_0 - L\varphi_0/2$ on the left board of the platform is a perturbation for the semi-infinite domain of the string. On the left of the inclusion we have:

$$w(s) = (W_0 - L\varphi_0/2)\, e^{\lambda(s+L/2)}, \qquad \lambda = \sqrt{\frac{\omega_{bw}^2 - \omega^2}{c_w^2}}, \qquad c_w^2 = \frac{T_0}{\rho}. \qquad (16.8)$$

The boundary conditions for $L/2 \le s \le s_\star$ are

$$u(L/2)=U_0, \quad P_0=Au'(s_\star), \quad w(L/2) = W_0+L\varphi_0/2, \quad w'(s_\star)=0.$$

The solution of the problem of longitudinal oscillations for the right finite domain of the string can be represented as a sum of two absolutely convergent series,

$$\begin{aligned}
u(s) &= U_0\left(1 + \frac{2\omega^2}{L_\star}\sum_{k=1}^{\infty}\frac{\sin\gamma_k(s-L/2)}{\gamma_k(\gamma_k^2 c_u^2 - \omega^2)}\right) \\
&\quad + \frac{P_0}{A}\left(s - \frac{L}{2} - \frac{2\omega^2}{L_\star}\sum_{k=1}^{\infty}\frac{(-1)^k \sin\gamma_k(s-L/2)}{\gamma_k^2(\gamma_k^2 c_u^2 - \omega^2)}\right),
\end{aligned}$$

$$\gamma_n(L_\star) = \tfrac{1}{2}\pi + \pi(n-1), \quad n = 1, 2, \ldots, \qquad L_\star = s_\star - L/2.$$

Differentiating this solution with respect to s, and substituting $s = L/2$, we obtain

$$u'(L/2) = G_U^u U_0 + G_P^u P_0/A. \qquad (16.9)$$

Coefficients G_U^u and G_P^u are functions of the source's frequency,

$$G_U^u = \frac{2}{L_\star} \sum_{k=1}^{\infty} \frac{\omega^2}{\gamma_k^2 c_u^2 - \omega^2}, \quad G_P^u = 1 - \frac{2}{L_\star} \sum_{k=1}^{\infty} \frac{(-1)^k \omega^2}{\gamma_k(\gamma_k^2 c_u^2 - \omega^2)}.$$

On the right of the inclusion the boundary-value problem is satisfied by the function

$$w(s) = \frac{e^{-\lambda(s-L/2)} + e^{-2\lambda s_\star + \lambda(s+L/2)}}{1 + e^{-2\lambda L_\star}} \, (W_0 + L\varphi_0/2) .$$

Taking into account the localization phenomenon, i.e. $\exp(-\lambda L_\star) \ll 1$, we can get the following approximation for $w(s)$,

$$w(s) \approx (W_0 + L\varphi_0/2) \, e^{-\lambda(s-L/2)}. \tag{16.10}$$

Using solutions (16.8), and (16.10), we find $w'(s)$ at the points where the elastic line and the platform are in contact. Those functions are involved in the equations of multi-element inclusion motion. Hence

$$w'(-L/2) = \lambda(W_0 - L\varphi_0/2), \quad w'(L/2) = -\lambda(W_0 + L\varphi_0/2). \tag{16.11}$$

The structure of (16.4) allows us to get expressions for the amplitudes of the interior oscillator vibrations U_{i0}. We have

$$U_{i0} = \frac{m_i(\omega^2 U_0 \cos\alpha_i + \omega^2 W_0 \sin\alpha_i - H_i^0 \omega^2 \varphi_0)}{\mu_i - m_i\omega^2}, \quad i = 1,\ldots,N. \tag{16.12}$$

Substituting (16.7), (16.9), (16.11), and (16.12) into (16.1)–(16.3), we get a system of equations for the platform displacements

$$\left(m_\Sigma \omega^2 - \mu_0 - iA\omega/c_u + G_{uu} + AG_U^u\right) U_0 + G_{uw} \, W_0$$
$$+ \left(G_{u\varphi} - m_{1,N} H_{1,N}\omega^2\right) \varphi_0 = -G_P^u P_0,$$
$$G_{wu} U_0 + \left(G_{ww} - 2T_0\lambda + m_\Sigma\omega^2\right) W_0$$
$$+ \left(G_{w\varphi} + m_{1,N} S_{1,N}\omega^2\right) \varphi_0 = 0, \tag{16.13}$$
$$\left(G_{\varphi u} - m_{1,N} H_{1,N}\omega^2\right) U_0 + \left(G_{\varphi w} + m_{1,N} S_{1,N}\omega^2\right) W_0$$
$$+ \left(J_{0\Sigma}\omega^2 - LT_0 + G_{\varphi\varphi} - \tfrac{1}{2}\lambda L^2 T_0\right) \varphi_0 = 0,$$

where $G_{uw} = G_{wu}$, $G_{u\varphi} = G_{\varphi u}$, $G_{w\varphi} = G_{\varphi w}$,

$$G_{uu} = \sum_{i=1}^{N} \frac{m_i^2 \omega^4 \cos^2\alpha_i}{\mu_i - m_i\omega^2}, \quad G_{uw} = \sum_{i=1}^{N} \frac{m_i^2 \omega^4 \cos\alpha_i \sin\alpha_i}{\mu_i - m_i\omega^2},$$

$$G_{u\varphi} = -\sum_{i=1}^{N} \frac{m_i^2 \omega^4 H_i^0 \cos \alpha_i}{\mu_i - m_i \omega^2}, \qquad G_{ww} = \sum_{i=1}^{N} \frac{m_i^2 \omega^4 \sin^2 \alpha_i}{\mu_i - m_i \omega^2},$$

$$G_{w\varphi} = -\sum_{i=1}^{N} \frac{m_i^2 \omega^4 H_i^0 \sin \alpha_i}{\mu_i - m_i \omega^2}, \qquad G_{\varphi\varphi} = \sum_{i=1}^{N} \frac{m_i^2 H_i^{02} \omega^4}{\mu_i - m_i \omega^2}.$$

4.   SUPPRESSING 'HARMFUL' VIBRATIONS

Considering our compound multi-element system from the applied viewpoint, we intend to show why such a system may be used as a selective filter of mechanical perturbations, or a so-called 'vector' filter. Therefore, suppose some trapped oscillation forms are 'harmful' or 'undesirable' in the system, and we are in need of an idea to suppress each form. This suppression is to be effected by transformation of the 'undesirable' motions into the 'relatively safe' inclusion's rigid box oscillations. By the terminolgy 'relatively safe' the following is meant: oscillations are said to be of a 'relatively safe' type for a concrete system if they occur in orthogonal directions to the 'undesirable' ones. For example, the transverse oscillations of a train body induce vibrations of the surrounding earth surface and a noise in the ambient air, while the longitudinal oscillations influence only the vehicle dynamics and are safe for environment.

Rewrite (16.13) in the generalized notation

$$\begin{aligned}
a_{uu}(\omega)\, U_0 + a_{uw}(\omega)\, W_0 + a_{u\varphi}(\omega)\, \varphi_0 &= b_1\, P_0 \\
a_{wu}(\omega)\, U_0 + a_{ww}(\omega)\, W_0 + a_{w\varphi}(\omega)\, \varphi_0 &= 0 \qquad (16.14) \\
a_{\varphi u}(\omega)\, U_0 + a_{\varphi w}(\omega)\, W_0 + a_{\varphi\varphi}(\omega)\, \varphi_0 &= 0.
\end{aligned}$$

Let us consider the case

$$a_{uw}(\omega) = a_{wu}(\omega) = 0, \qquad a_{u\varphi}(\omega) = a_{\varphi u}(\omega) = 0.$$

Separating the longitudinal motion, we get the homogeneous system for the transverse oscillations. One can easily make the 'transverse' system's determinant nonzero. According to the theorem, which is well known in linear algebra, we claim that our system has no 'transverse' solution except $W_0 = 0$ and $\varphi_0 = 0$. Thus, in order to suppress the transverse trapped vibrations of the string it is enough to arrange the special packing of the interior oscillators into the rigid box depending on the source's frequency, or change some eigen-frequencies for a number of internal oscillators to fulfill

$$\sum_{i=1}^{N} \frac{m_i \cos \alpha_i \sin \alpha_i}{\omega_i^2 - \omega^2} = 0, \qquad \omega_i^2 = \mu_i / m_i,$$

$$-\sum_{i=1}^{N} \frac{m_i \omega^2 H_i^0 \cos \alpha_i}{\omega_i^2 - \omega^2} = \sum_{i=1}^{N} m_i \left(S_i^0 \sin \alpha_i + H_i^0 \cos \alpha_i \right).$$

$$(16.15)$$

Conditions (16.15) give relations for elastic, inertia, and geometric parameters of the interior oscillators, which are to be used for complete suppressing of the transverse vibrations in the continual part of the object by means of box's interior dynamics. Using (16.5) in the case $N = 2$ we have the following condition to be fulfilled

$$\frac{m_1 \sin 2\alpha_1}{\omega_1^2 - \omega^2} + \frac{m_2 \sin 2\alpha_2}{\omega_2^2 - \omega^2} = 0,$$

$$\frac{m_1 \omega^2 H_1^0 \cos \alpha_1}{\omega_1^2 - \omega^2} + \frac{m_2 \omega^2 H_2^0 \cos \alpha_2}{\omega_2^2 - \omega^2} = -m_{1,2} H_{1,2}.$$

Combining these equations, we get

$$\frac{m_1 \omega^2}{\omega_1^2 - \omega^2} \frac{\cos \alpha_1}{\sin \alpha_2} = \frac{m_{1,2} H_{1,2}}{H_2^0 \sin \alpha_1 - H_1^0 \sin \alpha_2}.$$

Let us require $U(0) = 0$. Obviously, we have to arrange the following

$$\begin{cases} a_{ww} W_0 + a_{w\varphi} \varphi_0 = 0, \\ a_{\varphi w} W_0 + a_{\varphi\varphi} \varphi_0 = 0, \end{cases} \Rightarrow \begin{cases} U_0 = 0, \quad \mathrm{Det}(\omega) = 0, \Rightarrow \\ a_{ww} a_{\varphi\varphi} - a_{w\varphi} a_{\varphi w} = 0. \end{cases} \quad (16.16)$$

The total number of equation roots is not more than $2(2 + 2N)$, where N is the number of the elementary oscillators. We block the perturbation energy of frequencies ω_{β_1}, ω_{β_2}, etc in the right finite part of the structure if relation (16.16) is made to be fulfilled for those frequencies. This expression contains all parameters, which may be varied to control the transmission of a longitudinal harmonic perturbation through the inclusion.

5.   CONCLUSION

Note, there is no string parameter in conditions (16.15), hence we may claim that conditions (16.15) are fulfilled not only for an elastic-and-inertia string, but for an arbitrary simple one-dimensional elastic-and-inertia continuum. The case when $U_0 = 0$ satisfies (16.14) for longitudinal source's frequency is very remarkable. The rigid box oscillates only in the transverse direction and rotates about the axis orthogonal to the drawing plane on the induced frequency of a source. In linear approximation there is no longitudinal displacement in the left semi-infinite part of the string. Hence, this means that the multi-element inclusion filtrates a longitudinal signal of corresponding frequency.

Although we have talked only about synthetically made devices and protective coats, the results are relevant to be taken into account when work is under execution to determine the interior structure of naturally formed systems of molecular or atomic scale.

Suppose we carry out test experiments and we know exactly everything about the interior structure of the system. In order to investigate theoretically the responses on system parameters varying, we have to be sure that the mathematical description of the system composite elements reproduces their properties in an exhaustive way. In our consideration we have had this case: the equations of elastic line, Winkler's basis, etc are definitely known. And we have seen the real possibility to get different responses of the system at the same loading as a consequence of just a little varying of the oscillators allocation or eigen-frequences. Analyzing such an absolutely 'transparent' object, we have to state that any uncertainty or obscurity in describing individual elements leads to difficulties in predicting the response even in our very simple compound multi-element model, not to mention a system having inclusions and oscillators of more complicated nature.

IV

DIFFRACTION AND PROPAGATION OF ACOUSTIC AND ELECTROMAGNETIC WAVES

DIFFRACTION COEFFICIENTS FOR A CIRCULAR IMPEDANCE CONE

Y. A. Antipov

Department of Mathematical Sciences

University of Bath, Bath BA2 7AY, UK

masya@maths.bath.ac.uk

Abstract   The axially symmetric problem of scattering of an incident plane wave
by a circular cone with an impedance boundary condition on its face is
considered. The scheme of solution includes applying the Kontorovich–
Lebedev transform, derivation of a second-order difference equation in
a strip of a complex variable and its reduction to an integral equation of
the convolution type with variable coefficients. An approximate solution
by a collocation method is constructed. The diffraction coefficients are
found in terms of the solution of the integral equation. Numerical results
for the diffraction coefficients are reported.

1.   INTRODUCTION

The problem of diffraction by an acoustically soft and hard circular
cone admits an exact solution by applying either the Sommerfeld integral representation or the Kontorovich–Lebedev integral transformation.
If the face of a cone is not completely hard, but reacts to an incident
pressure wave with a normal surface impedance z_s, then the wave field
satisfies a boundary condition of the third kind. If the quantity z_s is infinite, then the surface is hard (rigid). If the surface impedance is zero,
then the obstacle is acoustically soft with the pressure-release surface.
Bernard [1] considered the problem of diffraction by an impedance circular cone using the Sommerfeld integral representation, the theory of
the Maliuzhinets difference equation and reduced the problem to an integral equation on infinite imaginary axis. He analysed the kernel of the
equation, showed that the characteristic part of the kernel is the Dixon
kernel. In addition, Bernard studied the solvability of the equation in
the space L_1. The leading asymptotic term for the diffraction coefficients
in the case of a narrow impedance cone was found in [2]. However, there
are some open questions: (i) what is the behaviour of the solution of the

I.D. Abrahams et al. (eds.),

IUTAM Symposium on Diffraction and Scattering in Fluid Mechanics and Elasticity, 151–158.
© 2002 *Kluwer Academic Publishers. Printed in the Netherlands.*

equation at the ends? (ii) how to solve the integral equation? and (iii) what are the formulae for the diffraction coefficients?

In Section 2, we formulate the problem and discuss the physical sense of the impedance boundary condition. Section 3 provides an alternative derivation of the integral equation. The advantage of this technique is in its simplicity in comparison with [1]. A numerical scheme of solution is also presented. An exact formula (in terms of the solution of the integral equation) and an approximate formula sufficient for numerical computations for the diffraction coefficients are given in Section 4. Numerical results and new physical effects induced by the impedance condition with those of acoustically soft and hard cones are discussed.

2.   FORMULATION OF THE PROBLEM

Let us consider a wave field U satisfying the Helmholtz equation in the exterior of a conical obstacle, namely in the domain $C = \{0 < r < \infty, -\pi \le \phi \le \pi, 0 \le \theta < \alpha\}$, where $\pi/2 < \alpha < \pi$. The primary source is a plane scalar wave incident from the direction $\theta = 0 : u^{\mathrm{inc}} = e^{ikr\cos\theta}$, k is a wave number and $-\pi < \arg k < 0$. We aim to analyse scattering of the incident wave u^{inc} from the surface of the cone, along which the ratio of the θ-component of the velocity vector $\mathbf{v}$ and the pressure p is constant $p/v_\theta = -z_s$, where the constant z_s is termed the surface impedance. The last condition can be rewritten in terms of the velocity potential U ($\mathbf{v} = e^{-i\omega t}\nabla U$) as follows: $r^{-1}\partial U/\partial\theta + ik\beta U = 0$ at $\theta = \alpha$, $0 < r < \infty$. This is the impedance boundary condition. The parameter β is the acoustic admittance and it is expressible in terms of the surface impedance $z_s : \beta = \rho c/z_s$, c is the wave velocity, $\omega = kc$. The acoustic admittance is complex and its real part is positive in the case of the absorbing boundary. When $\beta = 0$ the boundary is rigid and the cone is acoustically hard. If $\beta = \infty$ we arrive at the Dirichlet boundary condition that corresponds to an acoustically soft cone or a perfectly conducting cone. Let $U = u^{\mathrm{inc}} + u^{\mathrm{diff}}$, where u^{diff} is the scattered velocity potential. Then the field $u = u^{\mathrm{diff}}$ is the solution of the following boundary value problem for the Helmholtz equation

$$\Delta u + k^2 u = 0, \quad 0 < r < \infty, \quad 0 < \theta < \alpha, \tag{17.1}$$

$$\left(u + \frac{1}{ik\beta r}\frac{\partial u}{\partial \theta}\right)_{\theta=\alpha} = g(r), \quad 0 < r < \infty, \tag{17.2}$$

where $g(r) = -(1 - \beta^{-1}\sin\alpha)\exp\{ikr\cos\alpha\}$.

3.   DERIVATION OF AN INTEGRAL EQUATION AND ITS SOLUTION

In this section we aim to reduce the boundary-value problem for the Helmholtz equation to an integral equation. Applying the Kontorovich–Lebedev transform [3]

$$u_s(\theta) = \int_0^\infty u(r,\theta) H_s^{(2)}(kr) \frac{dr}{\sqrt{r}}, \quad u(r,\theta) = -\frac{1}{2\sqrt{r}} \int_{-i\infty}^{i\infty} s u_s(\theta) J_s(kr) ds \tag{17.3}$$

to equation (17.1) and the boundary condition (17.2) yields the following ordinary differential equation

$$\frac{1}{\sin\theta} \frac{d}{d\theta} \left(\sin\theta \frac{d}{d\theta} u_s(\theta) \right) + \left(s^2 - \frac{1}{4} \right) u_s(\theta) = 0, \tag{17.4}$$

subject to the boundary condition

$$u_s(\alpha) + \frac{1}{2i\beta s} \frac{d}{d\theta} [u_{s-1}(\alpha) + u_{s+1}(\alpha)] = \frac{h(s)}{2i\beta k^{1/2} s}, \tag{17.5}$$

where

$$h(s) = \frac{2i\beta(1 - \beta^{-1}\sin\alpha)s\sqrt{2\pi}}{\cos\pi s} e^{i\pi(s/2 - 3/4)} P_{s-1/2}(-\cos\alpha). \tag{17.6}$$

The general solution, bounded at $\theta = 0$, is written down in terms of an unknown function $\Phi(s)$

$$u_s(\theta) = \frac{P_{s-1/2}(\cos\theta)}{k^{1/2}(d/d\theta)P_{s-1/2}(\cos\alpha)} \Phi(s-1). \tag{17.7}$$

Here $P_{s-1/2}(\cos\theta)$ is the Legendre function of the first kind. The boundary condition (17.5) gives rise to the following Carleman type boundary value problem of the theory of analytic functions for a strip:

Find the function $\Phi(s)$ that is analytic in the strip $\Pi = \{-2 < \Re(s) < 0\}$, satisfies the boundary condition

$$\Phi(\sigma) + K(\sigma)\Phi(\sigma - 1) + \Phi(\sigma - 2) = h(\sigma), \quad \sigma \in \Omega = \{-i\infty, i\infty\}, \tag{17.8}$$

where

$$K(s) = \frac{2i\beta s P_{s-1/2}^m(\cos\alpha)}{(d/d\theta)P_{s-1/2}^m(\cos\alpha)} \tag{17.9}$$

and the following auxiliary conditions:

$$\int_{-\infty}^\infty |\Phi(x + it)|^2 dt < \infty \tag{17.10}$$

uniformly with respect to $x \in [-2, 0]$ *and*

$$\Phi(\sigma - 1) = e^{i\pi\sigma}\Phi(-\sigma - 1), \quad \sigma \in \Omega. \tag{17.11}$$

The formulated problem can be considered as a difference equation of the second order in the strip Π with the two additional conditions (17.10) and (17.11).

The exact solution of problem (17.8), (17.10) and (17.11) can be found only for particular cases of the coefficient $K(\sigma)$. In general, we can rewrite equation (17.8) in such a way

$$\Phi(\sigma) + \Phi(\sigma - 2) = h^*(\sigma), \quad \sigma \in \Omega, \tag{17.12}$$

where $h^*(\sigma) = h(\sigma) - K(\sigma)\Phi(\sigma - 1)$. The function $\Phi(\sigma)$ is analytic in the strip $\Pi = \{-2 < \Re(s) < 0\}$, satisfies the boundary condition (17.12) and the auxiliary condition (17.10). Therefore, due to the Sokhotski–Plemelj formula modified for a strip, we write

$$\Phi(s) = \frac{1}{4i} \int_\Omega \frac{h^*(\tau)d\tau}{\sin\frac{\pi}{2}(\tau - s)}, \quad s \in \Pi, \tag{17.13}$$

$$\Phi(\sigma - 1 \pm 1) = \frac{1}{2}h^*(\sigma) \pm \frac{1}{4i} \int_\Omega \frac{h^*(\tau)d\tau}{\sin\frac{\pi}{2}(\tau - \sigma)}, \quad \sigma \in \Omega. \tag{17.14}$$

Formula (17.13) defines $\Phi(s)$ in the interior of the strip Π in terms of its values on the contour $\{-1 - i\infty, -1 + i\infty\}$. The relation (17.14) describes the boundary values of $\Phi(s)$ on the contour Ω and on the contour $\{-2 - i\infty, -2 + i\infty\}$. Now letting $s = \sigma - 1$ in (17.13) we get a convolution integral equation with respect to the function $\Phi(\sigma - 1)$

$$\Phi(\sigma - 1) + \frac{1}{4i} \int_{-i\infty}^{i\infty} \frac{K(\tau)\Phi(\tau - 1) - h(\tau)}{\cos\frac{\pi}{2}(\tau - \sigma)} d\tau = 0, \quad \sigma \in \Omega. \tag{17.15}$$

From a numerical point of view, it is convenient to transform this equation to a new one on the semi-infinite interval $(-\infty, 0)$,

$$\chi(\xi) + \frac{K(i\xi)}{4} \int_{-\infty}^{0} \chi(\eta) \left(\frac{1}{\cosh\frac{\pi}{2}(\eta - \xi)} - \frac{e^{\pi\eta}}{\cosh\frac{\pi}{2}(\eta + \xi)} \right) d\eta = -h(i\xi). \tag{17.16}$$

Here we take into account that $\Phi(-\sigma - 1) = \exp\{-i\pi\sigma\}\Phi(\sigma - 1)$, $K(-\tau) = -K(\tau)$, $h(-\tau) = -\exp\{-i\pi\tau\}h(\tau)$ and introduced the new function $\chi(\xi) = K(i\xi)\Phi(i\xi - 1) - h(i\xi)$. Analysis of the Cauchy integrals shows that the function $\chi(\xi)$ decays exponentially as $\xi \to -\infty$:

$$\chi(\xi) = \sqrt{\pi}A_0 e^{(-\pi/2 + \alpha)\xi} \left\{ \sqrt{-\xi} + O\left(\frac{1}{\sqrt{-\xi}}\right) \right\}, \quad \xi \to -\infty,$$

where

$$A_0 = \frac{2(i+1)(\beta - \sin\alpha)}{\beta + \sin\alpha}\sqrt{\frac{2\sin\alpha}{\pi}}.$$

Since the function $h(\xi)$ possesses the same behaviour as $\xi \to -\infty$, even the following simple interpolation should provide a good rate of convergence of an approximate solution to the exact one. Let us consider a set of points

$$0 = x_0 > x_1 > x_2 > \ldots > x_N \to -\infty, \quad N \to \infty,$$

$$x_n \sim -\nu n^\delta, \quad n \to \infty, \quad \delta > 0, \quad \nu > 0, \qquad (17.17)$$

introduce notations

$$\chi(\xi_m) = X_m \quad \text{as} \quad \xi_m = (x_{m-1} + x_m)/2, \quad m = 1, 2, \ldots, N \qquad (17.18)$$

and approximate the function $\chi(\xi)$ by a piecewise constant function: $\chi(\xi) = X_m$, $\xi \in (x_m, x_{m-1})$. Then we discretise (17.16) and arrive at the following system of algebraic equations with respect to the coefficients X_k,

$$X_n + \sum_{m=1}^{N} c_{nm} X_m = b_n, \quad n = 1, 2, \ldots, N, \qquad (17.19)$$

where $4b_n = -h(i\xi_n)$ and

$$c_{nm} = \frac{K(i\xi_n)}{4} \int_{x_m}^{x_{m-1}} \left(\frac{1}{\cosh\frac{\pi}{2}(\eta - \xi_n)} - \frac{e^{\pi\eta}}{\cosh\frac{\pi}{2}(\eta + \xi_n)} \right) d\eta,$$

The integrals in the representation of the coefficients c_{nk} can be estimated explicitly

$$c_{nm} = \frac{K(i\xi_n)}{\pi}(t_{nm}^{(1)} + t_{nm}^{(2)}), \quad t_{nm}^{(1)} = \tan^{-1}\frac{\lambda_{m-1}}{\mu_n} - \tan^{-1}\frac{\lambda_m}{\mu_n},$$

$$t_{nm}^{(2)} = \frac{\lambda_m - \lambda_{m-1}}{\mu_n} + \frac{1}{\mu_n^2}[\tan^{-1}(\lambda_{m-1}\mu_n) - \tan^{-1}(\lambda_m\mu_n)],$$

$$\lambda_m = e^{\pi x_m/2}, \quad \mu_n = \sqrt{\lambda_n\lambda_{n-1}}, \qquad (17.20)$$

Thus, the coefficients c_{nm}, b_n decay exponentially as $n \to \infty$ (m is fixed) or in the case of the c_{nm} if $m \to \infty$ (n is fixed), namely

$$c_{nm} = O\left(e^{\pi x_n/2}\right), \quad n \to \infty \ (x_n \to -\infty), \quad m \text{ is fixed},$$

$$c_{nm} = O\left(e^{\pi x_m/2}\right), \quad m \to \infty, \quad n \text{ is fixed},$$

$$b_n = O\left(e^{(\pi/2 - \alpha_0)x_n}\sqrt{-x_n}\right), \quad n \to \infty. \qquad (17.21)$$

4.   DIFFRACTION COEFFICIENTS

We now turn to the wave field u, the solution of problem (17.1) and (17.2). Applying the inverse Kontorovich–Lebedev transform (17.3) and taking into account that $u_s(\theta) = e^{i\pi s} u_{-s}(\theta)$ we get the expression for the scattering wave field

$$u^{\text{diff}}(r,\theta) = -\frac{1}{4i\beta\sqrt{r}k^{1/2}} \int_{-i\infty}^{i\infty} \frac{P_{s-1/2}(\cos\theta)[\chi(-is) + h(s)]}{P_{s-1/2}(\cos\alpha)} J_s(kr)ds.$$

$$(17.22)$$

Let us find the diffraction coefficients $D(\theta)$ (the scattering diagram) an important characteristic in the geometric theory of diffraction. They are introduced in such a way

$$u(r,\theta) = \frac{D(\theta)}{kr} e^{-ikr} + O\left((kr)^{-2}\right), \quad kr \to \infty \qquad (17.23)$$

and describe the high-frequency asymptotics of the solution. To find the diffraction coefficients, we use the behaviour of the Bessel function at infinity and from formula (17.22) we obtain the expression

$$D(\theta) = \frac{i+1}{4\beta\sqrt{\pi}} \int_{-\infty}^{0} \frac{P_{i\xi-1/2}(\cos\theta)}{P_{i\xi-1/2}(\cos\alpha)} [\chi(\xi) + h(i\xi)] e^{\pi\xi/2} \sinh\pi\xi d\xi. \quad (17.24)$$

The integral is improper, it converges uniformly in the domain $0 \le \theta < 2\alpha - \pi$ namely in the region where no reflected waves are observed. The rate of the convergence is exponential. The direction $\theta = 2\alpha - \pi$ is singular for the diffraction coefficients: they are infinite. In addition we write down an approximate formula for the diffraction coefficients, based on an approximate solution of equation (17.16),

$$D^{(N)}(\theta) = \frac{i+1}{4\beta\sqrt{\pi}} \sum_{m=1}^{N} (X_m - b_m)\Lambda(\xi_m,\theta)(x_{m-1} - x_m), \qquad (17.25)$$

where $D^{(N)}(\theta)$ is an approximation to $D(\theta)$ and

$$\Lambda(\xi,\theta) = \frac{P_{i\xi-1/2}(\cos\theta)}{P_{i\xi-1/2}(\cos\alpha)} e^{\pi\xi/2} \sinh\pi\xi. \qquad (17.26)$$

In Table 1, we write down the values of the diffraction coefficients for $\alpha = 5\pi/6$ for the perfectly conducting cone (the Dirichlet problem: $\beta = \infty$), the acoustically hard cone (the Neumann problem: $\beta = 0$) and the two cases of the impedance condition: $\beta = 1$ and $\beta = 1 + i$.

The coefficients $\Im I(f)$ were calculated in [4] for the same angle $\alpha = 5\pi/6$ and $\beta = \infty$ (the perfectly conducting cone), and we adduce them

Table 1   Diffraction coefficients

θ	$\beta = \infty$ [4] $2\pi i f$	$\beta = \infty$ $-iD$	$\beta = 1$ $-iD$	$\beta = 1 + i$ $\Re(D)$	$\beta = 1 + i$ $\Im(D)$	$\beta = 0$ $-iD$
0	0.2399	0.2399	0.03486	-0.05287	0.05152	-0.0522
$3\pi/24$	0.2576	0.2577	0.03839	-0.05776	0.05697	-0.0582
$6\pi/24$	0.3231	0.3232	0.05196	-0.07620	0.07812	-0.0812
$9\pi/24$	0.4966	0.4967	0.09110	-0.1269	0.1399	-0.1520
$12\pi/24$	1.0860	1.0862	0.2423	-0.3063	0.3843	-0.4590
$15\pi/24$	8.3073	8.2856	2.4441	-2.5373	3.9465	-6.0754

to compare with our own results (the third column of the table). The results of calculations for the acoustically soft cone ($\beta \to \infty$) are in good agreement with the corresponding results of [4]. Additionally, it is seen that as θ approaches the singular line $2\alpha - \pi$ (in this case it is $2\pi/3$) the diffraction coefficients grow to infinity for all three types of the boundary conditions.

The numerical results show that the imaginary part of the diffraction coefficients ($\Re(D) = 0$ if $\Im I(\beta) = 0$) grows as $\theta \to 2\alpha - \pi$: $-iD(\theta) \to -\infty$ as $\beta < \sin\alpha$, $-iD(\theta) \to +\infty$ as $\beta > \sin\alpha$. In this case $\sin\alpha = 1/2$ and $2\alpha - \pi = 2\pi/3$. At the point $\beta = \sin\alpha$ in a neighbourhood of the point $\theta = 2\alpha - \pi$ the curves $-iD(\theta)$ are unstable: for $\beta \to \sin\alpha + 0$ we have $-iD(\theta) \to +\infty$ and for $\beta \to \sin\alpha + 0$ we get $-iD(\theta) \to +\infty$ as $\theta \to 2\alpha - \pi - 0$. The point $\theta = 0$ is the global minimum for the modulus of the diffraction coefficients $D(\theta)$.

The value $\beta = \sin\alpha$ is critical for the coefficients $D(\theta)$: $D(\theta) = 0$ for all θ. If $\beta < \sin\alpha$ then $-iD(\theta) < 0$ and for $\beta > \sin\alpha$ the coefficients $-iD(\theta)$ are positive for all values of $\theta \in [0, 2\alpha - \pi)$.

The coefficients $-iD(\theta)$ for "blunt" cones are greater then those for narrow cones. For a narrow cone the diffraction coefficients vanish. This fact is in a good agreement with the approximate formula found in [2]. The situations $\alpha = \pi$ and $\alpha = \pi/2$ are both singular: if $\alpha \to \pi$ then the cone disappears and if $\alpha \to \pi/2$ then the problem turns into the problem on diffraction of a plane incident wave by a half-space. It is known that the leading term of the far scattering field for the two-dimensional problem differs from (17.23).

There always exists a bounded domain in the β-complex plane for which the scattering cross section $\sigma(\theta) = \lambda^2\pi^{-1}|D(\theta)|^2$ ($\lambda = 2\pi k^{-1}$) is less than that for the the hard cone. We note that the scattering cross section $\sigma(\theta)$ for the impedance cone is always less than $\sigma(\theta)$ for the perfectly conducting cone. Moreover, for each set of parameters of the problem on the impedance cone there exists a unique value of the acoustic admittance that minimises the scattering cross section.

Acknowledgments

The author is grateful to I. D. Abrahams, V. M. Babich and V. P. Smyshlyaev for discussions.

References

[1] Bernard, J M L (1997) *Méthode analytique et transformées functionnelles pour la diffraction d'ondes par une singularité conique,* Rapport CEA-R-5764 Editions Dist/Saslay.

[2] Bernard, J M L and Lyalinov, M A (1999) *The leading asymptotic term for the scattering diagram in the problem of diffraction by a narrow circular impedance cone,* J. Phys. A: Math. Gen. **32**, L43–L48.

[3] Kontorovich, M I and Lebedev, N N (1938) *On a method of solution of some problems of diffraction theory and related problems,* J. Exp. & Theor. Phys. **8**(10-11), 1192–1206.

[4] Babich, V, Dement'ev, D and Samokish, B (1995) *On the diffraction of high-frequency waves by a cone of arbitrary shape,* Wave Motion **21**, 203–207.

ON THE ACOUSTICS OF A THICK SHEAR LAYER

L. M. B. C. Campos, M. H. Kobayashi

Secção de Mecânica Aeroespacial, ISR

Instituto Superior Técnico, Av. Rovisco Pais, 1

1049-001 Lisbon, Portugal

lmbcampos.aero@alfa.ist.utl.pt, marcelo@popsrv.ist.utl.pt

Abstract   The transmission and reflection of sound in a thick shear layer with a hyperbolic tangent velocity profile is discussed, for arbitrary ratio of the wavelength λ to the lengthscale L of the shear flow. The analysis includes the opposite limits of the vortex sheet $\lambda/L \gg 1$ and ray theory $\lambda^2 \ll L^2$ and concentrates in the more interesting case of wavelength comparable to the lengthscale of the shear flow $\lambda \sim L$.

1.   INTRODUCTION

There is an extensive literature on the acoustics of shear flows, dating back more than half a century [1, 2, 3, 4], with particular emphasis on boundary layers [5, 6] and shear layers [7, 8]. In most of the literature the acoustic wave equation in a shear flow is solved approximately, except for two exact solutions, in the case of an homentropic linear shear [9, 10, 11, 12, 13, 14] and an exponential shear [15]. The present paper includes a third exact solution, for a hyperbolic tangent shear flow, which is used to model sound scattering by a thick shear layer [16].

2.   SOUND PROPAGATION IN THE HYPERBOLIC TANGENT SHEAR FLOW PROFILE

Consider a unidirectional shear flow with hyperbolic tangent profile given by $\mathbf{U}(y) = V \tanh(y/L)\mathbf{e}_x$. Since the mean state neither depends on time t nor on horizontal coordinate x, it is convenient to use the Fourier representation of the two-dimensional acoustic pressure field

$$p(x, y, t) = \int_{\mathbb{R}^2} P(y; k, \omega)\, e^{i(kx - \omega t)} dk\, d\omega,$$

159

I.D. Abrahams et al. (eds.),

IUTAM Symposium on Diffraction and Scattering in Fluid Mechanics and Elasticity, 159–168.

© 2002 *Kluwer Academic Publishers. Printed in the Netherlands.*

where $P(y; k, w)$ is the acoustic pressure perturbation spectrum, for a wave of frequency ω and horizontal wavenumber k, at transverse position y. It satisfies the acoustic wave equation in a unidirectional shear flow [17]

$$P'' + 2\left[c'/c + kU'/(\omega - kU)\right]P' + \left[(\omega - kU)^2/c^2 - k^2\right]P = 0, \quad (18.1)$$

where prime denotes derivative with regard to y and c is the adiabatic sound speed of the mean flow. In a homentropic flow the sound speed is constant, and the acoustic wave equation (18.1) simplifies to

$$(1 - kU/\omega)\,P'' + 2\,(k/\omega)\,U'P'$$
$$+\, k^2\,(1 - kU/\omega)\left[(1 - kU/\omega)^2\,(\omega/kc)^2 - 1\right]P \;=\; 0; \quad (18.2)$$

this is the usual form in the literature [1, 2, 3, 4, 15]. When introducing the hyperbolic tangent shear flow profile in (18.2), it is convenient to change the independent variable,

$$\xi \equiv \tanh(y/L), \quad \Phi(\xi) \equiv P(y; k, \omega),$$

leading to a linear second-order differential equation with polynomial coefficients,

$$(1 - \Lambda\xi)\left(1 - \xi^2\right)^2\Phi'' + 2\left(1 - \xi^2\right)(\Lambda - \xi)\,\Phi'$$
$$+\,(1 - \Lambda\xi)\left[\Omega^2(1 - \Lambda\xi)^2 - K^2\right]\Phi \;=\; 0, \quad (18.3)$$

where prime denotes derivative with regard to ξ, and three dimensionless parameters appear:

$$\Omega \equiv \omega L/c, \quad K \equiv kL, \quad \Lambda \equiv kV/L.$$

The first parameter appears as a dimensionless 'frequency',

$$\Omega \equiv 2\pi L/\tau c = 2\pi L/\lambda = 2\pi\delta, \quad \delta \equiv L/\lambda,$$

where the wave period $\tau = 2\pi/\omega$, and wavelength $\lambda = \tau c$ have been introduced; this parameter specifies the ratio of the lengthscale of the shear flow L to the wavelength of sound in the free stream λ, so $\Omega \ll 1$ or $\lambda \gg L$ corresponds to sound scattering by a 'vortex sheet', whereas $\Omega^2 \gg 1$ corresponds to the 'ray limit' of a short wave in a slowly varying mean flow, with $\Omega \sim 1$ leading to the more interesting case of interaction of sound with a shear layer of 'finite' thickness. Taking the horizontal wavenumber for a medium at rest, $k = (\omega/c)\cos\theta$, where θ is the angle of the direction of propagation with the mean flow, the second parameter K, which appears as 'dimensionless wavenumber',

$$K = (\omega L/c)\cos\theta = \Omega\cos\theta,$$

specifies the direction of propagation, namely oblique for $K < \Omega$, horizontal for $K = \Omega$, with evanescent waves corresponding to $K > \Omega$. The last parameter,

$$\Lambda = (V/c)\cos\theta = M\cos\theta, \quad M = V/c$$

involves the Mach number of the free stream M, and is related to the Doppler factor, as will be seen next.

3. EXISTENCE OF CRITICAL LAYER AND MATCHING OF THE WAVE FIELDS

The preceding account makes clear that the scattering of sound by a shear layer depends on three parameters Ω, K, Λ, which are combinations of the Mach number M, angle of incidence θ and shear layer thickness δ. The former two apply to a 'vortex sheet', and the latter distinguishes a shear layer of 'finite thickness', allowing for new flow-acoustic interaction effects. The acoustic wave equation (18.2) has a singularity where the Doppler shifted frequency

$$\omega_*(y) = \omega - kU(y) = \omega - kV\tanh(y/L),$$

vanishes, and this determines the position of the critical layer, which is specified by:

$$\omega_*(y_c) = 0: \ \ \xi_c \equiv \tanh(y_c/L) = \omega/kV = 1/\Lambda = 1/(M\cos\theta).$$

If the Doppler shifted frequency in the free stream $\omega_*(\infty) = \omega - kV$ is negative, then since it is positive at the mid-line $\omega_*(0) = \omega$, it must vanish in between, at the critical layer, i.e. y_c is real.

If the Doppler shifted frequency is positive in the free stream, it is positive everywhere, and no critical layer exists, i.e. y_c is imaginary for $\Lambda < 1$. The intermediate case $\Lambda = 1$ corresponds to a critical layer in the free stream $y_c = \infty$.

It is clear that the differential equation (18.3) has three regular singularities, $\xi = \pm 1, 1/\Lambda \equiv \xi_c \Leftrightarrow y = \pm\infty, y_c$, corresponding to the free streams and critical layer, respectively. In the particular case when the critical layer lies in the upper free stream $\Lambda = 1 = \xi_c$, the wave equation (18.3) simplifies to:

$$\Lambda = 1: \ \ \left(1 - \xi^2\right)^2 \Phi'' + 2\left(1 - \xi^2\right)\Phi' + \left[\Omega^2\left(1 - \xi\right)^2 - K^2\right]\Phi = 0,$$

which has only two singularities.

4.　　ASYMPTOTIC WAVE FIELDS

The asymptotic wave fields in the free streams can be readily determined because the wave equation (18.2) then has constant coefficients:

$$U\left(\pm\infty\right) = \pm V : \quad P'' + \left[\left(\omega \mp kV\right)^2/c^2 - k^2\right] P = 0.$$

Introducing the vertical wavenumber, respectively above k_+ and below k_- the shear layer:

$$k_\pm \equiv \left|\left(\omega \mp kV\right)^2/c^2 - k^2\right|^{\frac{1}{2}} = \left(\omega/c\right)\left|\left(1 \mp M\cos\theta\right)^2 - \cos^2\theta\right|^{\frac{1}{2}} \quad (18.4)$$

the asymptotic wave fields are given by:

$$P\left(y \to \pm\infty\right) \sim \begin{cases} A_\pm \exp\left(ik_\pm y\right) + B_\pm \exp\left(-ik_\pm y\right) & \text{if } \Delta > 0 \\ A_\pm y + B_\pm & \text{if } \Delta = 0 \\ A_\pm \exp\left(k_\pm y\right) + B_\pm \exp\left(-k_\pm y\right) & \text{if } \Delta < 0 \end{cases} \quad (18.5)$$

where $\Delta \equiv \left(\omega \mp kV\right)^2 - k^2c^2$ and $A_\pm$ and $B_\pm$ are arbitrary constants of integration, to be determined from boundary or radiation conditions. It is possible to have propagating, evanescent or marginal wave fields, above (resp. below) the shear layer depending on whether $\omega - kV > kc$ (resp. $\omega + kV > kc$) or $\omega - kV < kc$ (resp. $\omega + kV < kc$) or $\omega - kV = kc$ (resp. $\omega + kV = kc$), respectively.

5.　　EXACT WAVE FIELDS

As an example of the calculation of an exact wave field, near the upper free stream, consider the expansion about $y = \infty$ or $\xi = 1$, using the variable $\zeta \equiv 1 - \xi$, $F(\zeta) \equiv \Phi(\xi)$, which transforms the differential equation (18.3) to

$$\zeta^2 \left(2 - \zeta\right)^2 \left(1 - \Lambda + \Lambda\zeta\right) F'' + 2\zeta\left(2 - \zeta\right)\left(1 - \Lambda - \zeta\right)F'$$
$$+ \left(1 - \Lambda + \Lambda\zeta\right)\left[\Omega^2\left(1 - \Lambda + \Lambda\zeta\right)^2 - K^2\right]F = 0 \quad (18.6)$$

Since $\zeta = 0$, $\xi = 1$, $y = +\infty$ is a regular singularity, solutions exist as Frobenius–Fuchs series,

$$F(\zeta) = \sum_{n=0}^{\infty} a_n(\sigma)\zeta^{n+\sigma}, \quad (18.7)$$

with index σ and coefficients a_n to be determined. Substitution of (18.7) into (18.6) leads to the following recurrence formula for the coefficients:

$$\left(1 - \Lambda\right)\left[4\left(n + \sigma\right)^2 + k_+^2 L^2\right] a_n(\sigma) = Aa_{n-1} - Ba_{n-2} - Ca_{n-3} \quad (18.8)$$

where

$$
\begin{aligned}
A &= 2(n+\sigma-1)\left[2(1-2\Lambda)(n+\sigma-2)+3-\Lambda\right] \\
&\quad - \Lambda\left[k_+^2 L^2 + 2\Omega^2(1-\Lambda)^2\right], \\
B &= (n+\sigma-2)\left[(1-5\Lambda)(n+\sigma-3)+2\right] + 3\Omega^2\Lambda^2(1-\Lambda), \\
C &= \Lambda\left\{(n+\sigma-3)(n+\sigma-4)+\Omega^2\Lambda^2\right\};
\end{aligned}
$$

setting $n = 0$ yields the indicial equation,

$$
n = 0: \ \ 0 = 4\sigma^2 + k_+^2 L^2 = 4\left(\sigma - \sigma_+\right)\left(\sigma - \sigma_-\right). \tag{18.9}
$$

The indices, which are the roots of (18.9), are specified by (18.4) the vertical wavenumber in the free stream: $\sigma_\pm = \mp i k_+ L/2$. The variable ζ is given asymptotically by:

$$
\zeta = 1 - \tanh(y/L) = 2/\left(1 + e^{2y/L}\right) = 2e^{-2y/L}\left[1 + \mathrm{o}\left(e^{-2y/L}\right)\right],
$$

Thus the wavefields (18.7),

$$
a_0\left(\sigma_\pm\right) \equiv 1: \ \ F_\pm(\zeta) = \sum_{n=0}^{\infty} a_n(\sigma_\pm)\zeta^{n+\sigma_\pm},
$$

correspond to

$$
P_\pm(y) = \sum_{n=0}^{\infty} a_n(\mp i k_+ L/2)\left[1 - \tanh(y/L)\right]^{n \mp i k_+ L/2}, \tag{18.10}
$$

and have leading asymptotic terms

$$
P_\pm(y \to \infty) \sim \zeta_\pm^\sigma \sim e^{-2\sigma_\pm y/L} \sim e^{\pm k_+ y}, \tag{18.11}
$$

respectively upward P_+ and downward P_- propagating waves, in agreement with (18.5). The total acoustic field is thus given by

$$
P(y) = A_+ P_+(y) + B_+ P_-(y), \tag{18.12}
$$

where A_+, B_+ are arbitrary constants. In the free stream (18.12) splits into upward and downward propagating waves (18.11); in the shear layer, each of $P_\pm$ has upward and downward propagating components, as would result from taking a spatial Fourier spectrum of (18.10); the reason is that there are multiple partial reflections of sound in the shear layer, and since (18.10) is exact, it includes multiple internal scattering of all orders. Thus $P_\pm$ are the exact (18.10) wave fields in the shear layer which match to respectively upward and downward propagating waves (18.5) in the upper free stream.

6.   MATCHING WAVE FIELDS ON TWO SIDES OF A SHEAR LAYER

The exact solution, (18.12) and (18.10), applies to the sound field in the shear layer and in the absence of a critical layer it holds for $-\infty < y \leq +\infty$, i.e. it excludes the lower stream, which is another singularity of the wave equation. Thus it is necessary to match to the solution of the wave equation around the singularity $y = -\infty$, $\xi = -1$ or $\eta = -2$, where η is the new variable $\eta = 1 + \xi$. A calculation similar to that in section 5 shows that the wavefields below the shear layer are given by

$$P(y) = A_- P^+(y) + B_- P^-(y), \tag{18.13}$$

where A_- and B_- are arbitrary constants, and

$$P^\pm(y) = \sum_{n=0}^{\infty} b_n(\mp ik_- L/2)\left[1 + \tanh(y/L)\right]^{n \mp ik_- L/2}, \tag{18.14}$$

with the coefficients b_n satisfying a recurrence formula similar to (18.8).

For example, a wave of unit amplitude incident from below corresponds to P^+,

$$-\infty \leq y < +\infty : \quad P(y) = P^+(y) + RP^-(y), \tag{18.15}$$

and the reflection coefficient R affects the downward propagating wave P^-; the wave field (18.15) corresponds to (18.13) with $A_- = 1$ and $B_- = R$. Above the shear layer (18.12) there should be upward propagating waves $B_+ = 0$, with amplitude $A_+ = T$ equal to the transmission coefficient:

$$-\infty < y \leq +\infty : \quad P(y) = TP_+(y). \tag{18.16}$$

The wave fields (18.15) and (18.16) are valid in overlapping regions, and their matching specifies the reflection R and transmission T coefficients.

7.   CALCULATION OF REFLECTION AND TRANSMISSION COEFFICIENTS

The matching of the wave fields ensures the continuity of pressure and vertical displacement, implying the continuity of the acoustic pressure and its normal gradient. Continuity of P and dP/dy at $y = 0$ yields

$$TF_+(1) = G_+(1) + RG_-(1), \quad -TF'_+(1) = G'_+(1) + RG'_-(1),$$

which can be solved for the reflection and transmission coefficients:

$$\left[F_+(1)G'_-(1) + F'_+(1)G_-(1)\right]R = -F_+(1)G'_+(1) - F'_+(1)G_+(1),$$
$$\left[F_+(1)G'_-(1) + F'_+(1)G_-(1)\right]T = G_+(1)G'_-(1) - G'_+(1)G_-(1).$$

The vanishing of the square brackets on the left-hand sides of these equations is related to the triggering of instabilities of the shear flow by sound.

8.   MODULUS AND PHASE OF THE REFLECTION AND TRANSMISSION COEFFICIENTS

The effect of shear layer thickness on the scattering coefficients is shown first. The modulus of the reflection factor (Fig. 1A) is unity in the zone of silence $\theta < 56°$ or $\theta > 124°$ for a free stream Mach number $M = 0.8$. Propagation above and below the shear layer is possible only for the range of angles of incidence $56° < \theta < 124°$. In this range the modulus of the reflection coefficient for a vortex sheet $\delta = 0$ is zero for normal incidence $\theta = 90°$ and then increases away from the vertical, being maximum at $\theta = 65°, 115°$, before decaying towards the edge of the zone of silence $\theta \leq 56°$ or $\theta \geq 124°$, where it becomes unity. For a thin shear layer, e.g. thickness one-tenth of the wavelength $\delta = 0.1$, there is still noticeable wave reflection, but for thicker shear layers $\delta = 1, 10$ the modulus of the reflection coefficient is practically zero in the range of angle of incidence corresponding to propagation on both sides of the shear layer. The phase of the reflection coefficient (Fig. 1B) is either 0 or π for a vortex sheet $\delta = 0$, and is more rounded-off for a thin shear layer $\delta = 0.1$. For a shear layer of thickness equal to the wavelength $\delta = 1$ the phase of the reflection coefficient is significant, and it becomes small for a thick shear layer $\delta = 10$. The modulus of the transmission coefficient (Fig. 1C) differs little between a vortex sheet and a shear layer, and is weakly dependent on the thickness of the latter. The modulus of the transmission coefficient is unity $|T| = 1$ for normal incidence $\theta = 90°$, and is smaller $|T| < 1$ in the forward arc $\theta < 90°$, larger $|T| > 1$ in the rear are $\theta > 90°$, except near to the zone of silence; $|T| \to \infty$ for $\theta \to \theta_+$ because the wave field below the shear layer is evanescent, and $|T| \to 0$ for $\theta \to \pi - \theta_+$ because above the shear layer the sound field is evanescent. The phase of the transmission coefficient (Fig. 1D) is small over most of the range of angles of incidence for propagating waves on both sides of the shear layer, and diverges as evanescence conditions are approached at the edge of the 'zones of silence'.

9.   MODULUS AND PHASE OF ACOUSTIC PRESSURE IN SHEAR LAYER

The modulus and phase of the acoustic pressure $P = |P| \exp\left[i \arg(P)\right]$, are plotted next versus distance across the shear layer, made dimension-

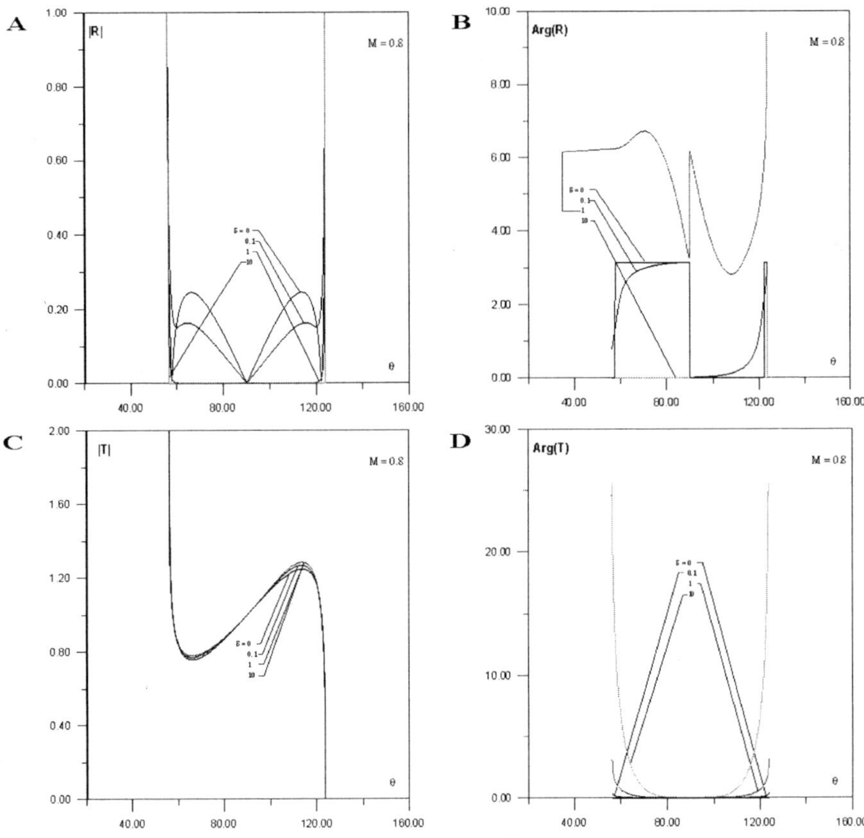

Figure 1   Modulus (A) and phase (B) of the reflection and (C, D) transmission coefficient, versus angle of incidence θ.

less by dividing by the shear layer thickness L, $-4 \leq Y \equiv y/L \leq 4$. Four free stream Mach numbers and a single shear layer thickness are considered (Fig. 2), for an angle of incidence $\theta = 75°$ corresponding to propagation on both sides of the shear layer in all cases. The modulus of the acoustic pressure (Fig. 2A) decays faster as it crosses the shear layer for larger Mach number; in the case of supersonic free streams there are visible amplitude oscillations at the lower but not in the upper stream. The phase of the acoustic pressure (Fig. 2B) has a comparable total variation across the shear layer for all Mach numbers, but is close to a linear function of thickness for a low subsonic free streams, and has a noticeable 'kink' for the supersonic free streams, i.e. it varies more rapidly in the lower half than in the upper half of the shear layer.

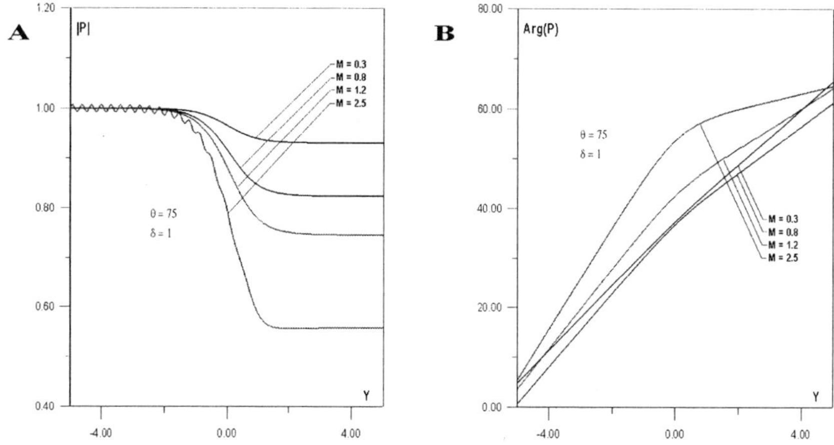

Figure 2   Modulus (A) and phase (B) of acoustic pressure versus distance across the shear layer normalised to shear layer thickness.

10.   DISCUSSION

The properties of the acoustic fields in a shear layer which have been demonstrated include the following: (i) a vortex sheet reflects sound waves much more strongly than a shear layer of moderate thickness $\delta \geq 1$ on a wavelength scale; (ii) the difference between the vortex sheet and shear layer is less marked as concerns the transmission coefficient; (iii) the interaction between sound and flow in a shear layer is much more complex, and richer in physical phenomena, than for a vortex sheet. Thus the vortex sheet [18] is an acceptable idealisation of a shear layer only for thickness less than about a tenth of the wavelength. The other simple idealisation is the opposite limit of sound 'rays', is the wavelength is much smaller than the thickness of the shear layer [19]. Outside these opposite extremes, and most notably for wavelengths comparable to the thickness of the shear layer, the acoustic fields in the latter are specified by the solution of the wave equation. The method of exact solution of the wave equation applies not only to the propagation of sound in a shear layer (as the main subject of the paper), but also to the triggering of its instabilities.

References

[1] Haurwitz, W (1932) *Zur theorie der wellenbewegungen in luft und wasser*, Veroff. Geophys. Inst. Univ. Leipz. **6**, 334–364.

[2] Kücheman, D (1938) *Störungebewegungen in einer gasströmung mit grenzschicht*, Zeit. für Angewandte Math. und Mech. **30**, 79–84.

[3] Pridmore-Brown, D (1958) *Sound propagation in a fluid flowing through an attenuating duct*, J. Fluid Mech. **4**, 393–406.

[4] Möhring, W, Müller, E and Obermeier, F (1983) *Problems in flow acoustics*, Rev. Modern Phys. **55**, 707–724.

[5] Almgren, G (1976) *Acoustic boundary layer influence in scale model simulation of sound propagation*, J. Sound Vib. **105**, 321–327.

[6] Hanson, D (1984) *Shielding of propfan cabin noise by the fuselage boundary layer*, J. Sound Vib. **92**, 591–598.

[7] Balsa, T (1976) *The far-field of high-frequency convected, singularities in shear flows with application to jet noise prediction*, J. Fluid Mech. **74**, 193–208.

[8] Balsa, T (1976) *Refraction and shielding of sound from a source in a jet*, J. Fluid Mech. **76**, 443–456.

[9] Goldstein, M E and Rice, E (1973) *Effect of shear on duct wall impedence*, J. Sound Vib. **30**, 79–84.

[10] Jones, D S (1977) *The scattering of sound by a simple shear layer*, Phil. Trans. Roy. Soc. Lond. A **284**, 287–328.

[11] Jones, D S (1978) *Acoustics of a splitter plate*, J. Inst. Maths & its Applic. **21**, 197–209.

[12] Scott, J N (1979) *Propagation of sound waves through a linear shear layer*, AIAA J. **17**, 237–245.

[13] Koutsoyannis, S P (1980) *Characterization of acoustic disturbance in linearly flows*, J. Sound Vib. **68**, 187–202.

[14] Campos, L M B C, Oliveira, J M G S and Kobayashi, M H (1999) *On sound propagation in a linear shear flow*, J. Sound Vib. **219**, 739–770.

[15] Campos, L M B C and Serrão, P G T Á (1998) *On the acoustics of an exponential boundary layer*, Phil. Trans. Roy. Soc. Lond. A **356**, 2335–2378.

[16] Campos, L M B C and Kobayashi, M H (2000) *On the reflection and transmission of sound in a thick shear layer*, J. Fluid Mech. **424**, 303–326.

[17] Campos, L M B C and Kobayashi, M H (2000) *On the propagation of sound in a non-isothermal shear flow*, preprint SMA.

[18] Miles, J W (1958) *On the disturbed motion of a vortex sheet*, J. Fluid Mech. **4**, 538–554.

[19] Amiet, R K (1978) *Refraction of sound by a shear layer*, J. Sound Vib. **58**, 467–482.

THE WAVENUMBER SURFACE IN BLADE-VORTEX INTERACTION

C. J. Chapman

Department of Mathematics

Keele University, Keele, Staffordshire ST5 5BG, UK

c.j.chapman@maths.keele.ac.uk

Abstract   This paper determines the wavenumber surface describing the production of sound when convected vorticity strikes the leading edge of an aircraft wing or fan blade. The acoustic part of the wavenumber surface consists of four sheets, namely an ellipsoid, half an elliptic cylinder, and two parallel planes. The half-cylinder, which has two infinite straight edges, is wrapped around the ellipsoid, i.e. is tangential to it along a half-ellipse. Each of the parallel planes contains an edge of the half-cylinder, and is tangential both to the half-cylinder and the ellipsoid; and each plane contains a special point at which there is a triple tangency of the plane, the half-cylinder, and the ellipsoid. The wavenumber surface also contains a fifth sheet, describing convection of vorticity with the mean flow; this sheet is a plane, perpendicular to the two acoustic planes. The four acoustic sheets determine the asymptotic properties of the radiated sound field; in particular the half-ellipse, two straight edges, and two triple points determine the special functions describing the sound radiated in certain special directions. The author believes that the wavenumber surface has not previously been determined for a diffraction problem. Intriguingly, the knife-edge in physical space leads to knife-edges in wavenumber space, a result which can be expected in general. The method in the paper for determining the wavenumber surface applies to any half-plane diffraction problem, e.g. the scattering of sound by a half-plane, the scattering of an elastic wave by a crack, or the scattering of a Rossby wave by a vertical barrier.

1.   INTRODUCTION

When a half-plane diffraction problem is tackled by the Wiener–Hopf technique, the customary approach is to Fourier transform all the space variables except one, and solve the resulting ordinary differential equation in this space variable. In consequence, the solution involves a Fourier inversion integral containing, typically, square roots of polynomials. Determination of the Riemann surface corresponding to the

I.D. Abrahams et al. (eds.),
IUTAM Symposium on Diffraction and Scattering in Fluid Mechanics and Elasticity, 169–178.
© 2002 *Kluwer Academic Publishers. Printed in the Netherlands.*

square root is then an important part of solving the problem, especially in asymptotic analysis.

In this paper a method is suggested which gives directly an integral representation with no square roots in the exponential term. The method is to solve the ordinary differential equation, referred to above, by a Fourier transform. At first sight this appears to be taking a backward step, because the equation has constant coefficients and its solution can be written directly in terms of exponentials. But the method gives an inversion integral in which the exponential term is simple, and in which all three space variables are treated on an equal footing. The asymptotic behaviour of the solution is then determined by the surface in three-dimensional wavenumber space on which a denominator vanishes. This is the wavenumber surface.

Although the wavenumber surface has been extensively analysed in non-diffraction problems of wave motion, particularly following Lighthill [1], the author has not come across any previous determination of a wavenumber surface in a diffraction problem. This paper determines the wavenumber surface for the half-plane diffraction problem appropriate to calculation of the acoustic field produced when convected vorticity, e.g. a gust, strikes the leading edge of an aircraft wing or fan blade. The acoustic part of the wavenumber surface is shown to consist of four sheets, namely an ellipsoid, half an elliptic cylinder, and two planes, between which there are certain tangency relations; the wavenumber surface has a fifth sheet, a plane, representing convection of vorticity with the mean flow. This description of the wavenumber surface applies when the flow is subsonic, which is the only case we shall consider; for supersonic flow, the ellipsoid would be replaced by a hyperboloid of two sheets, and the elliptic half-cylinder would be replaced by the whole of a hyperbolic cylinder (i.e. both of its sheets). The equations of some of the sheets of the wavenumber surface correspond to the factors in the Wiener–Hopf factorisation by which the diffraction problem is solved. This wholly three-dimensional interpretation of Wiener–Hopf factors appears to be new. A highly developed asymptotic theory is available to relate mathematical properties of the sheets, e.g. their Gaussian and mean curvature, to physical properties of the sound field, e.g. the far-field directivity pattern, as expounded in detail in [2].

The next section gives briefly the technological background, in studies of aircraft noise, to a blade-vortex interaction problem. The rest of the paper consists of a mathematical analysis of an idealised version of this problem.

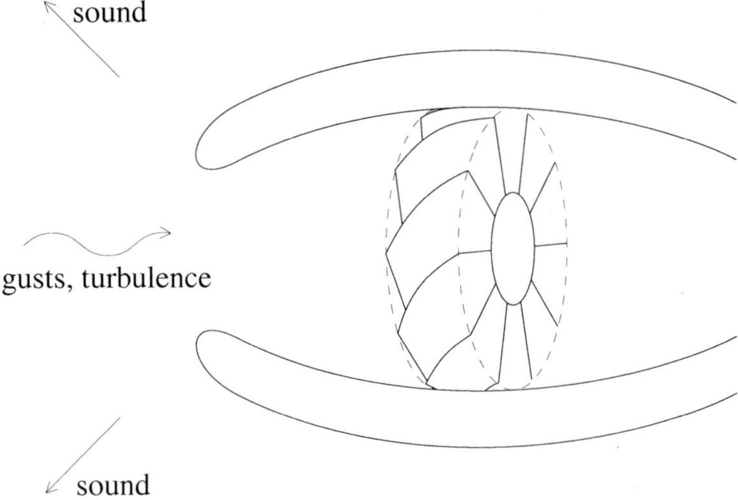

Figure 1   Blade-vortex interaction in a turbofan aeroengine. Incoming gusts and turbulence are chopped by the fan blades, and sound is produced.

2.   AEROENGINE NOISE

One source of noise from a turbofan aeroengine is the chopping, by the fan blades, of incoming gusts and turbulence. This mechanism of sound production, illustrated schematically in Fig. 1, is blade-vortex interaction. For sound of wavelength smaller than a chord, the conversion from vorticity to sound takes place largely at the leading edge of the fan blades. If the wavelength is not so small as to be comparable with the radius of curvature of the leading edge, a fan blade may be idealised as a half-plane, and the conversion of energy treated as a half-plane diffraction problem. This approach has been adopted by Amiet [3]–[6], Martinez and Widnall [7, 8, 9], Ffowcs Williams and Guo [10], and Guo [11]–[14], and has been extended, for example to quarter-planes and to wavelengths comparable with the leading-edge radius of curvature, by Peake [15]–[19], Majumdar and Peake [20], Peake and Kerschen [21], and Myers and Kerschen [22, 23].

We consider the simplest version of the problem, but allow the incoming gust to be arbitrary. The boundary-value problem is stated precisely in Section 3 and its solution given in Section 4. The wavenumber surface is constructed and its main features described in Section 5. Conclusions are drawn, and further work indicated, in Section 6.

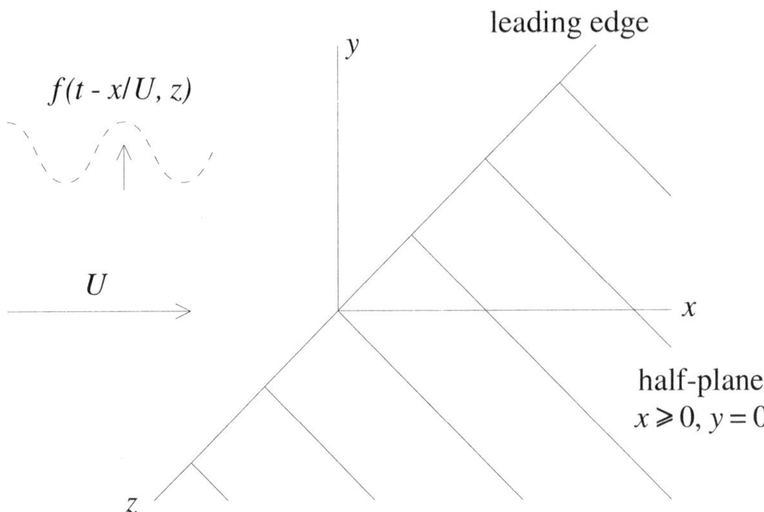

Figure 2   Gust convected at speed U past a stationary rigid half-plane $x \geq 0$, $y = 0$. In the upstream surface $x \leq 0$, $y = 0$ the vertical component of gust velocity is $f(t - x/U, z)$. Although this velocity component is blocked everywhere on the half-plane, the sound production, on linear theory, is confined entirely to the leading edge, i.e. to the z-axis.

3.   STATEMENT OF THE HALF-PLANE DIFFRACTION PROBLEM

An infinitesimally thin rigid half-plane is assumed to be at rest in a uniform subsonic mean flow of speed U. The mean flow is oriented relative to the half-plane as shown in Fig. 2. We use a Cartesian co-ordinate system (x, y, z) in which the leading edge of the half-plane is $x = y = 0$, i.e. the z-axis, and the half-plane is $x \geq 0$, $y = 0$. Thus x is the stream coordinate and z is the span coordinate. The coordinate directions (x, y, z) are the (downstream, upwards, span) directions. The velocity field of a gust which is convected with the mean flow can depend on time t and on the stream coordinate x only in the combination $t - x/U$. On the surface $x \leq 0$, $y = 0$ in the fluid the vertical velocity component of an arbitrary convected gust is therefore expressible in terms of an arbitrary function f of two variables as $f(t - x/U, z)$.

The acoustic velocity field $\mathbf{u} = (u, v, w)$ may be written in terms of a velocity potential ϕ as $\mathbf{u} = \nabla\phi$. If the undisturbed density of the fluid is ρ_0, and the convective derivative $\partial/\partial t + U\partial/\partial x$ is denoted D/Dt, the acoustic pressure p is given by $p = -\rho_0 D\phi/Dt$. The boundary conditions for the acoustic problem are that on the rigid half-plane the vertical component of acoustic velocity is minus that of the gust velocity, so that the total vertical component of velocity is zero; and on the upstream surface $x < 0$, $y = 0$ the acoustic pressure is zero. Thus the boundary conditions on $y = 0$ are

$$\frac{\partial \phi}{\partial y} = -f\left(t - \frac{x}{U}, z\right) \quad (x > 0) \quad \text{and} \quad \frac{D\phi}{Dt} = 0 \quad (x < 0). \tag{19.1}$$

The acoustic pressure p is discontinuous across the rigid half-plane and is odd in y. Hence the same may be assumed of ϕ, so that $\partial\phi/\partial x$ and $\partial\phi/\partial z$ are odd, and $\partial\phi/\partial y$ is even, in y, and $\phi = 0$ on $x < 0, y = 0$. The governing equation for ϕ is the convected wave equation, which in terms of the undisturbed sound speed c_0 is

$$\left\{ \frac{1}{c_0^2}\left(\frac{\partial}{\partial t} + U\frac{\partial}{\partial x}\right)^2 - \left(\frac{\partial^2}{\partial x^2} + \frac{\partial^2}{\partial y^2} + \frac{\partial^2}{\partial x^2}\right) \right\} \phi = 0. \tag{19.2}$$

Near the leading edge, ϕ is at most of order the square root of distance from the edge, and in the far field ϕ satisfies a radiation condition. The function f may be thought of as the 'data' for the problem: the matter of primary interest is the dependence of ϕ and other variables on f.

4. SOLUTION OF THE HALF-PLANE DIFFRACTION PROBLEM

In Section 1 we described a variant of the Wiener–Hopf technique in which the final ordinary differential equation is solved not directly but by a Fourier transform. When this technique is applied to the boundary-value problem stated in Section 3, the result is that the acoustic velocity potential ϕ may be written in the form

$$\phi = \frac{-1}{(2\pi)^4} \int\!\!\int\!\!\int\!\!\int \frac{2c_0^2 U l F(\omega, m)}{\tilde{D}(\omega, k, l, m)} e^{-i(\omega t - kx - ly - mz)}\, d\omega\, dk\, dl\, dm. \tag{19.3}$$

Here the integration region is the whole of (ω, k, l, m) space, and

$$F(\omega, m) = \int\!\!\int f(t', z) e^{i(\omega t' - mz)}\, dt'\, dz,$$

$$\tilde{D}(\omega, k, l, m) = \gamma_+(\omega, \omega/U, m)\gamma_-(\omega, k, m)(\omega - Uk)D(\omega, k, l, m),$$

$$D(\omega, k, l, m) = (\omega - Uk)^2 - c_0^2(k^2 + l^2 + m^2),$$

$$\gamma_-(\omega, k, m) = \left(1 - U^2/c_0^2\right)^{1/4}\left(k - k_+(\omega, m)\right)^{1/2},$$

$$\gamma_+(\omega, \omega/U, m) = \left(1 - U^2/c_0^2\right)^{1/4}\left((\omega/U) - k_-(\omega, m)\right)^{1/2},$$

$$k_\pm(\omega, m) = -\frac{U\omega}{c_0^2 - U^2} \pm \left\{ \frac{c_0^2\omega^2}{(c_0^2 - U^2)^2} - \frac{c_0^2 m^2}{c_0^2 - U^2} \right\}^{1/2}. \tag{19.4}$$

Thus ω is the frequency of a Fourier component of the acoustic field, and (k, l, m) is the wavevector in wavenumber space; $D(\omega, k, l, m)$ is the dispersion function for the convected wave equation; and γ_- and γ_+ are Wiener–Hopf factors. The acoustic pressure and velocity are obtained from the derivatives of (19.3) with respect to t, x, y and z.

The conventional use of the Wiener–Hopf technique gives the result of performing explicitly the l integration in (19.3) and thereby introduces in the exponential a square-root term multiplying y. The idea of our approach is that (19.3) is preferable for geometrical analysis, because of the symmetry of treatment of the position-vector components (x, y, z) and of the wavevector components (k, l, m). In fact, the k integration in (19.3) can also be performed explicitly, to leave a double integral over ω and m. The author has analysed this double integral in some detail, but the results will not be presented here.

5.   THE WAVENUMBER SURFACE

In the integral (19.3) for the acoustic velocity potential ϕ, a crucial role is played by those values of (ω, k, l, m) for which the denominator of the integrand vanishes, i.e. for which

$$\tilde{D}(\omega, k, l, m) = 0. \tag{19.5}$$

A suitable name for this equation is 'the diffraction dispersion relation'; and the wavenumber surface, for a given value of ω, consists of all the points (k, l, m) in wavenumber space for which the equation holds. To find these points, we consider separately the four factors of $\tilde{D}$ given in (19.4) for real values of k, l, and m, allowing for the possibility that a single factor can give rise to more than one sheet. The result is that for $\omega > 0$ the wavenumber surface is as shown in Fig. 3. The ellipsoid is the surface $D(\omega, k, l, m) = 0$, which is also found in non-diffraction problems of acoustic propagation in a mean flow; the plane perpendicular to the k-axis is the surface $\omega - Uk = 0$, i.e. $k = \omega/U$, which also is found in non-diffraction problems and describes convection of vorticity with the mean flow; the vertical half-cylinder is the surface $\gamma_-(\omega, k, m) = 0$; and the two parallel planes perpendicular to the m-axis are the two sheets of the surface $\gamma_+(\omega, \omega/U, m) = 0$. The half-cylinder and two parallel planes, corresponding as they do to Wiener–Hopf factors, are characteristic of diffraction. Fig. 3 makes evident the tangency relations of the ellipsoid, half-cylinder and two parallel planes, and shows how the edges of the half-cylinder lie in these two planes, forming two vertical lines of abutment. The two most special points on the wavenumber surface are where the two edges of the half-cylinder touch the ellipsoid, or equivalently they are the two end-points of the half-ellipse on which

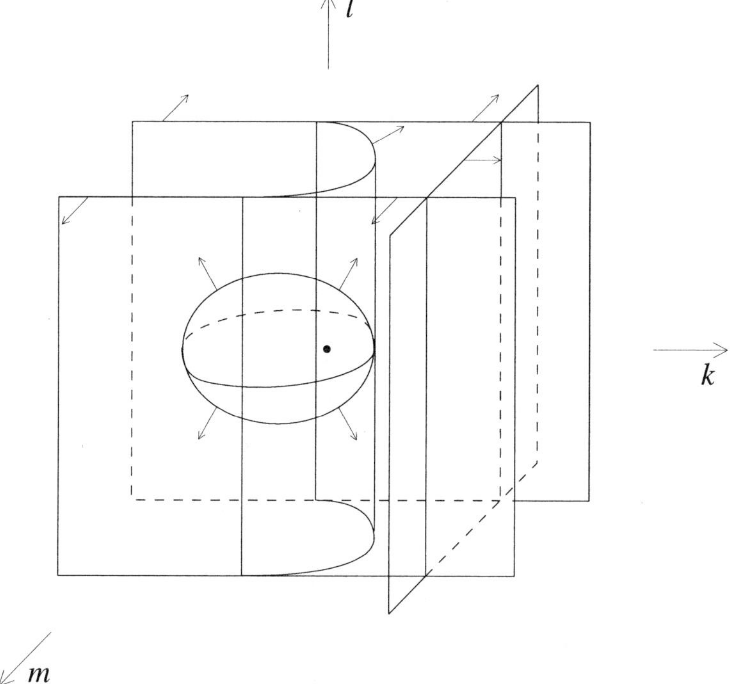

Figure 3 The wavenumber surface for the blade-vortex interaction problem shown in Fig. 2. A dot marks the origin of wavenumber space (k, l, m). The five sheets of the wavenumber surface are (i) the ellipsoid $D(\omega, k, l, m) = 0$; (ii) the elliptical half-cylinder $\gamma_-(\omega, k, m) = 0$, tangential to (i) along a half-ellipse; (iii), (iv) the two parallel planes, represented by the single equation $\gamma_+(\omega, \omega/U, m) = 0$, tangential to (i) and also to (ii) at its edges; and (v) the vertical plane $k = \omega/U$. Arrows show the direction of the group-velocity $\partial\omega/\partial\mathbf{k}$ on each sheet. The diagram is for $\omega > 0$. When $\omega < 0$, the half-cylinder (ii) and the vertical plane (v) are in the region $k < 0$, but their group-velocity arrows still have a positive component in the k direction.

the ellipsoid is tangent to the half-cylinder. These are triple points of tangency of the ellipsoid, the half-cylinder, and a plane.

Let us put $\mathbf{k} = (k, l, m)$ and write the equation of a sheet of the wavenumber surface as $\omega = \omega(\mathbf{k})$. For each point on the sheet, the corresponding far-field energy velocity is the group velocity $\partial\omega/\partial\mathbf{k}$, normal to the sheet. The factors of $\tilde{D}$ show that for $\omega > 0$ the group velocity for the ellipsoid, half-cylinder, two parallel planes, and the plane perpendicular to the k-axis are

$$\left(U + \frac{c_0 k}{|\mathbf{k}|}, \frac{c_0 l}{|\mathbf{k}|}, \frac{c_0 m}{|\mathbf{k}|}\right), \quad \left(U + \frac{c_0 k}{(k^2 + m^2)^{1/2}}, 0, \frac{c_0 m}{(k^2 + m^2)^{1/2}}\right),$$

$(0, 0, \pm(c_0^2 - U^2)^{1/2})$ and $(U, 0, 0)$. Arrows representing these velocities are shown on Fig. 3.

We have seen that the wavenumber surface admits a complete analytical description. The vorticity part of the surface is less complicated than the acoustical part, because vorticity can propagate only in the direction of the fluid flow. In other diffraction problems there may be several complicated parts of the wavenumber surface. For example, in diffraction of elastic waves by a crack the longitudinal-wave and transverse-wave parts of the wavenumber surface will both be complicated.

6.   CONCLUSIONS AND FURTHER WORK

The advantage of determining the wavenumber surface for blade-vortex interaction is that this surface makes intelligible the complicated three-dimensional anisotropic sound field produced by an arbitrary incoming gust, for example by a localised gust which strikes only a narrow segment of the leading edge of a blade. Once the wavenumber surface is known, a set of well-understood techniques, expounded by Lighthill [1, 2] and others, may be used to relate the geometry of the surface to the radiated sound. The theory has been extended to include special directions corresponding to tangency of two or more sheets of the surface. These directions invariably have physical significance. For example, propagation along the leading edge of the blade corresponds in Fig. 3 to the triple points where the half-cylinder, the ellipsoid, and the plane are tangential; and propagation on the surfaces of the blade corresponds to the half-ellipse of tangency of the half-cylinder and ellipsoid. This preliminary paper gives only the shape and the sheet structure of the wavenumber surface; the task of applying the asymptotic theory remains for further work.

An alternative approach, mentioned at the end of Section 4, is to perform explicitly the k and l integrations in (19.3), to leave a double integral over ω and m. This approach retains a geometrical flavour, and leads to a canonical diffraction integral for which the Riemann surface has the topology shown in Fig. 4. It would be interesting to know if the wavenumber surface determines the Riemann surface.

Acknowledgments

The early stages of this work were carried out with the support of DTI (CARAD) through the Defence Research Agency, Pyestock. The author is grateful to A. B. Parry, S. J. Perkins and other members of the aeroacoustics group at Rolls-Royce, Derby for their comments and assistance with the project.

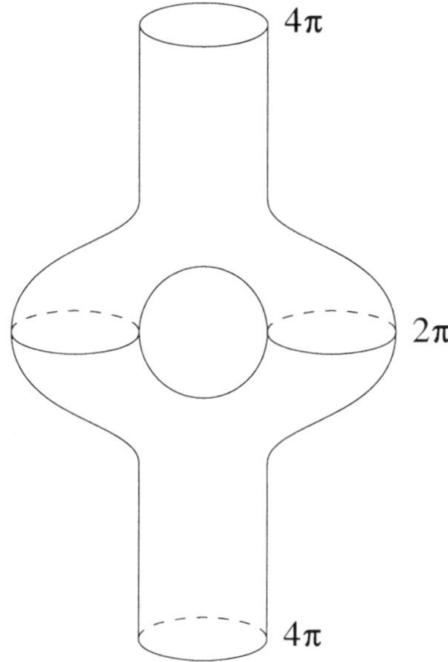

Figure 4   Riemann surface for the canonical diffraction integral in the theory of blade-vortex interaction. The circumference of the tubes is 2π or 4π as shown.

References

[1] Lighthill, M J (1960) *Studies on magneto-hydrodynamic waves and other anisotropic wave motions*, Phil. Trans. Roy. Soc. Lond. A **252**, 397–430.

[2] Lighthill, J (1978) *Waves in Fluids.* Cambridge: University Press.

[3] Amiet, R K (1976) *High frequency thin-airfoil theory for subsonic flow*, AIAA J. **14**, 1076–1082.

[4] Amiet, R K (1986) *Airfoil gust response and the sound produced by airfoil-vortex interaction*, J. Sound Vib. **107**, 487–506.

[5] Amiet, R K (1986) *Intersection of a jet by an infinite span airfoil*, J. Sound Vib. **111**, 409–503.

[6] Amiet, R K (1986) *Gust response of a flat-plate aerofoil in the time domain*, Quart. J. Mech. Appl. Mech. **39**, 485–505.

[7] Martinez, R and Widnall, S E (1980) *Unified aerodynamic-acoustic theory for a thin rectangular wing encountering a gust*, AIAA J. **18**, 636–645.

[8] Martinez, R and Widnall, S E (1983) *Aerodynamic theory for wing with side edge passing subsonically through a gust*, AIAA J. **21**, 808–815.

[9] Martinez, R and Widnall, S E (1983) *An aeroacoustic model for high-speed, unsteady blade-vortex interaction,* AIAA J. **21**, 1225–1231.

[10] Ffowcs Williams, J E and Guo, Y P (1988) *Sound generated from the interruption of a steady flow by a supersonically moving aerofoil,* J. Fluid Mech. **195**, 113–135.

[11] Guo, Y P (1989) *A note on sound from the interruption of a cylindrical flow by a semi-infinite aerofoil of subsonic speed,* J. Sound Vib. **128**, 275–286.

[12] Guo, Y P (1989) *On sound generation by a jet flow passing a semi-infinite aerofoil,* AIAA Paper 89–1070.

[13] Guo, Y P (1990) *Sound generation by a supersonic aerofoil cutting through a steady jet flow,* J. Fluid Mech. **216**, 193–212.

[14] Guo, Y P (1991) *Energetics of sound radiation from flow-aerofoil interaction,* J. Sound Vib. **151**, 247–262.

[15] Peake, N (1992) *Unsteady transonic flow past a quarter-plane,* J. Fluid Mech. **244**, 377–404.

[16] Peake, N (1993) *The interaction between a steady jet flow and a supersonic blade tip,* J. Fluid Mech. **248**, 543–566.

[17] Peake, N (1994) *The unsteady lift on a swept blade tip,* J. Fluid Mech. **271**, 87–101.

[18] Peake, N (1996) *Sound radiation from sources close to a corner in supersonic flow,* Wave Motion **24**, 197–210.

[19] Peake, N (1997) *The scattering of vorticity waves by a supersonic rectangular wing,* Wave Motion **25**, 369–383.

[20] Majumdar, S J and Peake, N (1996) *Three-dimensional effects in cascade-gust interaction,* Wave Motion **23**, 321–337.

[21] Peake, N and Kerschen, E J (1997) *Influence of mean loading on noise generated by the interaction of gusts with a flat-plate cascade: upstream radiation,* J. Fluid Mech. **347**, 315–346.

[22] Myers, M R and Kerschen, E J (1995) *Influence of incidence angle on sound generation by airfoils interacting with high-frequency gusts,* J. Fluid Mech. **292**, 271–304.

[23] Myers, M R and Kerschen, E J (1997) *Influence of a camber on sound generation by airfoils interacting with high-frequency gusts,* J. Fluid Mech. **353**, 221–259.

DIFFRACTION OF CREEPING WAVES BY CONICAL POINTS

V. P. Smyshlyaev, V. M. Babich[*], D. B. Dementiev[†], B. A. Samokish[‡]
Department of Mathematical Sciences, University of Bath, Bath BA2 7AY, UK
vps@maths.bath.ac.uk

Abstract   We briefly review our recent results on evaluation of the diffracted wave for electromagnetic creeping waves scattered by a conical point at a perfectly conducting surface. The theory uses matched asymptotic expansions and the reciprocity principle, and reduces the problem to the need to evaluate the "canonical" conical diffraction coefficients at the boundary. The latter is a special case of the theory developed and implemented by us before. Additional technical difficulty comes from the need to evaluate the diffraction coefficients *at the boundary* which within the application of our "spherical" boundary integral equation method leads to the need to evaluate appropriate *singular integrals*. The latter was resolved using the Discrete Fourier Transform and the whole strategy has been implemented numerically. We report sample numerical results demonstrating convergence of the algorithm.

1. INTRODUCTION

Creeping waves propagate in a high frequency regime along "shadow" parts of convex obstacles. They were detected experimentally long ago and their analytic theory was proposed by J. B. Keller in the 1950s. Since then Keller's formulae both inspired the development of the mathematical theory of high frequency diffraction, including their rigorous justification (V. B. Filippov, G. Lebeau), and proved to be extremely useful in applications.

This now classical theory of creeping waves is applicable only to *smooth* convex bodies. In a number of practically important problems, however, the obstacles contain *singularities*, a typical example of which

---

[*]Permanently at: St. Petersburg Branch of Steklov Mathematical Institute, Fontanka 27, St. Petersburg 191011, Russia

[†]Permanently at: Department of Mathematics and Computer Science, Emory University, Atlanta, GA 30322, USA

[‡]Permanently at: Department of Mathematics and Mechanics, St. Petersburg State University, St. Petersburg 198904, Russia

I.D. Abrahams et al. (eds.),
IUTAM Symposium on Diffraction and Scattering in Fluid Mechanics and Elasticity, 179–187.
© 2002 *Kluwer Academic Publishers. Printed in the Netherlands.*

is displayed in Fig. 1. The scatterer B contains a sharp "conical" point O in the shadow. The incident wave does not initially "see" the point O and the grazing incidence at the "horizon" curve l generates creeping waves as in the classical Keller theory. The creeping waves propagate along the geodesics on the surface S of B and some of them may come close to O. When this happens, the creeping waves diffract by the conical point O. General postulates of Keller's Geometric Theory of Diffraction [1] suggest that the point O acts as a "secondary source" diffracting wave in all directions. The problem of major interest is to evaluate this diffracted wave. Further, the above diffraction in *tangent* directions is expected to generate *secondary* creeping waves propagating along the geodesics of S *away* from O. Knowing the intensities of these secondary creeping waves is also very desirable.

For the acoustic problem, S. J. Chapman and V. Saward considered diffraction by an axisymmetric body with a sharp point [2, 3]. They used matched asymptotic expansions and argued that the evaluation of the diffracted wave reduces to the need to evaluate the *diffraction coefficients* for the infinitesimal semi-infinite cone, for a particular incident wave. The latter "inner" problem appears to fall into the general context of evaluation of the conical diffraction coefficients which has been developed and numerically implemented by us over the years, e.g. [4, 5, 6]. A specific feature of the present inner problem, and a source of considerable complication in numerical evaluation of the corresponding diffraction coefficients, is that the incidence is *along the boundary* of the infinitesimal cone. If the observation direction is however *away* from the boundary, the problem is numerically tractable via the use of the *reciprocity principle*. It allows us to interchange the direction of incidence (along the boundary) and the direction of observation (away from the boundary) which leads to certain technical simplifications. After this is done, we are able to compute the associated diffraction coefficients adapting appropriately the previously developed methods [4, 5, 6]. In addition we need to evaluate numerically emerging *singular integrals*, which is resolved using the Discrete Fourier Transform. Taken together, this allows us to compute the diffracted field away from the boundary.

This short paper provides some abbreviated details for the above strategy and results.

2.   FORMULATION OF THE PROBLEM

We consider incidence of an electromagnetic wave at a perfectly conducting axisymmetric scatterer B with a conical point O (Fig. 1). The construction is easily adapted to acoustic problems with ideal (Dirichlet

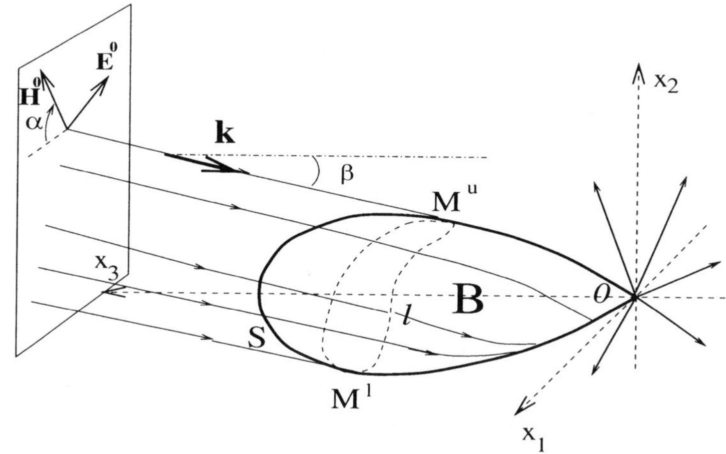

Figure 1   Geometry of the problem.

or Neumann) boundary conditions (cf. [2, 3]). The wave process outside B is described by the time-harmonic Maxwell equations

$$i\,k\,\mathbf{H} = \operatorname{curl}\mathbf{E}, \quad i\,k\,\mathbf{E} = -\operatorname{curl}\mathbf{H}, \tag{20.1}$$

with wave number k, electric intensity $\mathbf{E}$ and (normalized) magnetic intensity $\mathbf{H}$. Perfectly conducting boundary condition

$$\mathbf{E} \times \boldsymbol{n} = 0 \tag{20.2}$$

is assumed at the surface S of B ($\boldsymbol{n}$ is normal to S).

The incident plane wave $(\mathbf{E}^0, \mathbf{H}^0)$ is characterised by the angle β between the wave vector $\boldsymbol{k}$ and the axis of symmetry x_3 (Fig. 1). Choose the Cartesian coordinates so that $\boldsymbol{k}$ lies in the plane $x_1 = 0$. It is assumed that β is such that the conical point O is in the shadow; $\beta = 0$ corresponds to the axisymmetric incidence. The incident wave is also characterised by its polarisation angle α which is the angle between the axis x_1 and the vector $\mathbf{H}^0$ (Fig. 1). Therefore,

$$\mathbf{E}^0 = e^{i\boldsymbol{k}\cdot\boldsymbol{x}}\left(-\sin\alpha\,\boldsymbol{e}_1 + \cos\alpha\cos\beta\,\boldsymbol{e}_2 - \cos\alpha\sin\beta\,\boldsymbol{e}_3\right)$$

$$\mathbf{H}^0 = e^{i\boldsymbol{k}\cdot\boldsymbol{x}}\left(\cos\alpha\,\boldsymbol{e}_1 + \sin\alpha\cos\beta\,\boldsymbol{e}_2 - \sin\alpha\sin\beta\,\boldsymbol{e}_3\right), \tag{20.3}$$

where $\boldsymbol{e}_j$ are unit vectors along the axes x_j, $j = 1, 2, 3$. The boundary value problem (20.1)–(20.3) is supplemented by radiation conditions for the scattered field, as usual.

We seek the high frequency pattern of the total field, or, mathematically, the asymptotics of the solution to (20.1)–(20.3) with respect to small $\varepsilon = (kL)^{-1}$, where L is a characteristic length of B. This problem falls in the general context of the Geometric Theory of Diffraction [1]. The asymptotics of the scattered field is known to be "composed" of a number of components corresponding to waves of different types. In particular, the incident wave's rays (pointing in the direction of the wave vector **k**) generate the "creeping waves" at the points of their tangency with S.

The theory of creeping waves, propagating along geodesics of smooth convex surfaces, is well developed (see e.g. [7, Chapt. 13] and further references therein). The problem addressed here is what happens when the creeping waves come close to the "singular" (conical) point O.

In a typical configuration of non-axisymmetric incidence ($\beta \neq 0$, Fig. 1) there are only two small "bunches" of the creeping wave geodesics which come sufficiently close to O. Those are the creeping waves originating in the vicinities of the "upper" and the "lower" tangency points M^u and M^l, respectively (Fig. 1). This follows, for example, from Clairaut's theorem (see e.g. [8]): a geodesic curve at an axisymmetric surface satisfies the following property: $r \sin \gamma = $ constant, (cf. [2, Eqn. (8.25)]), where r is the distance to the axis and γ is the angle between the geodesics and the "meridian" of the figure of revolution. Therefore, it is sufficient to consider (separately) the above two small geodesic bunches. The Clairaut formula is an integral of the differential equation for geodesics. This equation is in fact explicitly integrable in quadratures for axisymmetric surfaces (see e.g. [8]).

3.   DIFFRACTION OF CREEPING WAVES BY THE TIP O

Creeping waves propagating along a convex perfectly conducting surface are known to be classified in two groups: the so called *magnetic* and *electric* creeping waves. General formulae for them are well known:

Magnetic creeping waves.

$$\begin{Bmatrix} \mathbf{E} \\ \mathbf{H} \end{Bmatrix} = \chi_h(\alpha)\, e^{ik\tau}\, e^{i\kappa_l' \Phi}\, J^{-1/2} \rho^{-\frac{1}{6}} w_1(\kappa_l' - \nu) \begin{Bmatrix} \mathbf{n} \\ -\mathbf{e}_\alpha \end{Bmatrix}. \qquad (20.4)$$

Electric creeping waves.

$$\begin{Bmatrix} \mathbf{E} \\ \mathbf{H} \end{Bmatrix} = \chi_e(\alpha)\, e^{ik\tau}\, e^{i\kappa_l \Phi}\, J^{-1/2} \rho^{-\frac{1}{6}} w_1(\kappa_l - \nu) \begin{Bmatrix} \mathbf{e}_\alpha \\ \mathbf{n} \end{Bmatrix}. \qquad (20.5)$$

Here w_1 is the Airy-type function $w_1(z) = e^{i\pi/6}\pi^{1/2}\mathrm{Ai}(\exp(2\pi i/3)z)$, κ_l and κ_l', $l = 1, 2, \ldots$ are zeros of w_1 and of the derivative of w_1 respectively,

J is the surface geometrical spreading of the creeping wave geodesics (surface "rays"), ρ is the radius of curvature of the normal section of S along the ray, $\nu = k^{2/3} n (2/\rho)^{1/3}$, n is the distance to the surface, $\boldsymbol{n}$ is the normal vector to S, α is the parameter labelling the geodesics, $\boldsymbol{e}_\alpha$ is the "tangent" unit vector orthogonal to the geodesics, τ is the "eikonal" of the creeping rays, $\tau_1 = \tau - p(\alpha)$ ($\tau = p(\alpha)$ is the boundary between the illuminated and the shadow zone on the surface S), $\chi_h(\alpha)$ and $\chi_e(\alpha)$ are the "intensities" of the creeping waves, which have to be found from matching with incident wave near the tangency points, and

$$\Phi = 2^{-1/3} k^{1/3} \int_0^{\tau_1} \rho^{-2/3} \, d\tau_1.$$

Direct specification of the formulae (20.4), (20.5) to the problem in hand is rather cumbersome. Saward (for the scalar problem) [2, §8.3] used perturbation methods to study the behaviour of geodesics close to the upper meridian and subsequently of the creeping waves on approaching the point O. For the present (electromagnetic) problem, the incident wave (20.3) can be decomposed into two other plane waves whose polarisations correspond to $\alpha = 0$ and to $\alpha = \pi/2$, respectively:

$$\mathbf{E}^0 = \cos\alpha \, \mathbf{E}^{01} + \sin\alpha \, \mathbf{E}^{02}, \quad \mathbf{H}^0 = \cos\alpha \, \mathbf{H}^{01} + \sin\alpha \, \mathbf{H}^{02},$$

where

$$\mathbf{E}^{01} = e^{i\boldsymbol{k}\cdot\boldsymbol{x}} \left(\cos\beta \boldsymbol{e}_2 - \sin\beta \boldsymbol{e}_3 \right), \quad \mathbf{H}^{01} = e^{i\boldsymbol{k}\cdot\boldsymbol{x}} \boldsymbol{e}_1, \tag{20.6}$$

$$\mathbf{E}^{02} = -e^{i\boldsymbol{k}\cdot\boldsymbol{x}} \boldsymbol{e}_1, \quad \mathbf{H}^{02} = e^{i\boldsymbol{k}\cdot\boldsymbol{x}} \left(\cos\beta \boldsymbol{e}_2 - \sin\beta \boldsymbol{e}_3 \right). \tag{20.7}$$

It is sufficient to consider separately these two plane incident waves.

Consider for definiteness the creeping waves generated at point M^u. They propagate along the meridian geodesic towards O. As long as we are sufficiently far away from O they are representable in the form (20.4), (20.5). [Note in passing (proof omitted here) that the "upper" creeping waves remain regular on the meridian, whereas the "lower" creeping waves (at the meridian corresponding to M^l) have a focussing point (away from O).] Direct matching with the incident waves (omitted here) derives explicit values for χ_h and χ_e and shows that it is *magnetic* wave (20.4) which dominates in the neighbourhood of the "upper" meridian for the incident wave (20.6), and it is *electric* wave (20.5) which dominates for (20.7).

The emerging problem is that of scattering of the creeping waves (20.4) and (20.5) by the conical point O. It can be approached by rescaling the neighbourhood of the point O and retaining the main order terms in (20.4) and (20.5) as "incident waves" for the resulting "inner" problem, Fig. 2 (cf. [2, §§8.3, 8.4]).

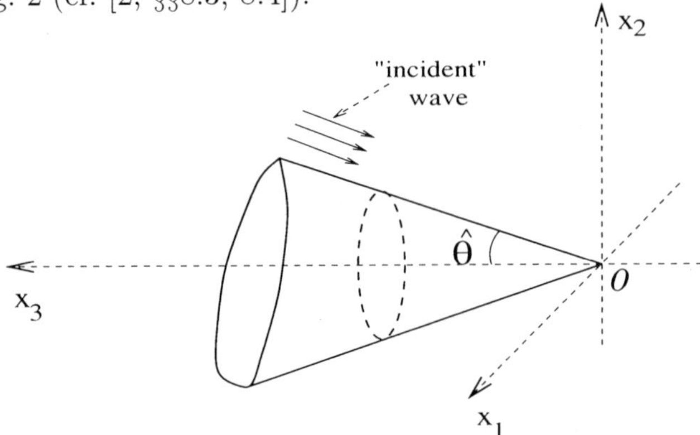

Figure 2   The "inner" problem.

On the technical side, we found that the use of the *reciprocity principle* (i.e. interchanging the directions of incidence and observation, cf. [2, §9.8], but *both* in the *original* and in the *inner* problems) allows us to simplify the analysis quite dramatically and to clarify the structure of the incident wave. In particular, the use of reciprocity establishes a posteriori that it is a *single plane-like wave* along the straight cone's generator which is the incident wave for the inner problem. In turn, the use of reciprocity for the *inner* problem allows us to express the pattern of the diffracted wave *away from the boundary* in terms of the diffracted field *at the boundary* for the reciprocal problem (with the incidence *away* from the boundary).

The latter problem falls into the context of evaluating the conical canonical diffraction coefficients [4, 5, 6]. A major technical complication for adapting our previously developed analytical and numerical routines for the problem in hand was the need to evaluate appropriate electromagnetic diffraction coefficients (and their derivatives) *at the boundary*. For this, the "spherical" integral equation method [5, 6] leads to the need to evaluate certain *singular* integrals. They had to be *regularised*, which had to be combined with the need to evaluate numerically slowly convergent integrals with respect to a complex parameter ν, via the Abel–Poisson regularisation [6].

The above rather extensive numerical strategy has been successfully implemented. As a result, convergence was observed for a trial example, although rather slow. We omit here the details of derivation and implementation, but give below a brief list of the eventual analytical and numerical results.

4. THE WAVE DIFFRACTED BY THE CONICAL POINT

For the original problem (20.1)–(20.3) the wave diffracted by the tip O is

$$\left\{ \begin{array}{c} \mathbf{E}_{\text{diff}} \\ \mathbf{H}_{\text{diff}} \end{array} \right\} \cong -2\pi \frac{e^{ikr}}{kr} \left\{ \begin{array}{c} \boldsymbol{\mathcal{E}}(\boldsymbol{\omega}) \\ \boldsymbol{\mathcal{H}}(\boldsymbol{\omega}) \end{array} \right\}. \tag{20.8}$$

The *diffraction coefficients* $\boldsymbol{\mathcal{E}}$ and $\boldsymbol{\mathcal{H}}$ depend on the observation direction $\boldsymbol{\omega} = (\theta, \varphi)$ and are to be found; (r, θ, φ) are spherical coordinates associated with (x_1, x_2, x_3): $(x_1 = r \sin\theta \cos\varphi$, $x_2 = r \sin\theta \sin\varphi$, $x_3 = -r \cos\theta)$. The diffraction coefficients have to be evaluated separately for the incident waves (20.6) and (20.7), and for the "upper" and "lower" creeping waves paths. Consider for definiteness the upper path and the incident wave (20.6).

Since $\boldsymbol{\mathcal{E}}$ and $\boldsymbol{\mathcal{H}}$ are mutually orthogonal, is it sufficient to evaluate $\boldsymbol{\mathcal{E}}$, which is in turn orthogonal to $\boldsymbol{\omega}$. For this, it is sufficient to evaluate the component $\mathcal{E}'$ of $\boldsymbol{\mathcal{E}}$ along a vector $\boldsymbol{e}'$ orthogonal to $\boldsymbol{\omega}$. Select $\boldsymbol{e}'$. The final result is listed below. (More precisely, we evaluate the contribution due to the "main" creeping wave ($l = 1$ in (20.4)); contributions for "higher" l are asymptotically smaller.) We find that $\mathcal{E}' = D E$, where

$$D = -i\pi^{1/2} \left(\frac{\rho_u}{\rho_0} \right)^{1/6} \left(\frac{\sin\hat{\theta}}{\sin\beta} \right)^{1/2} \times \tag{20.9}$$

$$\frac{e^{ik\tau(O)}}{\kappa_1' w_1(\kappa_1')} \exp\left\{ i\, 2^{-1/3} \kappa_1' k^{1/3} \int_O^{M^u} \rho^{-2/3}(s')\, ds' \right\},$$

$\tau(O)$ is the eikonal of the creeping wave at the tip O, $\hat{\theta}$ is the semi-angle of the "infinitesimal" cone at O, ρ_u and ρ_0 are the meridian's curvatures at M^u and O respectively.

Further,

$$E := \frac{i}{\pi} \lim_{\varepsilon \to 0+} \int_\gamma \frac{e^{-i\nu\pi - \varepsilon\nu}}{\nu^2 - \frac{1}{4}} G(\nu)\, \nu\, d\nu, \tag{20.10}$$

where γ is the contour "bending round" the positive real semi-axis of the complex plane ν (see e.g. [5, Fig. 3]);

$$G(\nu) := \frac{\partial}{\partial\theta} g_D(\theta, \pi/2)\big|_{\theta=\theta_1} - \frac{1}{\sin\theta_1} \frac{\partial}{\partial\varphi} g_N(\theta_1, \varphi)\big|_{\varphi=\pi/2}. \tag{20.11}$$

Finally, g_D and g_N are the solutions of the "spherical" boundary value problems at $M = \{\boldsymbol{\omega} = (\theta, \varphi) : 0 \le \theta \le \theta_1 := \pi - \hat{\theta}\}$ (see [4, 5]) for the "reciprocal" incident wave from the direction $\boldsymbol{\omega}$, whose electric field

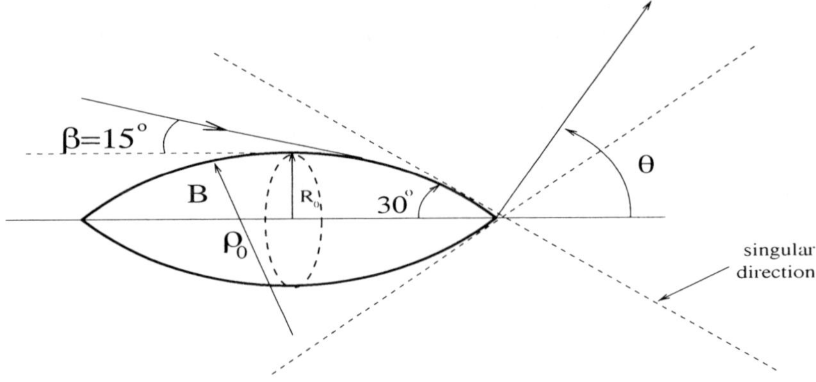

Figure 3   The scatterer for numerical implementation.

is polarised in direction e'. The latter are found numerically via the integral equation method on the sphere, by introducing a large number of "nodes" N (see [5, 6] for details).

As a result, $(\partial/\partial\theta)g_D$ and $(\partial/\partial\varphi)g_N$ entering (20.11) are expressible in terms of the solutions of the integral equations (the "densities" μ_D and μ_N [5]) via singular integrals. The latter have been regularised via the Discrete Fourier Transform.

5.   NUMERICAL RESULTS

We have chosen as a scatterer B the body formed by rotating a segment of the circle of radius ρ_0, with cone's semi-angle equal to $30°$ ($\theta_1 = 5\pi/6$). The wavenumber k was selected in such a way that the dimensionless parameter $k\rho_0$ equals 10. The angle of incidence β (see Section 2) was taken as $15°$ ($\beta = \pi/12$). The observation directions are taken: $\theta_0 = \pi/3$, $\varphi_0 = \pi/4$.

We calculate below $S := -i\pi E$ for $e' = e_\theta$ for selected values of N and ε (Table 1). One can observe convergence, although rather slow. There remains room for improvements both in the analytical part (e.g. in extracting the underlying singularity to "higher orders") and in improving the numerical convergence.

Note in conclusion that for the problem of evaluation of amplitudes of *secondary creeping waves* the underlying canonical diffraction coefficients have to be evaluated when *both* the source and the observation points lie *at the boundary* of the spherical domain. This introduces additional complications. Nevertheless, at present, we have been able to compute the secondary creeping wave amplitudes for *axisymmetric* incidence ($\beta = 0$). The symmetry ensures that the spherical boundary value problem can be solved explicitly, i.e. without the need to resort to

Table 1   Convergence of the numerical algorithm for $e' = e_\theta$.

	S_ε^N for $e' = e_\theta$	
N	$\varepsilon = 0.1$	$\varepsilon = 0.02$
$N = 240$	$0.40947 - 0.35519\,i$	$0.40257 - 0.31318\,i$
$N = 300$	$0.40675 - 0.34206\,i$	$0.40101 - 0.30819\,i$
$N = 360$	$0.40496 - 0.33327\,i$	$0.39999 - 0.30489\,i$

the integral equation and to the associated singular integrals. This (as well as a more detailed account of the above presented results) will be reported elsewhere.

Acknowledgments

The work has been supported by BAE Systems. The authors are grateful to Dr Jill Ogilvy (BAE Systems) for support and permanent interest in this work.

References

[1] Keller, J B (1962) *The geometrical theory of diffraction*, J. Opt. Soc. Amer. **52**, 116–130.

[2] Saward, V H (1997) *Some Problems in Diffraction theory*, Ph.D. Thesis: Oxford University.

[3] Chapman, S J (1997) personal communication.

[4] Smyshlyaev, V P (1993) *The high-frequency diffraction of electromagnetic waves by cones of arbitrary cross-sections*, SIAM J. Appl. Math. **53**, 670–688.

[5] Babich, V M, Smyshlyaev, V P, Dement'ev, D B and Samokish, B A (1996) *Numerical calculation of the diffraction coefficients for an arbitrarily shaped perfectly conducting cone*, IEEE Trans. Antenn. & Propag. **44**, 740–747.

[6] Babich, V M, Dement'ev, D B, Samokish, B A and Smyshlyaev, V P (2000), *On evaluation of the diffraction coefficients for arbitrary "non-singular" directions of a smooth convex cone*, SIAM J. Appl. Math. **60**, 536–573.

[7] Babič, V M and Buldyrev, V S (1991) *Short-Wave-Length Diffraction Theory*, Berlin: Springer.

[8] Dubrovin, B A, Fomenko, A T and Novikov, S P (1992) *Modern Geometry – Methods and Applications I*, New York: Springer.

EFFECTS OF TEMPERATURE GRADIENT ON THE PROPAGATION OF AN ACOUSTIC SOLITARY WAVE IN AN AIR-FILLED TUBE

N. Sugimoto, K. Tsujimoto

Department of Mechanical Science

Graduate School of Engineering Science

University of Osaka, Osaka 560-8531, Japan

sugimoto@me.es.osaka-u.ac.jp

Abstract   This paper examines effects of temperature gradient on the propagation of an acoustic solitary wave in an air-filled tube with an array of Helmholtz resonators. With neglect of viscosity and heat conduction, interest is focused on the effects due to axial nonuniformity of the temperature and density of air in undisturbed state. On the basis of the nonlinear wave equations for unidirectional propagation of plane waves, evolutions of the acoustic solitary wave are sought numerically in two cases where the temperature increases or decreases linearly along the tube. Discussions are given by comparing the present problem with the one of propagation of a shallow-water soliton over variable topography.

1.   INTRODUCTION

It has recently been shown that an acoustic solitary wave can be propagated in an air-filled tube if Helmholtz resonators are connected with the tube axially in periodic array [1, 2, 3]. Of course it is assumed that all lossy effects are negligibly small and undisturbed state is uniform throughout the tube. With the first assumption held, this paper examines effects of axial temperature gradient in undisturbed state on the propagation of the acoustic solitary wave.

A typical length of temperature variation is assumed much longer than the width of the solitary wave. Because the gradient is thus gentle and no boundary layer is assumed to develop on the tube wall, unidirectional propagation of plane waves is considered. The temperature gradient introduces an extra term into the evolution equations derived previously in the case of uniform temperature [2]. Since the equations are based

189

I.D. Abrahams et al. (eds.),

IUTAM Symposium on Diffraction and Scattering in Fluid Mechanics and Elasticity, 189–198.

© 2002 *Kluwer Academic Publishers. Printed in the Netherlands.*

on the approximation of geometrical acoustics [4], the total energy flux passing at some location is conserved.

Imposing the acoustic solitary wave as an initial condition, its spatial evolution is solved numerically in such a typical case that the temperature varies linearly with the axial distance, e.g., from a room temperature to about one thousand kelvin or a room temperature to the absolute zero. The present problem appears to resemble the one of the shallow-water solitary wave propagating over variable topography. In fact, in a special case, the acoustic solitary wave is reduced to a KdV soliton. Similarity and dissimilarity between the two problems are clarified from a viewpoint of the KdV equation.

2.   THE ACOUSTIC SOLITARY WAVE

We summarise of the acoustic solitary wave [2]. It is the steady propagation of a pressure pulse localized spatially and temporally. It consists of compression phase only where both the density and temperature rise adiabatically. When the wave is passing, a particle of air is pushed forward in the direction of propagation: there is no backward motion.

The propagation speed v is subsonic and limited in the range $a_0/(1 + \kappa/2) < v < a_0$, where a_0 is the linear sound speed and $\kappa = V/Ad \; (\ll 1)$ is a small parameter to measure the size of the array of resonators, V, A and d being, respectively, cavity's volume, tube's cross-sectional area and axial spacing between the neighbouring resonators. Given a value of κ and a natural angular frequency of the resonator ω_0, the solitary wave is determined uniquely in terms of its propagation speed $v(s) = a_0/(1 + \kappa s/2)$ for $0 < s < 1$.

As v approaches the upper bound ($s \to 0$), the peak pressure of the solitary wave increases monotonically. But there exists a limiting solitary wave, which is given for the excess pressure p' over the equilibrium pressure p_0 by

$$\frac{p'}{p_0} = \begin{cases} \frac{8}{3}\gamma(\gamma + 1)^{-1}\kappa \cos^2(\zeta/4) & \text{for } |\zeta| \leq 2\pi \\ 0 & \text{for } |\zeta| > 2\pi \end{cases} \tag{21.1}$$

with $\zeta = \omega_0(t - x/a_0) +$ const., where t and x are the time and the axial coordinate along the tube, respectively, γ being the ratio of specific heats. Note that the regularity is lost at $\zeta = \pm 2\pi$. As v approaches the lower bound ($s \to 1$), the profile tends to a KdV soliton given by

$$p'/p_0 = \alpha\gamma(\gamma + 1)^{-1}\kappa \text{sech}^2(\alpha/12)^{1/2}\zeta, \tag{21.2}$$

with $\zeta = \omega_0[t - (1 + \kappa s/2)x/a_0] +$ const., where $s = 1 - \alpha/3 \; (0 < \alpha \ll 1)$.

3.　NONLINEAR WAVE EQUATIONS

Evolution equations for nonlinear acoustic waves propagating in the positive direction of x under the temperature gradient are given in the dimensionless form as follows [5]:

$$\frac{\partial f}{\partial X} - f\frac{\partial f}{\partial \theta} + \frac{1}{4H_e}\frac{\mathrm{d}H_e}{\mathrm{d}X}f = -K\frac{\partial g}{\partial \theta}, \qquad \frac{\partial^2 g}{\partial \theta^2} + \Omega_e g = \Omega_e f. \qquad (21.3)$$

Here the variables are defined as follows: $\varepsilon f = \frac{1}{2}(1+\gamma)\gamma^{-1}p'/p_0 = \frac{1}{2}(\gamma+1)u/a_e$, $\varepsilon g = \frac{1}{2}(1+\gamma)\gamma^{-1}p'_c/p_0$, $K = \kappa/2\varepsilon$, $\Omega_e = (\omega_e/\omega)^2$,

$$\theta = \omega\left(t - \int_0^x \frac{\mathrm{d}x}{a_e(x)}\right), \qquad X = \varepsilon\omega\int_0^x \frac{\mathrm{d}x}{a_e(x)}, \qquad H_e(X) = \frac{T_e}{T_0};$$

p' and p'_c denote the excess pressures over p_0 in the tube and in the cavity of the Helmholtz resonator, respectively; u denotes the axial velocity of air; θ and X represent, respectively, the retarded time and the far-field axial coordinate; ε is a parameter much smaller than unity to measure the order of magnitude of excess pressure relative to p_0, and ω is its typical angular frequency; $T_e(x)$, $a_e(x) = \sqrt{\gamma \mathcal{R} T_e}$ and $\omega_e(x) = \sqrt{Ba_e^2/LV}$ denote, respectively, local values of the temperature of air in equilibrium, the linear sound speed and the natural angular frequency of the resonator, $\mathcal{R}$ being the gas constant, and B and L throat's cross-sectional area of the resonator and throat length, respectively; the suffix 0 specifies respective values at $X = 0$, e.g., $T_0 = T_e(0)$, but note that p_0 is constant everywhere.

When T_e is specified, a_e is determined and the relation between x and X is established. In the following, we assume a typical case in which the temperature increases or decreases linearly along the tube: $T_e/T_0 = 1 \pm x/l$; l is a typical axial length of temperature variation. Then

$$X = \varepsilon\omega\int_0^x \frac{\mathrm{d}x}{a_e} = \pm\frac{2\varepsilon\omega l}{a_0}\left[\left(1 \pm \frac{x}{l}\right)^{1/2} - 1\right], \qquad (21.4)$$

since $a_e/a_0 = (1 \pm x/l)^{1/2}$. Hence H_e is expressed in terms of X as

$$H_e = \left(1 \pm \tfrac{1}{2}\chi X/\varepsilon\right)^2, \qquad (21.5)$$

where $\chi = a_0/\omega l$ represents the ratio of a typical wavelength a_0/ω to l and it is assumed to be much smaller than unity.

For localized waves in θ, the following two conservation equations are derived from (21.3):

$$\frac{\mathrm{d}}{\mathrm{d}X}\int_{-\infty}^{\infty} H_e^{1/4}f\,\mathrm{d}\theta = 0 \quad \text{and} \quad \frac{\mathrm{d}}{\mathrm{d}X}\int_{-\infty}^{\infty} H_e^{1/2}f^2\,\mathrm{d}\theta = 0. \qquad (21.6)$$

Although the conserved quantity in $(21.6)_1$ does not correspond to any physical quantities, $(21.6)_2$ states the conservation of the total energy flux $p'u$ ($\propto H_e^{1/2} f^2$) passing at a location X. The latter may be anticipated from the approximation of geometrical acoustics.

Here we digress to discuss the case where the array of resonators is absent ($K = 0$). Then (21.3) decouple and f is found easily by the method of characteristics, $f = H_e^{-1/4} F(\theta + fX)$, where $F(\theta)$ denotes an initial value of f at $X = 0$. The power law of $-1/4$ is simply the acoustic version of Green's law for shoaling process in shallow-water waves [6], if f and H_e are regarded as surface elevation and depth, respectively. As the temperature increases, it is found that p' decreases in proportion to $H_e^{-1/4}$ but u increases in proportion to $H_e^{1/4}$ because a_e is proportional to $T_e^{1/2}$. Conversely as the temperature decreases, p' increases while u decreases. In any case, the energy flux $p'u$ is conserved. This is always valid in the linear case where f is so small that f in the argument of F may be ignored. However, in the nonlinear case, there usually emerges a shock in the solution and Green's law will then becomes invalid.

4.   SOLITARY-WAVE SOLUTIONS

When the temperature gradient is absent in the tube with the array, i.e., $dH_e/dX = 0$ and $\Omega_e = \Omega_0$, (21.3) allow the solitary-wave solution. Since the solitary wave is determined uniquely by κ, ω_0 and s, ε and ω used in derivation of the equations may be regarded as formal parameters and chosen to be $\kappa/2$ and ω_0, respectively. Then K and Ω_0 can be set equal to unity. This is equivalent to the replacements

$$f \to Kf, \quad g \to Kg, \quad X \to X/K\sqrt{\Omega_0} \quad \text{and} \quad \theta \to \theta/\sqrt{\Omega_0}. \quad (21.7)$$

Putting $\zeta = \theta - sX = \omega_0(t - x/v)$ with $v = a_0/(1 + \kappa s/2)$ ($0 < s < 1$), the solitary-wave solution is given by $f(\zeta)$ expressed inversely as

$$4 \tan^{-1} \sqrt{\frac{f_+ - f}{f - f_-}} - \frac{2s}{\sqrt{-f_+ f_-}} \log \left| \frac{\Lambda^2}{(f_+ - f_-)f} \right| = |\zeta|, \quad (21.8)$$

with $\Lambda = \sqrt{-f_-(f_+ - f)} - \sqrt{f_+(f - f_-)}$, $f_\pm = -2\left(s - \frac{2}{3}\right) \pm \sqrt{-\frac{4}{3}s + \frac{16}{9}}$ and $g(\zeta)$ expressed in terms of f as

$$g = \tfrac{1}{2}f^2 + sf. \quad (21.9)$$

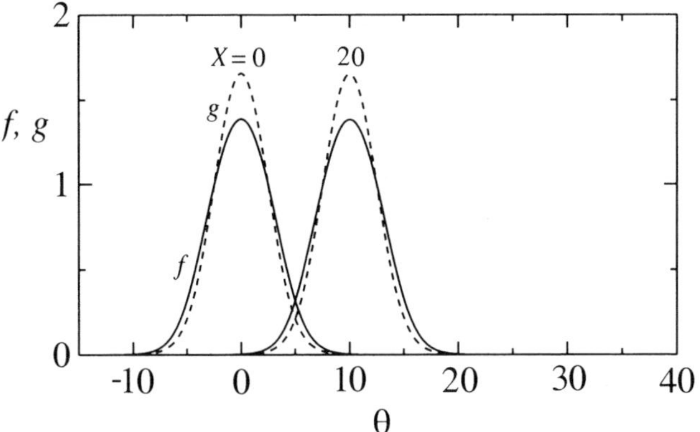

Figure 1   Profiles of the acoustic solitary wave for $s = 0.5$. The solid and broken lines represent f and g, respectively, and the profiles at $X = 20$ are obtained by solving (21.3) without the temperature gradient for the profiles at $X = 0$ as initial conditions.

Fig. 1 shows explicit profiles of f and g for $s = 0.5$ at $X = 0$ in the solid and broken lines, respectively. When the limits $s \to 0$ and $s \to 1$ are taken in (21.8), f and therefore p' is reduced to (21.1) and (21.2), respectively, by taking account of the scale factors between f and p'.

5.   TEMPERATURE GRADIENT EFFECTS

We now examine effects of temperature gradient. Suppose that the solitary wave be propagated steadily in a region without temperature gradient ($X \leq 0$) to enter a region with temperature gradient ($X > 0$). Then the initial conditions are taken as $f(\theta, X = 0) = F(\theta)$ and $g(\theta, X = 0) = G(\theta)$, where F and G represent the solutions (21.8) and (21.9) at $X = 0$.

In this case as well, we introduce (21.7) with $\Omega_0 = \Omega_e(0)$. Then K and Ω_e are set equal to unity and $H_e(X)$, respectively because $\Omega_e/\Omega_0 = T_e/T_0 = H_e(X)$. By this replacement, θ and X are redefined as

$$\theta = \omega_0 \left(t - \int_0^x \frac{\mathrm{d}x}{a_e} \right) \quad \text{and} \quad X = \frac{\kappa \omega_0}{2} \int_0^x \frac{\mathrm{d}x}{a_e}. \tag{21.10}$$

Effecting this replacement in (21.5), the term of temperature gradient in (21.3)$_1$ takes the form

$$\frac{1}{4H_e} \frac{\mathrm{d}H_e}{\mathrm{d}X} \to \pm \frac{\sigma/4K\sqrt{\Omega_0}}{1 \pm \sigma X/2K\sqrt{\Omega_0}}, \tag{21.11}$$

with the sign $\pm$ ordered vertically and $\sigma = \chi/\varepsilon$. In the following, we take $\sigma/2K\sqrt{\Omega_0} = 1/10$, i.e., $H_e = (1 \pm X/10)^2$.

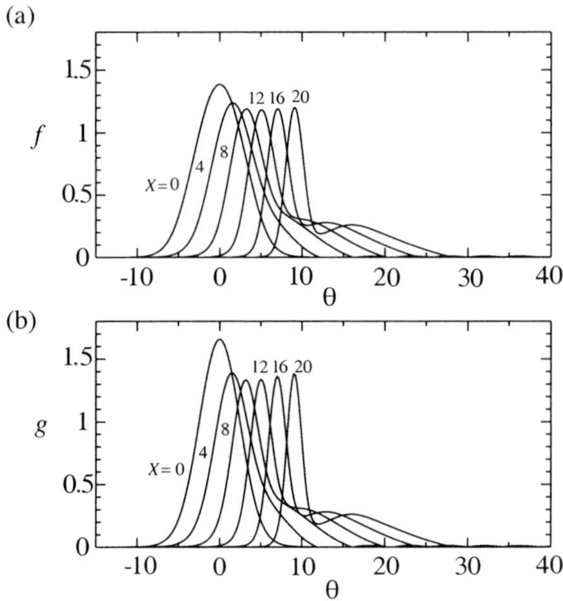

Figure 2   Evolution of the solitary wave for $s = 0.5$ at $X = 0$ to $X = 20$ by step 4 in the case of the temperature distribution $H_e = (1 + X/10)^2$: (a) and (b) represent the profiles of f and g in θ, respectively.

For the positive sign, the temperature increases from $15°C$ at $X = 0$ to $879°C$ at $X = 10$ while it decreases from $15°C$ to $-273°C$ at $X = 10$ for the negative sign. Using $\sigma/2K\sqrt{\Omega_0} = a_0/\kappa\omega_0 l$, l is given by $10a_0/\kappa\omega_0$ for this choice. For $\kappa = 0.2$ and $\omega_0/2\pi = 238$ Hz used in the experiment [3], l corresponds to 11.4 m and $x = l$ corresponds to $X = 4.14$ in the case of positive gradient.

Fig. 1 shows the steady propagation of the solitary wave when the temperature gradient is absent. Discarding $\mathrm{d}H_e/\mathrm{d}X$, (21.3) with $K = 1$ and $\Omega_e = 1$ are solved numerically by finite differences. In view of the definition of θ in (21.10), as ω_0 becomes large, i.e., at high temperature, the temporal half-value width of the solitary wave is found to be narrow.

Fig. 2 shows the evolution of the solitary wave with $s = 0.5$ in the case of the positive temperature gradient where (a) and (b) represent f and g, respectively. It is seen that the initial pulse decays for a moment ($X \leq 4$) but begins to grow very slowly and becomes narrow while producing a small pulse behind; f and g behave similarly. Even when the initial pulse has spread into two pulses, they consist of compression phase only. Because the linear temperature distribution is assumed, we cannot expect a steady state in which a new solitary wave will emerge

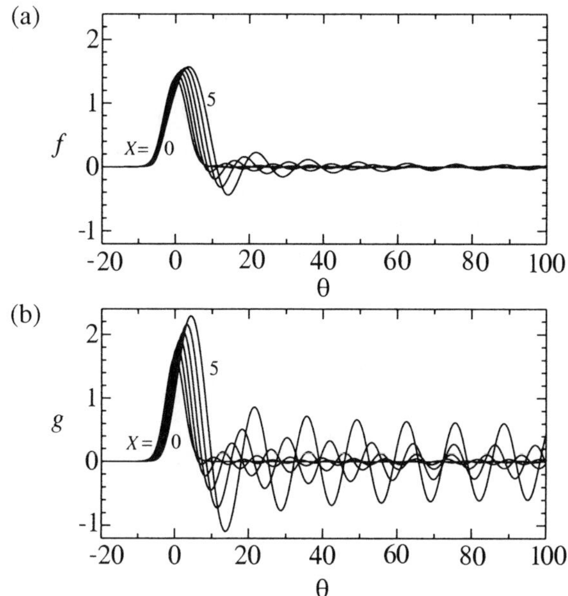

Figure 3   Evolution of the solitary wave for $s = 0.5$ at $X = 0$ to $X = 5$ by step 1 in the case of the temperature distribution $H_e = (1 - X/10)^2$: (a) and (b) represent the profiles of f and g in θ, respectively.

eventually. But the narrow leading pulse at $X = 20$ is consistent with the remark that the solitary wave becomes narrow as the equilibrium temperature increases. The total energy flux, i.e., the integral of $H_e^{1/2} f^2$, is always checked. The relative error remains within the order of 10^{-5}.

When the temperature decreases for the same initial condition, on the contrary, it is seen in Fig. 3 that the initial pulse grows but no fission occurs. Instead there tends to emerge an oscillatory tail consisting of compression and expansion phase. The tail tends to spread far down-stream and then the strong oscillation occurs in the cavity than in the tube. Because of this, a wide calculation domain should be prepared far downstream and the right end point is set at $\theta = 150$. But the accuracy in the check of the energy flux begins to worsen beyond 10^{-5} at $X = 2$ and the error becomes as large as 10^{-2} at $X = 5$. Radiation of the tail is a characteristic of a negative temperature gradient.

6.   DISCUSSIONS

We look at the numerical results from a viewpoint of evolution of a KdV soliton. Although the initial solitary wave differs from a KdV soliton (21.2), it might be instructive in understanding the results qualitatively. The acoustic soliton is the solitary wave in the limit of small height and wide width. Note, in passing, that the width of the acoustic solitary wave for a moderate value of s does not differ appreciably from that of the limiting one (21.1) as $s \to 0$ [2]. The wide width of the soliton suggests that its typical angular frequency ω defined by an inverse of a temporal half-value width, is much smaller than ω_e. If the limit $\Omega_e \to \infty$ is taken in $(21.3)_2$, then g may be approximated as

$$g = f - \frac{1}{\Omega_e}\frac{\partial^2 g}{\partial\theta^2} = f - \frac{1}{\Omega_e}\frac{\partial^2 f}{\partial\theta^2} + O(\Omega_e^{-2}), \qquad (21.12)$$

and g approaches f in the limit. This also holds generally in the case that the temperature increases, i.e., $\omega_e/\omega \to \infty$ (Fig. 2). Substituting this into $(21.3)_1$, we derive the KdV equation with the term of temperature gradient:

$$\frac{\partial f}{\partial X} + K\frac{\partial f}{\partial\theta} - f\frac{\partial f}{\partial\theta} + \frac{1}{4H_e}\frac{\mathrm{d}H_e}{\mathrm{d}X}f = \frac{K}{\Omega_e}\frac{\partial^3 f}{\partial\theta^3}. \qquad (21.13)$$

For the sake of comparison, we show the dimensionless KdV equation for the shallow-water waves over a varying depth $h(x)$:

$$\frac{\partial f}{\partial X} - \frac{3}{2h}f\frac{\partial f}{\partial\theta} + \frac{1}{4H}\frac{\mathrm{d}H}{\mathrm{d}X}f = \frac{\mu^2}{6\varepsilon}\frac{\partial^3 f}{\partial\theta^3}, \qquad (21.14)$$

with $H(X) = h(x)/h_0$. Here εf corresponds to a surface elevation normalized by a typical depth h_0, ε ($\ll 1$) and ω being a small parameter of weak nonlinearity, and a typical angular frequency; $\mu^2 = h\omega^2/g = (\omega/\omega_h)^2$ and $\omega_h = \sqrt{g/h(x)}$, g being acceleration due to gravity. Also X and θ are defined in a similar fashion as follows:

$$\theta = \omega\left(t - \int^x [gh(x)]^{-1/2}\,\mathrm{d}x\right) \quad \text{and} \quad X = \varepsilon\omega\int^x [gh(x)]^{-1/2}\,\mathrm{d}x.$$

Evolution of the initial acoustic soliton based on (21.13) may be solved by following the same procedure used for (21.14) [6, pp. 560–564]. In fact, (21.13) may be recast in the standard form:

$$\frac{\partial\psi}{\partial Z} - 6\psi\frac{\partial\psi}{\partial\tau} + \frac{\partial^3\psi}{\partial\tau^3} + \nu\psi = 0, \qquad (21.15)$$

with $\nu(Z) = -(3/4H_e)\mathrm{d}H_e(X(Z))/\mathrm{d}Z$, where the following replacement has been made:

$$f = -\frac{6K}{\Omega_e}\psi, \; Z = \int_0^X \frac{K}{\Omega_e(X)}\mathrm{d}X, \; \tau = KX - \theta. \qquad (21.16)$$

Table 1   Comparison between the shallow-water and acoustic solitary waves

	shallow-water	acoustic wave
dispersion relation	$\omega = \sqrt{gk\tanh kh}$	$k = \frac{\omega}{a_0}\sqrt{1 + \kappa/(1 - \omega^2/\omega_0^2)}$
reference speed	$c_0 = \sqrt{gh}\ (kh \to 0)$	$a_0 = \sqrt{\gamma \mathcal{R} T_0}\ (\kappa \to 0)$
propagation speed v	supersonic $(c_0 < v)$	subsonic $(a_0/(1 + \kappa/2) < v < a_0)$
nonuniformity	depth h	temperature T_0
dispersion	weaker as $h \to 0$	stronger as $T_0 \to 0$

But we do not go into details of solving (21.15) but compare (21.13) with (21.14) to find mutual correspondence.

The pressure corresponds to the elevation while the temperature corresponds to the depth. The second term on the left-hand side of (21.13) is immaterial because it can be removed by redefining θ as $\theta - KX$. But we notice a difference in the dispersion term (the coefficient of the third-order derivative). Since $K/\Omega_e = K/\Omega_0 H_e$, the dispersion becomes stronger as the temperature decreases, whereas as the depth decreases, the dispersion vanishes and the nonlinear term becomes pronounced.

The above observation may help to understand the numerical results. In the case of the positive gradient, f decreases initially so that the nonlinearity becomes weak, while the dispersion also becomes weak, contrary to the case of shallow-water waves. The evolution may substantially be described by the KdV equation (21.13) and the solitary wave gives rise to fission. On the other hand, in the case of the negative gradient, f increases so that the nonlinearity becomes strong. At the same time, the dispersion also becomes strong, which gives birth of a strong oscillatory tail and does not give rise to fission. The role of nonlinearity and dispersion should be contrasted with that in the shallow-water solitary waves. Table 1 summarizes these comparisons; the dispersion relation of the acoustic wave is derived from [7, Eq. (61)].

7.   CONCLUSION

The effects of the temperature gradient on the lossless propagation of the acoustic solitary wave have been examined for the two cases where the temperature increases or decreases linearly along the tube. It is appropriate to separate the evolutions into two stages, i.e., a short term just after the initial state and a long term later.

In the short-term evolution, the initial pulse decays or grows uniformly in accordance with Green's law and the dispersion does not play a primary role. Note that the pressure and the particle velocity behave reversely because the pressure is proportional to the local Mach number u/a_e, and also because the total energy flux must be conserved. For the positive gradient, hence, the pressure decreases but the velocity increases, while the reverse holds for the negative gradient.

In the long-term evolution, the dispersion comes into play and Green's law does not hold. For the positive gradient, the fission of the initial pulse begins. But the peak pressure tends to increase slightly in spite of increase of the temperature. This results in considerable increase of the peak velocity. As the temperature rises, the dispersion becomes small so that the further evolution may well be described by the KdV equation. For the negative gradient, the long-term evolution cannot be pursued because the temperature reaches the absolute zero. But as the temperature lowers, the dispersion becomes strong so that the oscillatory tail emerges while neither fission nor shock occurs.

The acoustic solitary wave has some similarity with the shallow-water solitary wave. But the remarkable dissimilarity lies in the action of dispersion. This characterizes evolution different from that of the shallow-water solitary wave.

References

[1] Sugimoto, N (1992) *Propagation of nonlinear acoustic waves in a tunnel with an array of Helmholtz resonators*, J. Fluid Mech. **244**, 55–78.

[2] Sugimoto, N (1996) *Acoustic solitary waves in a tunnel with an array of Helmholtz resonators*, J. Acoust. Soc. Am. **99**, 1971–1976.

[3] Sugimoto, N, Masuda, M, Ohno, J and Motoi, D (1999) *Experimental demonstration of generation and propagation of acoustic solitary waves in an air-filled tube*, Phys. Rev. Lett. **83**, 4053–4056.

[4] Pierce, A D (1991) *Acoustics*, Acoustical Society of America.

[5] Sugimoto, N and Tsujimoto, K, to be submitted.

[6] Mei, C C (1989) *The Applied Dynamics of Ocean Surface Waves*, Singapore: World Scientific.

[7] Sugimoto, N and Horioka, T (1995) *Dispersion characteristics of sound waves in a tunnel with an array of Helmholtz resonators*, J. Acoust. Soc. Am. **97**, 1446–1459.

SCATTERING BY BLUNT AND SHARP CONVEX OBSTACLES IN TWO DIMENSIONS

R. H. Tew

School of Mathematical Sciences, University of Nottingham

University Park, Nottingham NG7 2RD, UK

richard.tew@nottingham.ac.uk

Abstract   Following a brief description of the phenomenological differences between scattering by 'blunt' and 'sharp' bodies the bulk of this paper concerns the latter situation. In particular, we aim to show how modern asymptotic techniques can say more about the modal structure of the creeping field and also to provide a full description of the scattered field engendered by plane wave incidence upon a finite obstacle. This includes expressions for the far-field directivity associated with the scattered field at distances sufficiently large for the obstacle to appear point-like.

1.   INTRODUCTION

Amongst the diffraction problems involving high-frequency, time harmonic plane-wave incidence upon an obstructing boundary, two of the most important are the 'Sommerfeld' problem (scattering by the tip of a half-line) - PI - and the case of tangential ray incidence upon a bluntly curved, convex boundary - PII.

PI engenders an outgoing expansion fan of rays centred on the tip whilst PII gives rise to a forward-propagating creeping field that 'hugs' the boundary in the geometrical shadow region. This then continuously sheds further rays, of exponentially small amplitude, in the direction of the local tangent to the boundary as it propagates. Accounts of these much-studied problems can be found in [1] and [2], respectively.

However, what is less well known is that we cannot examine the solution of the creeping field problem in the limit as the boundary becomes less and less blunt, and eventually sharp, to obtain the Sommerfeld solution from it. For one thing, the creeping field depends sensitively upon the zeros of the Airy (or differentiated Airy) function whereas these do not feature at all in the Sommerfeld solution.

I.D. Abrahams et al. (eds.),
IUTAM Symposium on Diffraction and Scattering in Fluid Mechanics and Elasticity, 199–206.
© 2002 *Kluwer Academic Publishers. Printed in the Netherlands.*

A recent study [3] has identified an intermediate 'canonical' scattering geometry, the solution to which does indeed reproduce these two classic solutions in opposite limits. Referring to this work for the details, an interesting prediction from it is the existence of a new backward-propagating creeping field, exponentially subdominant to the forward-travelling field, which cannot be predicted by traditional theories [2]. Furthermore, the forward-propagating field is shown to be likened to propagation through a medium with a variable refractive index, the variation being supplied in this case by the boundary curvature. Thus this forward-travelling wave will undergo continuous exponentially small back-reflection and this partially explains the new backward-propagating field.

This does not occur if the curvature is constant, implying that diffraction by a circle is not generic in this class of scattering problems. Hence, descriptions of creeping field excitation and propagation on general bodies which depend in any way upon the solution to the circular case to extract diffraction coefficients or to generalise decay exponents, for example, do not necessarily give the full picture.

The aim of this paper – which is a shortened version of [4] – is to propose a general theory for scattering by a finite, convex scatterer. This will include descriptions of the creeping fields, the shadow zone structure and also the directivity pattern associated with the scattered field.

2.   THE SCATTERING PROBLEM

If an incident field $e^{ikx-i\omega t}$ propagates in the region D exterior to a vacuum bounded by the closed, convex curve $\partial D : (x, y) = (x_0(s), y_0(s))$, where s is arclength along ∂D, then the total field $\phi(x, y)e^{-i\omega t}$ satisfies the Helmholtz equation

$$\left(\frac{1}{1+n\kappa}\right)\frac{\partial}{\partial n}\left((1+n\kappa)\frac{\partial\phi}{\partial n}\right)+\left(\frac{1}{1+n\kappa}\right)\frac{\partial}{\partial s}\left(\frac{1}{(1+n\kappa)}\frac{\partial\phi}{\partial s}\right)+k^2\phi = 0,$$

$$(22.1)$$

in $n > 0$, where $k = \omega/c \gg 1$, n is normal distance from ∂D into D and $\kappa(s)$ is the boundary curvature. We impose a Dirichlet condition

$$\phi = 0, \quad n = 0 \qquad (22.2)$$

on ∂D and insist that the scattered field $\phi - e^{ikx-i\omega t}$ is outgoing at infinity.

Decomposing the incident field into a family of parallel rays there will be two points of ray tangency with ∂D (at the 'top' and 'bottom' of the scatterer); the continuation of these rays past the scatterer are then the upper and lower geometrical shadow boundaries, respectively. The diffraction regions to be studied associated with the former are I-V in Fig. 1; those corresponding to the lower shadow boundary can then be read off from these by a trivial re-definition of co-ordinates and local boundary properties.

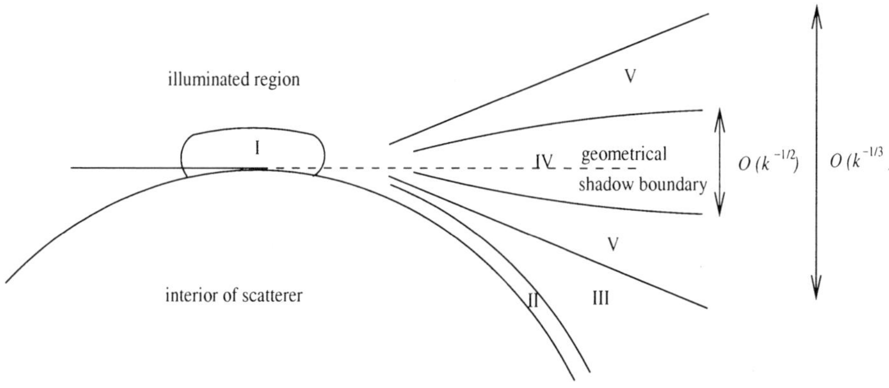

Figure 1   Diffraction zones and scalings that are analysed in the body of the paper.

THE GEOMETRICALLY REFLECTED FIELD AND RAY TANGENCY

For a typical non-tangent incident ray, the 'geometrical optics' reflected field can easily be determined in the form

$$\phi_r \sim -\left[\frac{\sin\psi}{2\kappa\tau + \sin\psi}\right]^{\frac{1}{2}} e^{ik(x_0(s)+\tau)} \tag{22.3}$$

where $\psi(s)$ is the angle of curvature on ∂D and τ is the distance along the reflected ray, whose direction is specified by Snell's Law.

This all breaks down near the point of ray tangency (region I in Fig. 1) and if we measure s in a clockwise sense from this point, the scalings are

$$s = k^{-\frac{1}{3}}\kappa_t^{-\frac{2}{3}}\hat{s}, \quad n = k^{-\frac{2}{3}}\kappa_t^{-\frac{1}{3}}\hat{n} \tag{22.4}$$

where κ_t is the curvature at the 'top' tangency point.

Writing

$$\phi = e^{ikx} + e^{iks}\hat{A}(\hat{s}, \hat{n}) \tag{22.5}$$

it follows that $\hat{A}$ satisfies the Fock–Leontovich equation

$$\frac{\partial^2 \hat{A}}{\partial \hat{n}^2} + 2i\frac{\partial \hat{A}}{\partial \hat{s}} + 2\hat{n}\hat{A} = 0, \quad \hat{n} > 0, \tag{22.6}$$

$$\hat{A} = -e^{-is^3/6}, \quad \hat{n} = 0, \tag{22.7}$$

with solution

$$\hat{A}(\hat{s},\hat{n}) = -2^{\frac{1}{3}} \int_{-\infty}^{\infty} \frac{\mathrm{Ai}\left(-2^{\frac{1}{3}}p\right) \mathrm{Ai}\left(-2^{\frac{1}{3}}(p+\hat{n})e^{2\pi i/3}\right)}{\mathrm{Ai}\left(-2^{\frac{1}{3}}e^{2\pi i/3}p\right)} e^{-isp} dp. \tag{22.8}$$

This fixes the solution in region I and it can be shown that this matches into the local form of ϕ_r [4].

NEAR-SURFACE SHADOW STRUCTURE

In region II it is appropriate to write

$$s = \mathrm{O}(1), \quad n = k^{-\frac{2}{3}}\overline{n} \tag{22.9}$$

and seek an ansatz for ϕ as

$$\phi \sim e^{iks + ik^{\frac{1}{3}}f(s,\overline{n})} \sum_{M=0}^{\infty} \frac{C_M(s,\overline{n})}{k^{M/3}}. \tag{22.10}$$

Substituting (22.10) into (22.1) and considering the various powers of k that arise gives, after some calculation, $f = f(s)$ and

$$C_0(s,\overline{n}) = \alpha_0 \kappa^{\frac{1}{6}} \mathrm{Ai}\left[-2^{\frac{1}{3}}\kappa^{\frac{1}{3}}\left(\overline{n} - \frac{f'}{\kappa}\right)e^{2\pi i/3}\right]. \tag{22.11}$$

The boundary condition now gives that f satisfies

$$f(s) = f_p(s) = -\lambda_p^2 \kappa^{\frac{2}{3}}(s)e^{-2\pi i/3}2^{-\frac{1}{3}}; \quad \mathrm{Ai}(-\lambda_p^2) = 0. \tag{22.12}$$

The constant α_0 is found by matching into (22.8) and is omitted here.

Thus each $C_M(s,\overline{n})$ is itself a sum of modes (with mode-number p), and

$$\phi \sim e^{iks} \sum_{p=0}^{\infty} e^{ik^{\frac{1}{3}}f_p(s)} \sum_{M=0}^{\infty} \frac{C_{Mp}(s,\overline{n})}{k^{M/3}}. \tag{22.13}$$

The $(p+1)$th mode is exponentially subdominant compared to the pth mode and so if we apply the ideas of optimal truncation of divergent asymptotic expansions [5] to (22.13) then we can see (after some detailed calculations, not presented here) that there will be 'mode conversion' between the higher-order modes. This is the backbone to the mathematical explanation for the backward-propagating creeping field referred to in the Introduction. We discuss this no further here.

THE SHED CREEPING FIELD

For region III we take $s, n = \mathrm{O}(1)$ and set

$$\phi \sim k^{\nu} D_0 e^{iku + ik^{\frac{1}{3}}v}, \tag{22.14}$$

from which it follows that

$$
\begin{aligned}
\nabla u \cdot \nabla u &= 1 \\
\nabla u \cdot \nabla v &= 0 \\
2\nabla D_0 \cdot \nabla u + D_0 \nabla^2 u &= 0,
\end{aligned}
$$

with the understanding that (22.14) should match into the leading order term from (22.13). This yields $\nu = -1/6$ and

$$u(\rho, \tau) = \rho + \tau \tag{22.15}$$

$$v(\rho) = -\lambda_0^2 e^{-2\pi i/3} 2^{-\frac{1}{3}} \int_0^{\rho} \kappa^{\frac{2}{3}}(t)\, dt \tag{22.16}$$

$$D_0(\rho, \tau) = -\tau^{-\frac{1}{2}} \left[\frac{\pi^{\frac{1}{2}} e^{-i\pi/12} 2^{\frac{1}{6}} \mathrm{Ai}\left(-\lambda_0^2 e^{-2\pi i/3}\right)}{\kappa_i^{\frac{1}{6}} \kappa^{\frac{1}{6}}(\rho) \mathrm{Ai}'\left(-\lambda_0^2\right)} \right]. \tag{22.17}$$

This is best interpreted as a ray solution which propagates an arclength ρ along the boundary from the point of ray tangency and is then shed along the boundary tangent, τ now being distance along this tangent.

THE SHADOW BOUNDARY

Along the geometrical shadow boundary, the traditional scalings are

$$x = \mathrm{O}(1), \quad y = k^{-\frac{1}{2}} \hat{y} \tag{22.18}$$

where (x, y) are local cartesian coordinates with origin at the point of ray tangency. Writing

$$\phi = e^{ikx} A_{SB}(x, \hat{y}) \tag{22.19}$$

we then find that

$$\frac{\partial^2 A_{SB}}{\partial \hat{y}^2} + 2i \frac{\partial A_{SB}}{\partial x} = 0, \quad x > 0, \tag{22.20}$$

$$A_{SB} = H(\hat{y}), \quad x = 0, \tag{22.21}$$

which has solution

$$A_{SB} = \frac{e^{-i\pi/4}}{\sqrt{\pi}} \int_{\eta}^{\infty} e^{it^2}\, dt; \quad \eta = \frac{\hat{y}}{2\sqrt{x}}. \tag{22.22}$$

However, this solution neither matches into the geometrical optics field in the 'lit' region nor into the shed field (22.14) in the shadow region. There must therefore be another layer adjacent to each of these parabolic shadow boundaries.

It turns out that the appropriate scalings are

$$x = O(1), \quad y = k^{-\frac{1}{3}} \tilde{y} \tag{22.23}$$

and once again we set

$$\phi \sim k^{\mu} e^{ikx + ik^{\frac{1}{3}} \tilde{u}(x,\tilde{y})} \tilde{A}_0(x, \tilde{y}), \tag{22.24}$$

which quickly yields

$$\left(\frac{\partial \tilde{u}}{\partial \tilde{y}}\right)^2 + 2\frac{\partial \tilde{u}}{\partial x} = 0 \tag{22.25}$$

$$2\frac{\partial \tilde{u}}{\partial \tilde{y}}\frac{\partial \tilde{A}_0}{\partial \tilde{y}} + 2\frac{\partial \tilde{A}_0}{\partial x} + \tilde{A}_0 \frac{\partial^2 \tilde{u}}{\partial \tilde{y}^2} = 0. \tag{22.26}$$

The appropriate solutions are found to be

$$\tilde{u}(x, \tilde{y}) = \frac{\tilde{y}^2}{2x}, \quad \tilde{A}_0(x, \tilde{y}) = x^{-\frac{1}{2}} g\left(\frac{\tilde{y}}{x}\right) \tag{22.27}$$

and matching into (22.8) finally yields that

$$g(\xi) = \frac{-1}{\sqrt{2\pi}} \kappa_t^{-\frac{1}{3}} e^{i\pi/12} e^{-i\xi^3/(6\kappa_t)} \int_{-\infty}^{\infty} \frac{\text{Ai}\left(-2^{\frac{1}{3}} p\right)}{\text{Ai}(-2^{\frac{1}{3}} e^{2\pi i/3} p)} e^{ip\xi/\kappa_t^{1/3}} \, dp. \tag{22.28}$$

The function $g(\xi)$ dictates the transitional solution at both the upper and lower branches of the parabolic 'Fresnel zone' given by (22.22) for the top shadow boundary; an exactly similar analysis (with κ_t replaced by κ_b, b for 'bottom') holding for the other shadow boundary. Thus, this transitional solution matches the reflected field as $\tilde{y}/x \to +\infty$ (though we must then remember to add in the incident field, corresponding to the exact solution $\mu = 0$, $\tilde{u}(x, \tilde{y}) \equiv 0$ and $\tilde{A}_0(x, \tilde{y}) \equiv 1$) and the shed creeping field in the opposite limit.

3.   FAR-FIELD STRUCTURE

The transitional and Fresnel zones associated with top and bottom ray tangencies diverge as we move away from the scatterer, and they must ultimately merge. Details of this interaction are given in [4] and are omitted here. Instead, we concentrate on the ultimate far-field where the incident field dominates everywhere and the form of the solution is then

$$\phi \sim e^{ikx} + e^{ikr} v(r, \theta; k). \tag{22.29}$$

Then to leading order in r, v satisfies

$$2\frac{\partial v}{\partial r} + \frac{v}{r} = 0, \tag{22.30}$$

from which we get

$$v(r, \theta; k) = \frac{G(\theta; k)}{\sqrt{r}}. \tag{22.31}$$

We therefore identify G as the diffraction coefficient for the far-field scattered radiation. Our plan is to find G in the limit $k \to \infty$ by matching (22.31) with the solutions found previously.

For $\theta = O(1)$, an examination of (22.3) as $r \to \infty$ yields that

$$G(\theta; k) \sim - \left[\frac{\sin \frac{1}{2}\theta}{2\kappa(s)}\right]^{\frac{1}{2}} e^{-ik(x_0(s)\cos\theta + y_0(s)\sin\theta)} \tag{22.32}$$

where $(x_0(s), y_0(s))$ is the point of emergence of the reflected ray, and $\psi \sim \frac{1}{2}\theta$.

For $\theta = k^{-\frac{1}{3}}\hat{\theta}$, we must take the limit of the superposed transition zones emerging from the top and bottom tangency points, (x_t, y_t) and (x_b, y_b), respectively. This gives

$$G(\theta; k) \sim k^{-\frac{1}{6}} \left[e^{-ik^{\frac{2}{3}}y_t\hat{\theta} + \frac{1}{2}ik^{\frac{1}{3}}x_t\hat{\theta}^2 + \frac{1}{6}iy_t\hat{\theta}^3} g(\hat{\theta}; \kappa_t) \right.$$
$$\left. + e^{-ik^{\frac{2}{3}}y_b\hat{\theta} + \frac{1}{2}ik^{\frac{1}{3}}x_b\hat{\theta}^2 + \frac{1}{6}iy_b\hat{\theta}^3} g(-\hat{\theta}; \kappa_b) \right],$$

where g is given by (22.28).

For smaller polar angles, the next feature is the merging Fresnel zones, the boundaries to which are $\theta = O\left((kr)^{-\frac{1}{2}}\right)$. These become vanishingly thin as $r \to \infty$ affecting the directivity for $\theta = 0$ only, when

$$\phi \sim e^{ikx} \left[1 + \frac{(y_t - y_b)e^{-i\pi/4}}{\sqrt{2\pi kx}} \right] \tag{22.33}$$

from which we deduce

$$G(0; k) \sim \frac{(y_t - y_b)e^{-i\pi/4}}{\sqrt{2\pi k}}. \tag{22.34}$$

4.   CONCLUSIONS

By constructing a full, leading-order asymptotic description of the diffracted field and shadow boundaries, we have obtained the high-frequency limit of the directivity of the scattered field when viewed at sufficiently large distances for the scatterer to appear point-like. Whilst explicit expressions are given in the body of this paper, it is worthwhile to note that as $k \to \infty$, $G = \mathrm{O}(1)$ for $\theta = \mathrm{O}(1)$, $G = \mathrm{O}(k^{-\frac{1}{6}})$ for $\theta = \mathrm{O}(k^{-\frac{1}{3}})$ and $G = \mathrm{O}(k^{-\frac{1}{2}})$ for $\theta = 0$. Thus, for $\theta \leq \mathrm{O}(k^{-\frac{1}{3}})$, the directivity is an order of magnitude lower than it is elsewhere and this feature persists for arbitrarily large distances.

Acknowledgments

The author expresses his gratitude to S. J. Chapman, J. R. King, J. R. Ockendon, B. J. Smith and I. Zafarullah for their significant contributions to this work.

References

[1] Jones, D S (1986) *Acoustic and Electromagnetic Waves*, Oxford: University Press.

[2] Keller, J B (1955) *Diffraction by a convex cylinder*, URSI Michigan Electromagnetic Wave Theory Symposium, 312–321.

[3] Engineer, J C, King, J R and Tew, R H (1998) *Diffraction by slender bodies*, Euro. J. Appl. Math. **9**, 129–158.

[4] Tew, R H, Chapman, S J, King, J R, Ockendon, J R, Smith, B J and Zafarullah, I (2000) *Scalar wave diffraction by slender bodies*, Wave Motion **32**, 363–380.

[5] Olde Daalhuis, A B, Chapman, S J, King, J R, Ockendon, J R and Tew, R H (1995) *Stokes phenomenon and matched asymptotic expansions*, SIAM J. Appl. Math **55**, 1469–1483.

V

WAVE SCATTERING AND PROPAGATION IN SOLID BODIES CONTAINING CRACKS

DIFFRACTION COEFFICIENTS FOR TILTED SURFACE-BREAKING CRACKS

V. M. Babich, V. A. Borovikov, L. Ju. Fradkin, D. Gridin, V. Kamotski
Centre for Waves and Fields, School of EEIE, South Bank University
103 Borough Road, London SE1 0AA, UK
fradkil@sbu.ac.uk

V. P. Smyshlyaev
School of Mathematical Sciences, University of Bath, Bath BA2 7AY, UK
V.P.Smyshlyaev@maths.bath.ac.uk

1.   INTRODUCTION

Many cracks of practical interest in ultrasonic non-destructive testing (NDT) are surface-breaking (Fig. 1). In order to model inspection of such cracks with the geometrical theory of diffraction (GTD), the diffraction coefficients for surface corners need to be calculated. The corresponding canonical problem is that of the diffraction of a plane wave by the vertex of an elastic wedge of less than 180°.

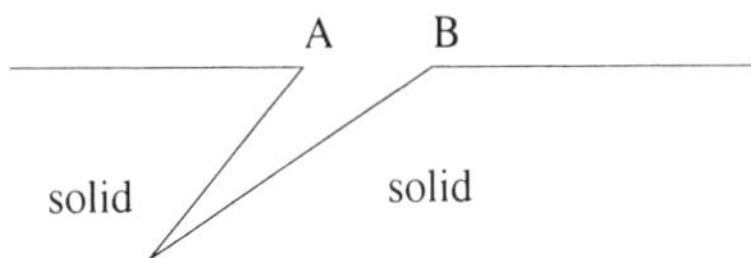

Figure 1   The tilted surface-breaking crack in a solid.

Study of diffraction by wedge-shaped regions has a long history. Here we follow the approach based on Sommerfeld transforms [1] and first applied in acoustics by Malyuzhinets [2, 3, 4] in his study of diffraction by a wedge with impedance boundary conditions. Malyuzhinets reformulated the conditions in the form of functional equations for Sommerfeld transforms, and thus reduced the problem to a Wiener–Hopf type equation: $F(\vartheta + \alpha) = K(\vartheta)F(\vartheta - \alpha)$. Budaev [5] and Budaev and Bogy [6, 7] have extended the approach to elastic wedges with zero-

I.D. Abrahams et al. (eds.),
IUTAM Symposium on Diffraction and Scattering in Fluid Mechanics and Elasticity, 209–216.
© 2002 *Kluwer Academic Publishers. Printed in the Netherlands.*

traction boundary conditions. They reduced the diffraction problem to two decoupled systems of two functional equations and then to two decoupled singular integral equations.

2.   PROBLEM STATEMENT

Let us consider the problem of diffraction of a time-harmonic plane compressional wave

$$\phi^{inc}(kr, \theta) = e^{i[\gamma kr \cos(\theta - \theta_P^{inc}) + \omega t]}, \qquad (23.1)$$

by an elastic wedge $\{r \geq 0, -\alpha \leq \theta \leq \alpha, -\infty < z < \infty\}, \alpha \leq \pi$ (Fig. 2). Incident shear waves may be treated in a similar manner. Here $k = \omega/c_S$, $\gamma = c_S/c_P$, c_P and c_S are the compressional and shear speeds respectively. The wave motion is two-dimensional and may be described

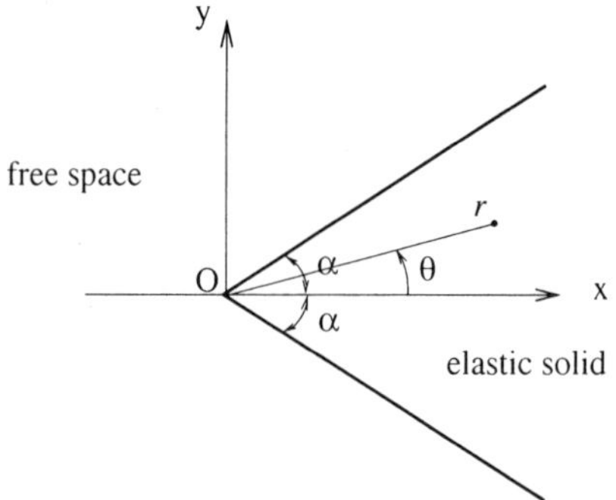

Figure 2   Geometry of the traction-free elastic wedge.

by elastic potentials which satisfy Helmholtz equations:

$$\nabla^2 \phi + \gamma^2 k^2 \phi = 0, \quad \nabla^2 \psi + k^2 \psi = 0. \qquad (23.2)$$

The wedge faces are assumed to be free of tractions:

$$\sigma_{r\theta}|_{\theta=\pm\alpha} = \sigma_{\theta\theta}|_{\theta=\pm\alpha} = 0. \qquad (23.3)$$

We impose the radiation condition in the form of the limiting absorption principle and the Meixner condition at the wedge vertex

$$\phi(kr, \theta) = O[(kr)^p], \quad \psi(kr, \theta) = O[(kr)^p], \quad \text{Re } p > -1, \quad kr \to 0. \qquad (23.4)$$

3.   MALYUZHINETS–BUDAEV APPROACH

Sommerfeld transforms Φ and Ψ.   Let us represent the elastic potentials as the Sommerfeld integrals:

$$\phi(kr, \theta) = \int_{C\cup\tilde{C}} \Phi(\vartheta)\, e^{i\gamma kr \cos(\vartheta-\theta)} d\vartheta = \int_{C\cup\tilde{C}} \Phi(\vartheta+\theta)\, e^{i\gamma kr \cos\vartheta} d\vartheta,$$

$$\psi(kr,\theta) \;=\; \int_{C\cup\tilde{C}} \Psi(\vartheta)\,e^{ikr\cos(\vartheta-\theta)}\,d\vartheta = \int_{C\cup\tilde{C}} \Psi(\vartheta+\theta)\,e^{ikr\cos\vartheta}\,d\vartheta,$$

with $\Phi(\vartheta)$ and $\Psi(\vartheta)$ are analytic for sufficiently large $|\mathrm{Im}\,\theta|$. The shaded areas in Fig. 3 indicate the regions where $\mathrm{Im}\{\cos\vartheta\} > 0$ and as $kr \to \infty$, the exponents $\to 0$.

If the above integrals converge they satisfy the Helmholtz equation, thus, we seek Φ, Ψ which in the shaded areas are $O(e^{c_1|\mathrm{Im}\,\vartheta|})$ as $|\mathrm{Im}\,\vartheta| \to \infty$. Further, the contours $C \cup \tilde{C}$ may be transformed into the steepest descent paths C_1 and C_2. The asymptotic approximation of the integrals over the paths produces the edge waves (and thus the diffraction coefficients), and the singularities crossed over during the path transformation give multiply reflected GE, Rayleigh and head waves. It follows that all the physically meaningful poles and branch points of $\Phi(\vartheta)$ and $\Psi(\vartheta)$ must be located between C_1 and C_2 shifted horizontally by $-\alpha$ and α respectively (*region I*). Furthermore, an asymptotic contribution of any singularity ϑ involves $\exp[ikr\cos(\vartheta - \theta)] = \exp[ikr\cos(\mathrm{Re}\,\vartheta - \theta)\cosh(\mathrm{Im}\,\vartheta)]\exp[kr\sin(\mathrm{Re}\,\vartheta - \theta)\sinh(\mathrm{Im}\,\vartheta)]$. Thus, singularities inside $|\mathrm{Re}\,\vartheta| \le \alpha$ (*strip II*) describe incoming waves, and the problem is reduced to seeking Φ and Ψ such that

$$\Phi(\vartheta) = -\frac{1}{4\alpha i}\frac{1}{\sin\frac{\pi}{2\alpha}(\vartheta - \theta_P^{inc})} + \bar{\Phi}(\vartheta), \quad \Psi(\vartheta) = \bar{\Psi}(\vartheta),$$

with $\bar{\Phi}(\vartheta)$ and $\bar{\Psi}(\vartheta)$ regular inside the strip II. Note that the contribution to the Sommerfeld integral of the pole $\vartheta = \theta - \theta_P^{inc}$ is the incident plane wave (the choice of the sine function happens to simplify calculations below).

Symmetry considerations. Consider $\phi_\pm^{inc}(kr,\theta) = \frac{1}{2}[\phi^{inc}(kr,\theta) \pm \phi^{inc}(kr,-\theta)]$. Then the original problem decouples into two, symmetric

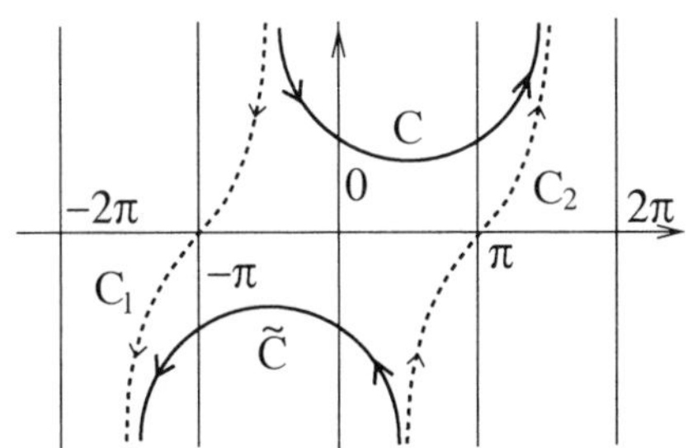

Figure 3   The contours in the complex ϑ-plane.

and antisymmetric, with the incident waves respectively ϕ_+^{inc} and ϕ_-^{inc}. The corresponding ϕ turns out to be even (odd) and ψ odd (even). We can now re-formulate the problem and seek

$$\Phi_\pm(\vartheta) = -\frac{1}{8\alpha i}\left[\frac{1}{\sin\frac{\pi}{2\alpha}(\vartheta - \theta_P^{inc})} \pm \frac{1}{\sin\frac{\pi}{2\alpha}(\vartheta + \theta_P^{inc})}\right] + \bar{\Phi}_\pm(\vartheta),$$

such that $\bar{\Phi}_\pm(\vartheta)$ are regular inside the strip II.

Functional equations for Φ and Ψ. Let us introduce a change of variable $g(\vartheta) = \cos^{-1}(\gamma^{-1}\cos\vartheta)$ in the integrals involving $\Phi_+(\vartheta)$. The deformed contour of integration may be transformed back to the original. Substituting the new representation into the boundary conditions and using the Nullification Theorem for the Sommerfeld integrals [6], we obtain the following functional equations

$$t_{11}\{\Phi_+[g(\vartheta) + \alpha] + \Phi_+[g(\vartheta) - \alpha]\} + t_{12}[\Psi_+(\vartheta + \alpha) + \Psi_+(\vartheta - \alpha)] = Q_1,$$
$$t_{21}\{\Phi_+[g(\vartheta) + \alpha] - \Phi_+[g(\vartheta) - \alpha]\} + t_{22}[\Psi_+(\vartheta + \alpha) - \Psi_+(\vartheta - \alpha)] = Q_2,$$

with $t_{11}(\vartheta) = \cos 2\vartheta \sin\vartheta\,(\gamma^2 - \cos^2\vartheta)^{-1/2}$, $t_{12}(\vartheta) = \sin 2\vartheta$, $t_{21}(\vartheta) = \sin 2\vartheta$, $t_{22}(\vartheta) = -\cos 2\vartheta$. It can be shown that $Q_1(\vartheta) = c_1\sin\vartheta$ and $Q_2(\vartheta) = -c_1\tan\alpha\sin\vartheta$, with c_1 an unknown constant (see e.g. [6]). The branch cuts for $g(\vartheta)$ are $[-\gamma_* + \pi n, \gamma_* + \pi n]$ where $\gamma_* = \cos^{-1}\gamma$, and the branch is chosen so that $g(\pi/2) = \pi/2$. Then $g(-\vartheta) = -g(\vartheta)$, $g(\vartheta + \pi n) = g(\vartheta) + \pi n$ and $g(\vartheta) \simeq \vartheta - i\log\gamma + O\left(\exp[-2|\mathrm{Im}\,\vartheta|]\right)$ as $\mathrm{Im}\,\vartheta \to \infty$. Below, we concentrate on the symmetric problem and omit the $+$-subscript.

Functional equations for the regular components of Φ and Ψ. Solving the above system with respect to $\Phi[g(\vartheta) + \alpha]$ and $\Psi(\vartheta + \alpha)$ gives

$$\begin{pmatrix} \Phi[g(\vartheta) + \alpha] \\ \Psi(\vartheta + \alpha) \end{pmatrix} = \begin{pmatrix} K_{11} & K_{12} \\ K_{21} & K_{22} \end{pmatrix}\begin{pmatrix} \Phi[g(\vartheta) - \alpha] \\ \Psi(\vartheta - \alpha) \end{pmatrix}$$
$$+ \begin{pmatrix} t_{11} & t_{12} \\ t_{21} & t_{22} \end{pmatrix}^{-1}\begin{pmatrix} Q_1 \\ Q_2 \end{pmatrix}, \qquad (23.5)$$

where K_{ij} are the reflection coefficients. If $\Phi(\vartheta)$ and $\Psi(\vartheta)$ are determined in strips of width 2α, (23.5) may be used to continue them into the right half-plane. A similar system may be obtained for analytical continuation into the left half-plane. The resulting systems may be used a finite number of times to determine all singularities of $\Phi(\vartheta)$ and $\Psi(\vartheta)$ located inside $\pi/2 - \alpha \le \mathrm{Re}\,\vartheta \le \pi/2 + \alpha$ (*strip III*) which correspond to (maybe, multiply) reflected P and S waves. The strip is of interest because it supports the machinery that is introduced below, in particular operators H and $\bar{H}$ as well as integrals (23.8) display the necessary properties. Φ and Ψ possess other singularities, poles and branch points giving rise to Rayleigh and head waves respectively – since so do coefficients K_{ij}. We can show that these singularities lie inside strip I but outside strip III. Thus, introducing the decomposition $\Phi(\vartheta) = \Phi_*(\vartheta) + \tilde{\Phi}(\vartheta)$, $\Psi(\vartheta) = \Psi_*(\vartheta) + \tilde{\Psi}(\vartheta)$, such that

$$\Phi_* = \frac{\pi}{2\alpha}\sum_k \mathrm{Res}\ \Phi(\vartheta_k^p)\frac{1}{\sin\frac{\pi}{2\alpha}(\vartheta - \vartheta_k^p)}, \quad \frac{\pi}{2} - \alpha \le \mathrm{Re}\,\vartheta_k^P \le \frac{\pi}{2} + \alpha,$$

$$\Psi_* = \frac{\pi}{2\alpha}\sum_k \mathrm{Res}\ \Psi(\vartheta_k^S)\frac{1}{\sin\frac{\pi}{2\alpha}(\vartheta - \vartheta_k^S)}, \quad \frac{\pi}{2} - \alpha \le \mathrm{Re}\,\vartheta_k^S \le \frac{\pi}{2} + \alpha,$$

and $\tilde{\Phi}(\vartheta)$, $\tilde{\Psi}(\vartheta)$ are regular in strip III, the functional equations to solve become

$$\tilde{\Phi}[g(\vartheta) + \alpha] + \tilde{\Phi}[g(\vartheta) - \alpha] + B(\vartheta)\Big\{\tilde{\Psi}(\vartheta + \alpha) + \tilde{\Psi}(\vartheta - \alpha)\Big\} = \tilde{Q}_1,$$

$$A(\vartheta)\Big\{\tilde{\Phi}[g(\vartheta) + \alpha] - \tilde{\Phi}[g(\vartheta) - \alpha]\Big\} + \tilde{\Psi}(\vartheta + \alpha) - \tilde{\Psi}(\vartheta - \alpha) = R_2 + \tilde{Q}_2,$$

where we use the following notations: $A(\vartheta) = -\tan 2\vartheta$,

$$B(\vartheta) = \frac{2\cos\vartheta\sqrt{\gamma^2 - \cos^2\vartheta}}{\cos 2\vartheta}, \quad \tilde{Q}_1(\vartheta) = Q_1(\vartheta)\frac{\sqrt{\gamma^2 - \cos^2\vartheta}}{\cos 2\vartheta \sin\vartheta},$$

$$\tilde{Q}_2(\vartheta) = Q_2(\vartheta)\sec 2\vartheta, \quad R_2 = -2\{A(\vartheta)\Phi_*[g(\vartheta) + \alpha] + \Psi_*(\vartheta + \alpha)\}.$$

Singular integral equations.   Changing unknowns to

$$\begin{aligned}
X(\vartheta) &= \tilde{\Phi}[g(\vartheta) + \alpha] + \tilde{\Phi}[g(\vartheta) - \alpha],\\
Y(\vartheta) &= \tilde{\Psi}(\vartheta + \alpha) + \tilde{\Psi}(\vartheta - \alpha),
\end{aligned}$$

the above system transforms into

$$\begin{aligned}
X(\vartheta) + B(\vartheta)Y(\vartheta) &= \tilde{Q}_1,\\
A(\vartheta)\bar{H}(\alpha)X(\vartheta) + H(\alpha)Y(\vartheta) &= R_2 + \tilde{Q}_2,
\end{aligned} \tag{23.6}$$

where $\bar{H} = G(\gamma)HG^{-1}(\gamma)$, $G(\gamma)F(\vartheta) = F[g(\vartheta)]$ and we use operator $H(\alpha)$, such that

$$H(\alpha)F(\vartheta) = \frac{1}{2\alpha\mathrm{i}}\mathrm{PV}\int_{\pi/2-\mathrm{i}\infty}^{\pi/2+\mathrm{i}\infty}\frac{F(\xi)\,d\xi}{\sin[(\pi/2\alpha)(\xi - \vartheta)]}, \quad \mathrm{Re}\,\vartheta = \frac{\pi}{2}. \tag{23.7}$$

For functions analytic in strip III, H and $\bar{H}$ have the following properties

$$H(\alpha)\Big[\tilde{\Psi}(\vartheta + \alpha) + \tilde{\Psi}(\vartheta - \alpha)\Big] = \tilde{\Psi}(\vartheta + \alpha) - \tilde{\Psi}(\vartheta - \alpha),$$

$$\bar{H}(\alpha)\left[\tilde{\Phi}(g(\vartheta)+\alpha)+\tilde{\Phi}(g(\vartheta)-\alpha)\right]=\tilde{\Phi}(g(\vartheta)+\alpha)-\tilde{\Phi}(g(\vartheta)-\alpha).$$

The singular integral equation in (23.6) is given on the line $\operatorname{Re}\vartheta=\pi/2$. $\tilde{\Phi}(\vartheta)$ and $\tilde{\Psi}(\vartheta)$ may be recovered inside strip III using the expressions

$$\tilde{\Phi}(\vartheta)\;=\;\frac{1}{4\alpha i}\int_{\pi/2-i\infty}^{\pi/2+i\infty}\frac{X(\xi)g'(\xi)\,d\xi}{\cos\{(\pi/2\alpha)[g(\xi)-\vartheta]\}},\qquad(23.8)$$

$$\tilde{\Psi}(\vartheta)\;=\;\frac{1}{4\alpha i}\int_{\pi/2-i\infty}^{\pi/2+i\infty}\frac{Y(\xi)\,d\xi}{\cos[(\pi/2\alpha)(\xi-\vartheta)]},\qquad|\operatorname{Re}\vartheta-\pi/2|\leq\alpha,$$

and outside strip III using an analytic continuation procedure like (23.5).

Introducing $\vartheta=\pi/2+i\eta$ and performing simple manipulations and trivial renaming of variables we finally arrive at the singular integral equation

$$H'y-a\bar{H}'by=r_2+c_1q,\qquad(23.9)$$

where the new unknown is $y(\eta)=Y(\pi/2+i\xi(\eta))$; $q=-q_3\tan\alpha-a\bar{H}'q_1$, with $q_1=-(\gamma^2+\sinh^2\eta)^{1/2}\operatorname{sech}2\eta$, $q_3=-\cosh\eta\operatorname{sech}2\eta$ and we use operators H' and $\bar{H}'$, such that

$$H'f(\eta)\;=\;\frac{1}{2\alpha i}\mathrm{PV}\int_{-\infty}^{\infty}\frac{f(t)\,dt}{\sinh\{(\pi/2\alpha)(t-\eta)\}},$$

$$\bar{H}'f(\eta)\;=\;\frac{1}{2\alpha i}\mathrm{PV}\int_{-\infty}^{\infty}\frac{f(t)\chi'(t)\,dt}{\sinh\{(\pi/2\alpha)[\chi(t)-\chi(\eta)]\}},\qquad(23.10)$$

with $g(\pi/2+i\eta)=\pi/2+i\chi(\eta)$. Note that $\int_{-\infty}^{\infty}H'f(\eta)\,d\eta=0$, and therefore, integrating (23.9) we get the equation

$$\int_{-\infty}^{\infty}(a\bar{H}'by+r_2+c_1q)\,d\eta=0.\qquad(23.11)$$

Solving equation (23.9).   First, we follow Budaev and Bogy [6] and re-write (23.9) as

$$H'(1-ab)y+Ky=r_2+c_1q,\qquad(23.12)$$

where the operator

$$Kf(\eta)=(H'a-a\bar{H}')bf(\eta)=\frac{1}{2\alpha i}\int_{-\infty}^{\infty}k(\eta,t)f(t)\,dt$$

has the regular kernel

$$k(\eta,t)=\frac{\tanh^2 2t}{\chi'(t)\sinh\frac{\pi}{2\alpha}(t-\eta)}-\frac{\tanh 2t\tanh 2\eta}{\sinh\frac{\pi}{2\alpha}[\chi(t)-\chi(\eta)]}.$$

Then we perform the following two steps:

Step 1:   Suppressing the kernel tail.

We multiply (23.11) by a function f_1, such that $\int_{-\infty}^{\infty} f_1(\eta)\, d\eta = 1$, and subtract the result from (23.12) to arrive at

$$[H'(1-ab)y](\eta) + \frac{1}{2\alpha\mathrm{i}} \int_{-\infty}^{\infty} \left[k(\eta,t) - f_1(\eta) \int_{-\infty}^{\infty} k(\eta',t)\, d\eta' \right] y(t)\, dt$$

$$= r_2(\eta) + c_1 q(\eta) - f_1(\eta) \int_{-\infty}^{\infty} [r_2(t) + c_1 q(t)]\, dt. \tag{23.13}$$

The kernel in the second term on the left-hand side is such that its integral over η is zero and this alleviates application of the second step.

Step 2:   Inversion.

Similarly to [6] we apply to (23.13) the operator $(H')^{-1}$, such that

$$(H')^{-1} f(\eta) = \frac{1}{2\alpha\mathrm{i}} \mathrm{PV} \int_{-\infty}^{\infty} \coth \frac{\pi}{2\alpha}(t - \eta) f(t)\, dt,$$

and obtain

$$(1-ab)y + Ly = (H')^{-1}\left[r_2 + c_1 q - f_1(\eta) \int_{-\infty}^{\infty} [r_2(t) + c_1 q(t)]\, dt \right],$$

where the operator L has a regular kernel $\ell(\eta,t)$. Finally, we introduce a new unknown function $\tilde{y} = \sqrt{1 - ab}\, y$ to obtain

$$\tilde{y}(\eta) + \int_{-\infty}^{\infty} \tilde{\ell}(\eta,t)\tilde{y}(t)\, dt = \tilde{r}(\eta) + c_1 \tilde{q}(\eta), \tag{23.14}$$

We obtain the system of two equations (23.11) and (23.14) with two unknowns c_1 and $\tilde{y}(\eta)$. For $\alpha \leq \pi/2$, it turns out to be amenable to a numerical solution.

Preliminary numerical results.   To test the code based on the above considerations we have chosen $\alpha = \pi/4$, $\theta_P^{inc} = 0$, and $f_1 = 2/(\pi \cosh 2\eta)$. We then computed the diffraction coefficient,

$$D_P(\theta) = \sqrt{2\pi} \mathrm{e}^{\mathrm{i}\pi/4} \left[\Phi(-\pi + \theta) - \Phi(\pi + \theta) \right]$$

(see Fig. 4). It has no pole and the symmetry of the problem implies that it should not. Thus our result is consistent with physical expectations. Further extensive testing is underway.

4.   CONCLUSIONS

We have produced a fuller theoretical justification of Budaev and Bogy's recipe for deriving singular integral equations, which describe

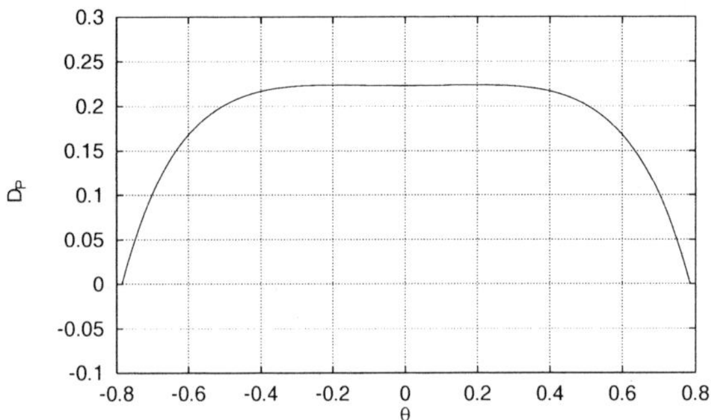

Figure 4   Modulus of diffraction coefficient $D_P(\theta)$.

the elastic wedge diffraction, and obtained a clearer understanding of the properties of the underlying operators. We have produced a novel scheme for solving the equations, which however does not work for $\alpha > \pi/2$.

Acknowledgments

The project was sponsored by the Industry Management Committee of the UK Nuclear Licensees under the IMC Contract PG/GNSR/5129. The first, second and fifth author are all affiliated with other Institutions but carried out the above work while visiting the Centre for Waves and Fields, SBU under conditions of the contract.

References

[1] Sommerfeld, A (1896) *Mathematische Theorie der Diffraction*, Math. Ann. **47**, 317–374.

[2] Malyuzhinetz, G D (1955) *Radiation of sound by oscillating faces of an arbitrary wedge*, Sov. Phys. Acoust. **1**, 152–174; 240–268.

[3] Malyuzhinets, G D (1957) *Inversion formula for the Sommerfeld integral*, Sov. Phys. Dokl. **3**, 52–56.

[4] Malyuzhinets, G D (1958) *Excitation, reflection and emission of surface waves from a wedge with given face impedances*, Sov. Phys. Dokl. **3**, 752–755.

[5] Budaev, B V (1995) *Diffraction by Wedges*, Pitman Research Notes in Mathematics series **322**, Harlow: Longman Scientific and Technical.

[6] Budaev, B V and Bogy, D B (1995) *Rayleigh wave scattering by a wedge*, Wave Motion **22**, 239–257; **24**, 307–314.

[7] Budaev, B V and Bogy, D B (1998) *Rayleigh wave scattering by two adhering elastic wedges*, Proc. Roy. Soc. Lond. A **454**, 2949–2996.

DOMAIN SENSITIVITY ANALYSIS OF THE ELASTIC FAR–FIELD PATTERNS

Scattering from non–smooth obstacles

M. Bochniak

Abteilung Mathematik I

Universität Ulm

89069 Ulm, Germany

bochniak@mathematik.uni-ulm.de

F. Cakoni

Department of Mathematical Sciences

University of Delaware

Newark, Delaware 19716-2553, USA

cakoni@math.udel.edu

Abstract   We consider the scattering of time harmonic elastic waves from a 2D bounded domain or from a crack taking into account the stress singularties of elastic waves near conical boundary points. The paper is concerned with the sensitivity of the far–field patterns with respect to small perturbations of the shape of the scatterer. Using the method of adjoint problems we obtain an integral representation for the corresponding shape derivative.

1.   FORMULATION OF THE PROBLEM

The mathematical modelling of the scattering of time–harmonic elastic waves from an obstacle Ω_i, surrounded by a homogeneous elastic medium with Lamé parameters μ and λ satisfying $\lambda + 2\mu > 0$ and $\mu > 0$, leads to an exterior boundary value problem for the Navier equations

$$\Delta^*\mathbf{u} + \omega^2\mathbf{u} := \mu\Delta\mathbf{u} + (\lambda+\mu)\nabla\nabla\cdot\mathbf{u} + \omega^2\mathbf{u} = 0 \quad \text{in} \quad \Omega := I\!\!R^2\backslash\Omega_i. \quad (24.1)$$

Here, the total elastic wave $\mathbf{u} = \mathbf{u}^i + \mathbf{u}^s$ is decomposed into a given time-harmonic incident wave $\mathbf{u}^i$ with the frequency ω and the unknown scattered wave $\mathbf{u}^s$.

The scatterer Ω_i can be one of the following two types:

217

I.D. Abrahams et al. (eds.),

IUTAM Symposium on Diffraction and Scattering in Fluid Mechanics and Elasticity, 217–225.

© 2002 *Kluwer Academic Publishers. Printed in the Netherlands.*

(B) Ω_i is a bounded domain with a piecewise smooth boundary Γ. Here $S = \{P_1, \ldots, P_Q\}$ is a finite set of boundary points, such that $\Gamma \setminus S$ is smooth. Furthermore, we assume that Ω is locally diffeomorphic in the neighbourhood of every corner point P_q to an infinite cone C_q with the opening angle $\varphi_q^0 \notin \{0, 2\pi\}$. The unit normal vector $\mathbf{n} = (n_1, n_2)$ on Γ is directed towards Ω_i.

(C) $\Omega_i = \Gamma$ is a crack, i.e. a piecewise smooth curve with a finite set $S = \{P_1, \ldots, P_Q\}$ consisting of both crack tips and of interior corner points which satisfy the angle condition $\omega_q^0 \neq \{0, 2\pi\}$. The direction of the unit normal vector $\mathbf{n}$ on Γ is chosen arbitrarily but is fixed along the crack.

The unknown scattered wave $\mathbf{u}^s$ is requested to satisfy the Dirichlet boundary conditions

$$\mathbf{u}^s = -\mathbf{u}^i \quad \text{on} \quad \Gamma. \tag{24.2}$$

For crack problems **(C)** the boundary conditions (24.2) have to be posed on both sides of the crack, i.e.

$$\mathbf{u}^s_\pm(\mathbf{x}) := \lim_{h \to 0+} \mathbf{u}^s(\mathbf{x} \pm h\mathbf{n}) = -\mathbf{u}^i(\mathbf{x}) \text{ for } \mathbf{x} \in \Gamma. \tag{24.3}$$

Moreover, the scattered field $\mathbf{u}^s = \mathbf{u}^s_p + \mathbf{u}^s_s$ is required to satisfy the Kupradze radiation condition for $r := |\mathbf{x}| \to \infty$ (see e.g. [1])

$$\lim_{r \to \infty} \sqrt{r}\left(\frac{\partial \mathbf{u}^s_p}{\partial r} - ik_p \mathbf{u}^s_p\right) = 0, \qquad \lim_{r \to \infty} \sqrt{r}\left(\frac{\partial \mathbf{u}^s_s}{\partial r} - ik_s \mathbf{u}^s_s\right) = 0 \tag{24.4}$$

uniformly in all directions $\hat{\mathbf{x}} = \mathbf{x}/|\mathbf{x}|$. Here, $\mathbf{u}^s_p$ and $\mathbf{u}^s_s$ are the longitudinal and the transversal part of the elastic scattered field, respectively, associated with the wave numbers $k_p, k_s > 0$ given by $k_p^2 = \omega^2/(\lambda + 2\mu)$, and $k_s^2 = \omega^2/\mu$. In the following, we will consider incident fields $\mathbf{u}^i := \mathbf{u}^i_p + \mathbf{u}^i_s$ in the form of a linear combination of longitudinal plane waves $\mathbf{u}^i_p(\mathbf{x}; \hat{\mathbf{d}}) = \hat{\mathbf{d}} e^{-ik_p \hat{\mathbf{x}} \cdot \hat{\mathbf{d}}}$ and transversal plane waves $\mathbf{u}^i_s(\mathbf{x}; \hat{\mathbf{d}}) = \hat{\mathbf{b}} e^{-ik_s \hat{\mathbf{x}} \cdot \hat{\mathbf{d}}}$, where the unit vector $\hat{\mathbf{d}} \in \mathbb{R}^2$ denotes the direction of propagation and the unit vector $\hat{\mathbf{b}} \in \mathbb{R}^2$ is a polarisation vector such that $\hat{\mathbf{b}} \perp \hat{\mathbf{d}}$.

The direct problem (24.1)–(24.4) has a unique solution in $(H^1_{\text{loc}}(\Omega))^3$. The scattered wave $\mathbf{u}^s$ behaves at infinity uniformly in $\hat{\mathbf{x}} = \mathbf{x}/|\mathbf{x}|$ as (see e.g. [1])

$$\mathbf{u}^s_p(\mathbf{x}) = \frac{1}{\lambda + 2\mu} \frac{e^{i\pi/4}}{\sqrt{8\pi k_p}} \frac{e^{ik_p r}}{\sqrt{r}} \mathcal{F}_p(\Gamma)(\hat{\mathbf{x}}) + o(\frac{1}{\sqrt{r}}), \tag{24.5}$$

$$\mathbf{u}^s_s(\mathbf{x}) = \frac{1}{\mu} \frac{e^{i\pi/4}}{\sqrt{8\pi k_s}} \frac{e^{ik_p r}}{\sqrt{r}} \mathcal{F}_s(\Gamma)(\hat{\mathbf{x}}) + o(\frac{1}{\sqrt{r}}). \tag{24.6}$$

The vector functions $\mathcal{F}_p(\Gamma)$, $\mathcal{F}_s(\Gamma)$, defined on the unit circle S_1, are known as the far-field patterns or the scattering amplitudes of the longitudinal part $\mathbf{u}_p^s$ and of the transversal part $\mathbf{u}_s^s$, respectively. They are uniquely determined by the shape of the boundary Γ. They can be expressed by the following integral formulae provided that R is big enough (see e.g. [1])

$$
\begin{aligned}
\mathcal{F}_p(\Gamma)(\hat{\mathbf{x}}) \;=\; & \int_{S_R} \left\{ \left[\mathbf{T}_{\mathbf{n(y)}} \hat{\mathbf{x}} \otimes \hat{\mathbf{x}} e^{-ik_p \hat{\mathbf{x}} \cdot \mathbf{y}} \right]^{\top} \mathbf{u}^s(\mathbf{y}) \right. \\
& \left. - \hat{\mathbf{x}} \otimes \hat{\mathbf{x}} e^{-ik_p \hat{\mathbf{x}} \cdot \mathbf{y}} \mathbf{T}_{\mathbf{n(y)}} \mathbf{u}^s(\mathbf{y}) \right\} ds_{\mathbf{y}},
\end{aligned}
\tag{24.7}
$$

$$
\begin{aligned}
\mathcal{F}_s(\Gamma)(\hat{\mathbf{x}}) \;=\; & \int_{S_R} \left\{ \left[\mathbf{T}_{\mathbf{n(y)}} \left[\mathbf{I} - \hat{\mathbf{x}} \otimes \hat{\mathbf{x}} \right] e^{-ik_s \hat{\mathbf{x}} \cdot \mathbf{y}} \right]^{\top} \mathbf{u}^s(\mathbf{y}) \right. \\
& \left. - \left[\mathbf{I} - \hat{\mathbf{x}} \otimes \hat{\mathbf{x}} \right] e^{-ik_s \hat{\mathbf{x}} \cdot \mathbf{y}} \mathbf{T}_{\mathbf{n(y)}} \mathbf{u}^s(\mathbf{y}) \right\} ds_{\mathbf{y}},
\end{aligned}
\tag{24.8}
$$

where $\mathbf{T_n}$ denotes the boundary traction operator

$$
(\mathbf{T_n})_{i,j} := \lambda \mathbf{n}_i(\mathbf{x}) \frac{\partial}{\partial x_j} + \mu \mathbf{n}_j(\mathbf{x}) \frac{\partial}{\partial x_i} + \mu \delta_{i,j} \nabla_x \cdot \mathbf{n}(\mathbf{x}) \qquad i,j = 1,2, \tag{24.9}
$$

and S_R is the boundary of the sphere B_R with radius R.

Our main concern in this work is to study how the perturbation of the domain influences the far-field operators $\Gamma \mapsto \mathcal{F}_p(\Gamma)$, $\Gamma \mapsto \mathcal{F}_s(\Gamma)$. This is performed using the material derivative approach (see e.g. [2]).

2.   BEHAVIOUR OF ELASTIC WAVES NEAR CORNER POINTS

Let $\mathbf{v}$ be an elastic wave which satisfies the homogeneous Navier equations (24.1) and homogeneous Dirichlet boundary conditions (24.2) or (24.3) in the vicinity of some conical boundary point $P_q \in S$. The behaviour of $\mathbf{v}$ near P_q can be characterised with the help of eigenvalues α_j^q and corresponding eigenfunctions of the operator pencil $\mathcal{A}(P_q)$ which is obtained by applying the Mellin transform to the principal part of the Navier operator in the infinite wedge C_q (see [3] for full details). Let $a_q := \min\{\Re \alpha\}$, where the minimum is taken over all eigenvalues α of $\mathcal{A}(P_q)$ with a positive real part. Note that $a_q > 1/2$ if $\varphi_q^0 \neq 2\pi$ and that $a_q = 1/2$ if $\varphi_q^0 = 2\pi$. Furthermore, let (r_q, φ_q) denote the local polar coordinates with origin in P_q. Then, according to [3] the elastic wave $\mathbf{v}$ behaves near P_q as

$$
\mathbf{v} = \sum_j K_q^j(\mathbf{v}) \mathbf{S}_j + O(r),
$$

where the so-called singular functions $\mathbf{S}_j$ have the form

$$\mathbf{S}_j = r_q^{\alpha_j^q} \Phi_j^q(\log r_q, \varphi_q)$$

with the smooth functions $\Phi_j^q(\log r_q, \varphi_q)$ depending polynomially on $\log r$, and $K_q^j(\cdot)$ are linear functionals.

3.   DOMAIN SENSITIVITY OF ELASTIC FIELDS

Description of the domain perturbation.   In order to describe the shape sensitivity of exterior boundary value problems, i.e. the influence of the shape of the domain on the solution, we introduce a family of perturbed domains Ω_ε, $\varepsilon \in [0, \varepsilon_0]$, as the image of a fixed domain Ω under a family of C^∞–diffeomorphisms

$$\Psi_\varepsilon = I + \varepsilon\Psi; \quad \varepsilon \in [0, \varepsilon_0], \tag{24.10}$$

i.e. $\Omega_\varepsilon := \Psi_\varepsilon(\Omega)$, $\Gamma_\varepsilon := \Psi_\varepsilon(\Gamma)$, $S_\varepsilon := \Psi_\varepsilon(S)$. Since we are interested in the perturbation of the boundary Γ we assume that $\exists \tilde{R} : \Psi_\varepsilon(\mathbf{x}) = \mathbf{x}$ for all $|\mathbf{x}| > \tilde{R}$.

The exterior Dirichlet boundary value problem for the scattered wave $\mathbf{u}_\varepsilon^s$ in Ω_ε reads as

$$\begin{aligned} \Delta^*\mathbf{u}_\varepsilon^s + \omega^2\mathbf{u}_\varepsilon^s &= 0 \quad \text{in} \quad \Omega_\varepsilon, \\ \mathbf{u}_\varepsilon^s &= -\mathbf{u}^i \quad \text{on} \quad \Gamma_\varepsilon, \end{aligned} \tag{24.11}$$

where $\mathbf{u}_\varepsilon^s$ satisfies the radiation condition (24.4).

Form sensitivity of the solution.   The form sensitivity of the perturbed scattered wave $\mathbf{u}_\varepsilon^s$ can be described with the help of the material derivative

$$\dot{\mathbf{u}}^s := \left.\frac{d(\mathbf{u}_\varepsilon^s \circ \Psi_\varepsilon)}{d\varepsilon}\right|_{\varepsilon=0} \tag{24.12}$$

and the shape derivative

$$\mathbf{u}^{s\prime} := \dot{\mathbf{u}}^s - \nabla\mathbf{u}_0^s \cdot \Psi. \tag{24.13}$$

The existence of the material derivative $\dot{\mathbf{u}}^s$ in appropriate weighted Sobolev spaces can be shown using the same method as in [4, 5] based on regular perturbation techniques for partial differential operators depending on a small parameter. The existence and the regularity of the shape derivative $\mathbf{u}^{s\prime}$ follows then directly from the definition (24.13) of $\mathbf{u}^{s\prime}$ and the existence of the material derivative $\dot{\mathbf{u}}^s$.

Let $\mathbf{U}_\varepsilon := \mathbf{u}^s_\varepsilon + \eta \mathbf{u}^i$, where η is a cut–off function with support in the vicinity of the boundary Γ. We note that $\mathbf{U}_\varepsilon$ satisfies the Navier equations with a right hand side having a compact support and vanishing in the neighbourhood of Γ_ε and satisfies homogeneous boundary conditions (24.2) or (24.3). Furthermore, $\mathbf{U}_\varepsilon = \mathbf{u}^s_\varepsilon$ outside some neighbourhood of Γ_ε.

According to [2, 6], the shape derivative $\mathbf{U}'$ satisfies the following exterior mixed boundary value problem

$$\begin{aligned} \Delta^*\mathbf{U}' + \omega^2\mathbf{U}' &= \mathbf{0} \quad \text{in} \quad \Omega, \\ \mathbf{U}' &= -\boldsymbol{\Psi}\cdot\mathbf{n}\frac{\partial\mathbf{U}_0}{\partial\mathbf{n_x}} \quad \text{on} \quad \Gamma. \end{aligned} \qquad (24.14)$$

Remember that in the case of cracks, the boundary condition must be imposed on both sides of the crack as in (24.3).

Form sensitivity of the far field pattern. Let Ω be an exterior domain with the scatterer Ω_i of type **(B)** or **(C)**. The perturbed scattered wave $\mathbf{u}^s_\varepsilon$ has at infinity the asymptotics (24.5) and (24.6) with the far–field pattern $\mathcal{F}(\Gamma_\varepsilon) = (\mathcal{F}_p(\Gamma_\varepsilon), \mathcal{F}_s(\Gamma_\varepsilon))$ given by an analogue of formulae (24.7) and (24.8). Let us calculate the Gâteaux derivative

$$d\mathcal{F}(\Gamma, \boldsymbol{\Psi}) = \lim_{\varepsilon \to 0} \frac{\mathcal{F}(\boldsymbol{\Psi}_\varepsilon(\Gamma)) - \mathcal{F}(\Gamma)}{\varepsilon} = \left.\frac{d\mathcal{F}(\Gamma_\varepsilon)}{d\varepsilon}\right|_{\varepsilon=0}. \qquad (24.15)$$

For big enough $R = |\mathbf{x}|$ we have $\boldsymbol{\Psi}_\varepsilon|_{S_R} = I$ and $\mathbf{u}^s_\varepsilon|_{S_R} = \mathbf{U}_\varepsilon \circ \boldsymbol{\Psi}_\varepsilon|_{S_R}$. Thus

$$\begin{aligned} \mathcal{F}_p(\Gamma_\varepsilon)(\hat{\mathbf{x}}) = \int_{S_R} \Big\{ &\left[\mathbf{T_{n(y)}}\hat{\mathbf{x}} \otimes \hat{\mathbf{x}}e^{-ik_p\hat{\mathbf{x}}\cdot\mathbf{y}}\right]^\top (\mathbf{U}_\varepsilon \circ \boldsymbol{\Psi}_\varepsilon)(\mathbf{y}) \\ &- \hat{\mathbf{x}} \otimes \hat{\mathbf{x}}e^{-ik_p\hat{\mathbf{x}}\cdot\mathbf{y}}\mathbf{T_{n(y)}}(\mathbf{U}_\varepsilon \circ \boldsymbol{\Psi}_\varepsilon)(\mathbf{y}) \Big\} ds_\mathbf{y}. \end{aligned} \qquad (24.16)$$

Differentiating both sides of the above equation at $\varepsilon = 0$ we obtain because of $\dot{\mathbf{U}}(\mathbf{x}) = \mathbf{U}'(\mathbf{x})$ for big enough $|\mathbf{x}|$

$$\begin{aligned} d\mathcal{F}_p(\Gamma)(\hat{\mathbf{x}}) = \int_{S_R} \Big\{ &\left[\mathbf{T_{n(y)}}\hat{\mathbf{x}} \otimes \hat{\mathbf{x}}e^{-ik_p\hat{\mathbf{x}}\cdot\mathbf{y}}\right]^\top \mathbf{U}'(\mathbf{y}) \\ &- \hat{\mathbf{x}} \otimes \hat{\mathbf{x}}e^{-ik_p\hat{\mathbf{x}}\cdot\mathbf{y}}\mathbf{T_{n(y)}}\mathbf{U}'(\mathbf{y}) \Big\} ds_\mathbf{y}. \end{aligned} \qquad (24.17)$$

A similar formula can be obtained for $d\mathcal{F}_s(\Gamma)(\hat{\mathbf{x}})$ with the only difference that $\hat{\mathbf{x}} \otimes \hat{\mathbf{x}}$ in (24.16) has to be replaced by $\mathbf{I} - \hat{\mathbf{x}} \otimes \hat{\mathbf{x}}$.

The representation (24.16) is not suitable for numerical realization because $\mathbf{U}'$, in general, can not be defined as a variational solution of the boundary value problem (24.14). According to Section 2, the solution $\mathbf{U}_0$ behaves in some neighbourhood $\mathcal{U}_q$ of the singular point $P_q \in S$ as $|\mathbf{x} - P_q|^{a_q}$ with $a_q \geq 1/2$. Therefore, $\mathbf{U}_0 \in H^{1+a_q}(\mathcal{U}_q \cap \Omega)$, $\nabla \mathbf{U}_0 \in H^{a_q}(\mathcal{U}_q \cap \Omega)$ and consequently $\mathbf{U}' \in H^{a_q}(\mathcal{U}_q \cap \Omega)$ due to (24.13). If the domain is not convex then we have $a_q < 1$ and thus $\mathbf{U}' \notin H^1_{\mathrm{loc}}(\Omega)$. Furthermore, we are interested in an expression for the derivative of the far field patterns depending only on the perturbation of the boundary.

In order to overcome this difficulty, we derive in the next section other representations for both $d\mathcal{F}_p(\Gamma, \Psi)(\hat{\mathbf{x}})$, $d\mathcal{F}_s(\Gamma, \Psi)(\hat{\mathbf{x}})$. As in [4, 5] we use the method of adjoint problems, which consists in applying Betti's formulae to the shape derivative $\mathbf{U}'$ and to the solution $\mathbf{w}$ of an appropriately defined adjoint problem. This leads to an expression in which only $\mathbf{U}_0$ and the adjoint field $\mathbf{w}$ appear.

4.   THE METHOD OF ADJOINT PROBLEMS

Exterior of a bounded domain.   Let us assume first that Ω is the exterior of bounded domain and consider the longitudinal far-field pattern $\mathcal{F}_p$. We define $\mathbf{w}$ as the solution of the following exterior boundary value problem

$$
\begin{aligned}
\Delta^*\mathbf{w}(\mathbf{y}) + \omega^2\mathbf{w}(\mathbf{y}) &= \mathbf{0} && \text{in} \quad \Omega, \\
\mathbf{w}(\mathbf{y}) &= \hat{\mathbf{x}}e^{-ik\hat{\mathbf{x}}\cdot\mathbf{y}} && \text{on} \quad \Gamma,
\end{aligned}
$$

which satisfies the Kupradze radiation condition (24.4) at infinity. Using Betti's formula for $\mathbf{U}'$ and $\mathbf{w}$ in $B_{R'} \cap B_R \cap \Omega$, $R' > R$, passing to the limit as $R' \to +\infty$ and taking into account that $\mathbf{U}', \mathbf{w}$ satisfy (24.4) we obtain

$$
0 = \hat{\mathbf{x}} \int_{S_R} \left[\mathbf{w}^\top(\mathbf{y})\mathbf{T}_{\mathbf{n}(\mathbf{y})}\mathbf{U}'(\mathbf{y}) - \mathbf{U}'^\top(\mathbf{y})\mathbf{T}_{\mathbf{n}(\mathbf{y})}\mathbf{w}(\mathbf{y}) \right] ds_{\mathbf{y}}. \tag{24.18}
$$

Summing up the expressions (24.18) and (24.16) we get

$$
d\mathcal{F}_p(\Gamma, \Psi)(\hat{\mathbf{x}}) = \hat{\mathbf{x}} \int_{S_R} \left[\mathbf{W}^\top(\mathbf{y})\mathbf{T}_{\mathbf{n}(\mathbf{y})}\mathbf{U}'(\mathbf{y}) - \mathbf{U}'^\top(\mathbf{y})\mathbf{T}_{\mathbf{n}(\mathbf{y})}\mathbf{W}(\mathbf{y}) \right] ds_{\mathbf{y}}, \tag{24.19}
$$

with $\mathbf{W}(\mathbf{y}) := \mathbf{w}(\mathbf{y}) - \hat{\mathbf{x}}e^{-ik\hat{\mathbf{x}}\cdot\mathbf{y}}$. Note that the normal vector $\mathbf{n}$ on S_R is directed outwards.

Let $B_\delta(P_q)$ be a ball with centre in P_q and radius δ. Inserting $\mathbf{U}'$ and $\mathbf{W}$ into Betti's formula in $\Omega \cap B_R \cap \bigcup\limits_{q=1}^{Q} B_\delta(P_q)$, we obtain from (24.19)

$$
d\mathcal{F}_p(\Gamma, \Psi)(\hat{\mathbf{x}}) \;=\; \hat{\mathbf{x}} \int\limits_{\Gamma \setminus \bigcup\limits_{q=1}^{Q} B_\delta(P_q)} \left[\mathbf{U}'^{\top} \mathbf{T}_{\mathbf{n}(\mathbf{y})} \mathbf{W} - \mathbf{W}^{\top} \mathbf{T}_{\mathbf{n}(\mathbf{y})} \mathbf{U}' \right] ds_{\mathbf{y}}
$$

$$
+\; \hat{\mathbf{x}} \int\limits_{\bigcup\limits_{q=1}^{Q} \partial B_\delta(P_q) \cap \Omega} \left[\mathbf{U}'^{\top} \mathbf{T}_{\mathbf{n}(\mathbf{y})} \mathbf{W} - \mathbf{W}^{\top} \mathbf{T}_{\mathbf{n}(\mathbf{y})} \mathbf{U}' \right] ds_{\mathbf{y}}.
$$

Let us formally pass to the limit as $\delta \to 0$ on both sides of this formula and rewrite it as

$$
d\mathcal{F}_p(\Gamma, \Psi)(\hat{\mathbf{x}}) \;=\; \hat{\mathbf{x}} \int\limits_{\Gamma} \left[\mathbf{U}'^{\top} \mathbf{T}_{\mathbf{n}(\mathbf{y})} \mathbf{W} - \mathbf{W}^{\top} \mathbf{T}_{\mathbf{n}(\mathbf{y})} \mathbf{U}' \right] ds_{\mathbf{y}} \qquad (24.20)
$$

$$
+\; \hat{\mathbf{x}} \lim_{\delta \to 0} \int\limits_{\bigcup\limits_{q=1}^{Q} \partial B_\delta(P_q) \cap \Omega} \left[\mathbf{U}'^{\top} \mathbf{T}_{\mathbf{n}(\mathbf{y})} \mathbf{W} - \mathbf{W}^{\top} \mathbf{T}_{\mathbf{n}(\mathbf{y})} \mathbf{U}' \right] ds_{\mathbf{y}}.
$$

In the following, we denote by L_q the limit in the above expression corresponding to the integral over $\partial B_\delta(P_q) \cap \Omega$. Finally, substituting the boundary values of $\mathbf{U}'$ and $\mathbf{W}$ in the first integral of (24.20), we obtain formally

$$
d\mathcal{F}_p(\Gamma, \Psi)(\hat{\mathbf{x}}) = -\hat{\mathbf{x}} \int\limits_{\Gamma} \Psi(\mathbf{y}) \cdot \mathbf{n}(\mathbf{y}) \frac{\partial \mathbf{U}_0(\mathbf{y})^{\top}}{\partial \mathbf{n}_{\mathbf{y}}} \mathbf{T}_{\mathbf{n}(\mathbf{y})} \mathbf{W}(\mathbf{y}) ds_{\mathbf{y}} + \hat{\mathbf{x}} \sum_{q=1}^{Q} L_q.
$$

$$
(24.21)
$$

In order to calculate the limits L_q we have to investigate the behaviour of their integrands as $\delta \to 0$ for every singular point $P_q \in S$. According to Section 2, the functions $\mathbf{U}_0$ and $\mathbf{W}$ behave in the neighbourhood of P_q as

$$
\mathbf{U}_0(\mathbf{x}) = O(r^{a_q}) \text{ and } \mathbf{W}(\mathbf{x}) = O(r^{a_q}) \qquad (24.22)
$$

with $\alpha_q > 1/2$. Consequently, $\mathbf{U}'(\mathbf{x}) = O(r^{a_q - 1})$ due to (24.13) and so the integrand of L_q behaves as $O(r^{2a_q - 2})$. Thus $L_q = 0$.

We note that a similar representation can be obtained for $d\mathcal{F}_s(\Gamma, \Psi)$. In this case the adjoint field has to be defined as the scattered elastic field produced by transversal incident plane wave of the form $-\hat{\mathbf{z}} e^{-ik\hat{\mathbf{x}} \cdot \mathbf{y}}$, where $\hat{\mathbf{z}}$ is perpendicular to the direction of observation $\hat{\mathbf{x}}$.

Exterior domains with cracks.   The above considerations can be repeated with some obvious changes for problems in the exterior of a crack. The adjoint backscattered wave has to be defined as the solution of the following boundary value problem in the exterior of the crack

$$\Delta^* \mathbf{w}(\mathbf{y}) + \omega^2 \mathbf{w}(\mathbf{y}) = \mathbf{0} \qquad \text{in} \quad \Omega,$$
$$\mathbf{w}_\pm(\mathbf{y}) = \hat{\mathbf{x}} e^{-ik\hat{\mathbf{x}}\cdot\mathbf{y}} \qquad \text{on} \quad \Gamma,$$

satisfying the Kupradze radiation condition (24.4) at infinity. Formula (24.21) reads now

$$d\mathcal{F}_p(\Gamma, \Psi)(\hat{\mathbf{x}}) = -\hat{\mathbf{x}} \int_\Gamma \Psi(\mathbf{y})\cdot\mathbf{n}(\mathbf{y}) \left[\!\left[\frac{\partial \mathbf{U}_0(\mathbf{y})^\top}{\partial \mathbf{n_y}} \mathbf{T}_{\mathbf{n}(\mathbf{y})} \mathbf{W}(\mathbf{y}) \right]\!\right] ds_{\mathbf{y}} + \hat{\mathbf{x}} \sum_{q=1}^{Q} L_q.$$

$$(24.23)$$

Here, $[\![\cdot]\!]$ denotes the jump across the crack. The limits L_q can be calculated similarly to those for problems in the exterior of a bounded domain. If P_q is an interior corner of the crack, then $\mathbf{U}' = O(r^{\alpha_q-1})$, $\mathbf{W} = O(r^{\alpha_q})$ with $\alpha_q > 1/2$ and therefore $L_q = 0$.

Let P_q, $q = 1, 2$ be the tips of the crack and let $\Psi_n(P_q)$ and $\Psi_t(P_q)$ be the normal and the tangential component of the perturbation at the tip P_q, respectively. According to [3] the eigenvalues corresponding to a crack tip are given by $\alpha_j = j/2$, with $j \in \mathbb{N}$ and have double geometric multiplicity. Therefore $\mathbf{U}_0$ and $\mathbf{W}$ behave near the tip P_q in local polar coordinates (r_q, ω_q) as

$$\mathbf{U}_0(r_q, \varphi_q) = \left[K_q^1(\mathbf{U}_0)\Phi_1(\varphi_q) + K_q^2(\mathbf{U}_0)\Phi_2(\varphi_q) \right] r_q^{1/2} + O(r_q),$$
$$\mathbf{W}(r_q, \varphi_q) = \left[K_q^1(\mathbf{W})\Phi_1(\varphi_q) + K_q^2(\mathbf{W})\Phi_2(\varphi_q) \right] r_q^{1/2} + O(r_q),$$

where

$$\Phi_1(\varphi_q) = \begin{bmatrix} -(2\kappa - 1)\cos\left(\frac{3}{2}\varphi_q\right) + (2\kappa - 1)\cos\left(\frac{1}{2}\varphi_q\right) \\ (2\kappa - 1)\sin\left(\frac{3}{2}\varphi_q\right) - (2\kappa + 1)\sin\left(\frac{1}{2}\varphi_q\right) \end{bmatrix},$$
$$\Phi_2(\varphi_q) = \begin{bmatrix} -(2\kappa + 1)\sin\left(\frac{3}{2}\varphi_q\right) + (2\kappa - 1)\sin\left(\frac{1}{2}\varphi_q\right) \\ -(2\kappa + 1)\cos\left(\frac{3}{2}\varphi_q\right) + (2\kappa + 1)\cos\left(\frac{1}{2}\varphi_q\right) \end{bmatrix}.$$

Here $\kappa = (\lambda + 3\mu)/(\lambda + \mu)$ is a material constant. The asymptotics of $\mathbf{U}'$ can be now calculated by means of (24.13) from the asymptotics of $\mathbf{U}_0$,

$$\mathbf{U}'(r_q, \varphi_q) = r_q^{-1/2} \left[\left(K_q^1(\mathbf{U}_0)\tilde{\Phi}_{1t}(\varphi_q) + K_q^2(\mathbf{U}_0)\tilde{\Phi}_{2t}(\varphi_q) \right) \Psi_t(P_q) + \right.$$
$$\left. \left(K_q^1(\mathbf{U}_0)\tilde{\Phi}_{1n}(\varphi_q) + K_q^2(\mathbf{U}_0)\tilde{\Phi}_{2n}(\varphi_q) \right) \Psi_n(P_q) \right] + O(1).$$

Here we omit for the sake of brevity explicit formulae for $\tilde{\Phi}_{1t}$, $\tilde{\Phi}_{2t}$, $\tilde{\Phi}_{1n}$ and $\tilde{\Phi}_{2n}$. Inserting the asymptotics of $\mathbf{U}'$ and $\mathbf{W}$ into L_q and performing the limit passage $\delta \to 0$ we obtain finally

$$
\begin{aligned}
L_q \;=\; & (K_q^1(\mathbf{U}_0)K_q^2(\mathbf{W}) + K_q^2(\mathbf{U}_0)K_q^1(\mathbf{W}))\frac{-16\pi\mu(\lambda + 2\mu)}{(\lambda + \mu)}\Psi_n(P_q) \\
& + (K_q^1(\mathbf{U}_0)K_q^1(\mathbf{W}) + K_q^2(\mathbf{U}_0)K_q^2(\mathbf{W}))\frac{-16\pi\mu(\lambda + 3\mu)(\lambda + 2\mu)}{(\lambda + \mu)^2}\Psi_t(P_q).
\end{aligned}
$$

Acknowledgments

The work of the first author was supported by the German Science Foundation in the framework of the Special Research Group SFB 404. The work of the second author was supported by the Alexander von Humboldt Stiftung during her stay at the University of Stuttgart, Germany.

References

[1] Dassios, G and Kiriaki, K (1984) *The low frequency theory of elastic waves*, Quart. Appl. Math. **42**, 225–248.

[2] Simon, J (1980) *Differentiation with respect to the domain in boundary value problems*, Num. Funct. Anal. Optim. **2**, 649–687.

[3] Nazarov, S A and Plamenevsky, B A (1994) *Elliptic Problems in Domains with Piecewise Smooth Boundaries*, Berlin: Walter de Gruyter.

[4] Bochniak, M and Cakoni, F (2000) *Domain Sensitivity Analysis for Acoustic Scattering Problems*, Mathematical and Numerical Aspects of Wave Propagation (eds. Bermudez et al.), 450–454, Philadelphia: SIAM Publications.

[5] Bochniak, M and Cakoni, F (2000) *The method of adjoint problems for the domain sensitivity analysis of the acoustic far-field pattern. Part I: 2D domains with corners*, Preprint 2000/07, SFB 404, University of Stuttgart.

[6] Masanao, T and Fujii, N (1992) *Second-order necessary conditions for domain optimisation problems in elastic structure I; II*, J. Opt. Th. Appl. **72**, 355–382; 383–401.

FACTORISATION OF A CERTAIN WIENER–HOPF MATRIX ARISING IN DIFFRACTION THEORY

P. A. Lewis

School of Computing, Staffordshire University
Stafford ST18 0DG, UK

P.A.Lewis@staffs.ac.uk

1. INTRODUCTION

Ultrasonic inspection is a widely used method for determining the existence of defects in components. It relies on specular reflection and diffraction of elastic waves at crack faces and edges to locate and size possible defects. The redistribution of energy into a range of diffracted angles is a purely local phenomenon and is determined by the properties of the material at the crack tip and its shape. As all materials are naturally anisotropic, it is clearly necessary to understand how diffraction is modified by anisotropy. An understanding of the effects of anisotropy could therefore help to optimise inspection techniques and justify procedures for detecting and sizing crack-like defects. The quantity of interest is the diffraction coefficient, which relates the vector amplitude of a ray incident on a crack edge to the amplitudes of the diffracted rays.

The geometrical theory of diffraction (GTD) is the most useful approach for determining diffraction coefficients [1]. However using GTD it is not possible to consider arbitrary angles of incidence, on arbitrarily oriented defects in materials with low degrees of symmetry. The only previously known results for diffraction coefficients are where there is a high degree of symmetry in the material. These include the three dimensional isotropic case [2] and the two dimensional transversely isotropic problem, considered in [3]. Temple and White [4] attempted to determine diffraction coefficients in more general materials numerically using time dependent finite difference simulations, with limited success.

The method we use is the Wiener–Hopf technique [5], which relies on being able to factorise a matrix into a product of two matrices, each analytic in certain overlapping complex half-planes. If this decomposi-

I.D. Abrahams et al. (eds.),
IUTAM Symposium on Diffraction and Scattering in Fluid Mechanics and Elasticity, 227–234.
© 2002 *Kluwer Academic Publishers. Printed in the Netherlands.*

tion is possible, then explicit expressions for the transform of the crack opening displacement, COD, and hence the scattered field, can be derived in terms of the factorised kernels. By considering the asymptotics of the scattered field, expressions for the diffraction coefficients can also be derived in terms of the transform of the COD.

As yet, there is no general method for determining such a factorisation. Some advance has been made by Meister and Speck [6], who derived an explicit formula for the three dimensional isotropic matrix. Our method relies on first being able to determine a scalar factorization of the diagonal elements of $\mathbf{K}$. These are found by taking logarithms and using Cauchy's integral formulae in the usual way. However, this is a non-trivial calculation as the kernel elements contain unknown singularities. In order to evaluate these integrals it is necessary to ensure that there are no singularities along the contour of integration. We note that the diffraction coefficient is, by definition, independent of frequency and so we introduce a small positive imaginary part to the frequency [7]. This has the effect of lifting the singularities in the kernel elements off the real axis, thus enabling the numerical evaluation of the scalar factorizations. The remaining decomposition can be reduced to the solution of a matrix system of Fredholm integral equations of the second kind. The formulation ensures that the integrands are small, smooth and non-singular along the real axis. The integral equations also provide an analytic continuation of the factorization into the whole complex plane. This system of coupled Fredholm integral equations is solved by collocation.

In this paper we are concerned with the implementation of the method to produce graphs of diffraction coefficients. Some of the results we present are previously unpublished and give an indication of the effects of anisotropy on diffraction coefficients. A more detailed description of the theory can be found in [8].

2.   THE SCATTERING PROBLEM

We consider a semi-infinite crack located in an unbounded linearly elastic body. The crack is supposed to be comprised of two planar non-contacting faces S^+ and S^- with normal ν, as shown in Fig. 1. There are two coordinate systems, one relative to the crack coordinates, denoted by $\mathbf{X}$ and one associated with the material coordinates $\mathbf{x}$.

The displacement field $\mathbf{u}(\mathbf{X})$ satisfies Navier's Equation

$$C_{ijkl}\frac{\partial^2 u_k}{\partial X_j \partial X_l}(\mathbf{X}) + \rho\omega^2 u_i(\mathbf{X}) = 0,$$

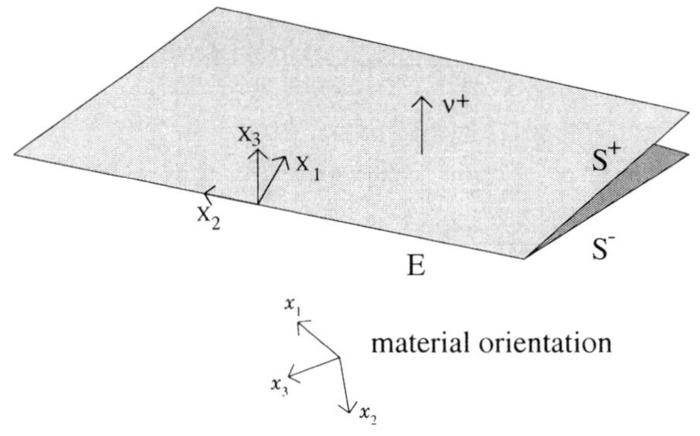

Figure 1   Geometry

where C_{ijkl} is the material stiffness tensor relative to the crack coordinate system. We also apply a stress-free boundary condition

$$\tau_{ij}\nu_j = C_{ijkl}\frac{\partial u_k}{\partial X_l}(\mathbf{X})\nu_j = 0 \quad \text{on} \quad S^+ \quad \text{and} \quad S^-$$

and insist that the scattered displacement be outgoing at infinity and bounded everywhere.

3.   THE FUNDAMENTAL SOLUTION

The Fundamental Solution, $\mathbf{U}^m(\mathbf{X}, \mathbf{X}')$, is a solution corresponding to a point source located at $\mathbf{X}'$ acting in a direction i_m in an unbounded material in the absence of the defect and satisfies the equation of motion

$$C_{ijkl}\frac{\partial^2 U_k}{\partial X_j\,\partial X_l}(\mathbf{X}, \mathbf{X}') + \rho\omega^2 U_i(\mathbf{X}, \mathbf{X}') + \delta_{im}\delta(\mathbf{X} - \mathbf{X}') = 0.$$

Taking Fourier transforms yields a solution

$$U_k^{(m)}(\mathbf{X}, \mathbf{X}') = \frac{1}{8\pi^3}\int_{-\infty}^{\infty}\int_{-\infty}^{\infty}\int_{-\infty}^{\infty}\frac{B_{km}(\alpha)}{S(\alpha)}e^{-i\alpha\cdot(\mathbf{X}'-\mathbf{X})}\,d\alpha, \qquad (25.1)$$

where $B_{km}(\alpha)$ and $S(\alpha)$ are the cofactors and determinant respectively of the Kelvin–Christoffel matrix

$$S_{ik}(\alpha) = C_{ijkl}\alpha_j\alpha_l - \rho\omega^2\delta_{ik}.$$

The only singularities are the zeros (real or complex) of $S(\alpha) = 0$, which is a homogeneous sixth order polynomial. The real zeros of S produce the slowness surface (see for example, [9]), which comprises three distinct sheets corresponding to the three modes of propagation in the material.

Closing the contour of integration in an appropriate half-plane enables one integral in (25.1) to be evaluated using the residue theorem.

A solution to the scattering problem is sought in the form

$$u_m^{(\mathrm{sc})}(\mathbf{X}') = \int_{S^+} \xi_i(\mathbf{X}) \Sigma_{ij}^{(m)}(\mathbf{X},\mathbf{X}') \, \nu_j^+ \, dS,$$

where $\xi_i(\mathbf{X})$ is the unknown jump in the displacement across the crack. Applying the boundary conditions yields the following integro-differential equation of the first kind with a difference kernel

$$\sigma_{r3}^{(\mathrm{inc})}(\mathbf{X}') = -C_{r3mt}C_{i3kl}\frac{\partial^2}{\partial X_t' \partial X_l'} \int_0^\infty \int_{-\infty}^\infty \xi_i(\mathbf{X}) U_k^{(m)}(\mathbf{X},\mathbf{X}') \, dS,$$

$$(25.2)$$

where $X_1 > 0$ and $X_3' = X_3 = 0$.

4.   THE FORMAL SOLUTION

For an incident plane wave, $\mathbf{u}(\mathbf{X}) = e^{\mathrm{i}\mathbf{k}\cdot\mathbf{X}}$, taking Fourier transforms of (25.2) yields the following matrix Wiener–Hopf equation

$$T_i^+(\alpha_1,\mathbf{k}) = \frac{A_i}{\alpha_1 - k_1} + U_j^-(\alpha_1,\mathbf{k}) \, K_{ij}(\alpha_1,k_2), \qquad (25.3)$$

where $U_i(\alpha_1,\mathbf{k})$ is the Fourier transform of the COD, $T^+(\alpha_1,\mathbf{k})$ is the Fourier transform of the unknown stresses on the complement to the crack and $\mathbf{A}$ is a constant vector dependent on the incident waveform. $\mathbf{K}(\alpha_1,k_2)$ is the kernel given by

$$K_{ij}(\alpha_1,k_2) = \mathrm{i}C_{i3kl}C_{j3mt} \sum_{s=1}^3 \frac{B_{km}(-\alpha_1,-k_2,p_s^+)\delta_{ts}^+\delta_{ls}^+}{S_3(-\alpha_1,-k_2,p_s^+)},$$

where p_s^+ are the roots of $S(\alpha) = 0$ lying in the upper half-plane and $\delta_s^+ = (-\alpha_1,-k_2,p_s^+)$. An assumption of translation invariance parallel to the edge of the crack has been made in the derivation of this equation. Next, we assume that the kernel $K_{ij}(\alpha_1,k_2)$ can be split as follows

$$[\mathbf{K}^+(\alpha_1,k_2)]^{-1} \cdot \mathbf{K}(\alpha_1,k_2) = \mathbf{K}^-(\alpha_1,k_2). \qquad (25.4)$$

Substituting (25.4) into (25.3) and using the asymptotic behaviour of the functions concerned produces the following explicit expression for the Fourier transform of the COD

$$U^-(\alpha_1,\mathbf{k}) = -\frac{1}{\alpha_1 - k_1}[\mathbf{K}(\alpha_1,\mathbf{k})]^{-1} \cdot [\mathbf{K}^+(\alpha_1,\mathbf{k})]^{-1} \cdot \mathbf{A}.$$

5.   THE FACTORISATION SCHEME

This factorisation scheme has been developed with this particular problem in mind. However, it may be possible to generalise the method to factorise Wiener–Hopf matrix kernels of other forms.

The first step in the factorisation scheme is to perform a product split of the diagonal elements. In the fully anisotropic problem, $S(\alpha)$ will not factorise and hence there are no explicit expressions for the kernel. In particular, there may be singularities on the real axis. To overcome this difficulty we introduce a complex component to the frequency. This has the effect of lifting any real singularities off the real axis. It is therefore possible to take logarithms and perform a sum split using the Cauchy Integral Formulae.

Next we introduce

$$\mathbf{K}(\alpha_1, k_2) = \mathbf{K}_d^+(\alpha_1, k_2) \cdot \mathbf{H}(\alpha_1, k_2) \cdot \mathbf{K}_d^-(\alpha_1, k_2).$$

We assume that the off-diagonal elements do not grow faster than the diagonal elements and seek a decomposition of the form

$$\mathbf{H}^+(\alpha_1, k_2) \cdot (\mathbf{I} - \mathbf{G}(\alpha_1, k_2)) = \mathbf{H}^-(\alpha_1, k_2),$$

where $\mathbf{H}(\alpha_1, k_2) = \mathbf{I} - \mathbf{G}(\alpha_1, k_2)$. Performing a sum split on $\mathbf{H}^+ \cdot \mathbf{G}$ and using the usual Wiener–Hopf arguments produces a coupled system of equations for the factorisations of $\mathbf{H}$

$$\mathbf{H}^+ - (\mathbf{H}^+ \cdot \mathbf{G})^+ = \mathbf{I} \quad \text{and} \quad \mathbf{H}^- + (\mathbf{H}^+ \cdot \mathbf{G})^- = \mathbf{I}.$$

These two equations can be combined in a single Fredholm integral equation of the second kind of the form

$$\mathbf{H}^-(\alpha_1, k_2) = \mathbf{I} - \mathbf{G}^-(\alpha_1, k_2) + \frac{1}{2\pi i} \int \mathbf{H}^-(\zeta, k_2) \cdot \mathcal{K}(\zeta, \alpha_1; k_2)\, d\zeta,$$

where

$$\mathcal{K}(\zeta, \alpha_1; k_2) = \left[\mathbf{I} - \mathbf{G}^-(\zeta, k_2)\right]^{-1} \cdot \mathbf{G}(\zeta, k_2) \cdot \frac{\mathbf{G}^-(\zeta, k_2) - \mathbf{G}^-(\alpha_1, k_2)}{\zeta - \alpha_1}.$$

We note that the kernel of this system of equations is smooth along the real axis and that by construction we have ensured that $\mathbf{G}$ is small and as a result the kernel is $O(\mathbf{G}^2)$. Further, the integral equation provides an analytic continuation of the solution into the whole complex plane. The unknown in this equation is a 3 by 3 matrix and the integral equations are solved by a simple collocation method.

6.   THE DIFFRACTION COEFFICIENTS

Using polar coordinates $X_1' = R\cos\Theta$, $X_3' = R\sin\Theta$, it may be shown, by applying Kelvin's method of stationary phase, that the diffraction coefficient, $D_m^s(\Theta)$, of component m and type s is given explicitly by

$$D_m^s(\Theta) = \frac{iC_{i3kl}B_{km}(\alpha_1, -k_2, p_s^+)\delta_{ls}^+ U_i^-(-\alpha_1, -k_2)\exp\left\{\frac{i\pi}{4}\mathrm{sgn}(\frac{d^2 p_s^+}{d\alpha_1^2})\right\}}{S_3(\alpha_1, -k_2, p_s^+)|\sin\Theta|^{\frac{1}{2}}|d^2 p_s^+/d\alpha_1^2|^{\frac{1}{2}}}.$$

The value of Θ is determined by the stationary points, which satisfy

$$\cos\Theta + \sin\Theta\frac{dp_s^+}{d\alpha_1} = 0.$$

For a general anisotropic material, for a given angle Θ this is an extremely difficult equation to solve for points lying on the slowness surface. It may be shown that on a slowness surface

$$\frac{\partial S}{\partial\alpha_1} + \frac{\partial S}{\partial\alpha_3}\frac{dp_s^+}{d\alpha_1} = 0 \quad \text{and so} \quad (\cos\Theta, \sin\Theta) \propto \mathrm{grad}S. \tag{25.5}$$

We therefore adopt an alternative approach. Rather than choose Θ, we select α_1, calculate the roots p_s^+ and then use (25.5) to determine the direction of the diffracted field.

7.   RESULTS

In all results we have removed the known singularity $\cos\theta - \cos\theta_0$, which arises in the term U_i^-. The first results we present are for a transversely isotropic material. These plots are for P-P backscatter for varying values of the parameter $p = c_{11}/c_{33}$. These results are in excellent agreement with those published by Norris and Achenbach [3].

The next set of results are for a material with increasing anisotropy. The anisotropy factor is $2c_{44}/(c_{11} - c_{12})$, as defined by Auld [10]. As can be seen from the figures, there is a gradual transition away from the corresponding isotropic results. As the anisotropy increases, the effects of the off-diagonal terms become greater. The final set of results we present is for a varying crack angle. Again, these cases have increasing off-diagonal kernel elements. The transition away from the corresponding isotropic results is again smooth and gradual.

Acknowledgments

This work was initiated under funding from the UK Health and Safety Executive as part of a programme of Nuclear Safety Research and AEA Technology. Other major contributors to this work are Dr E. J. Walker, Dr J. A. G. Temple and the late Prof. G. R. Wickham.

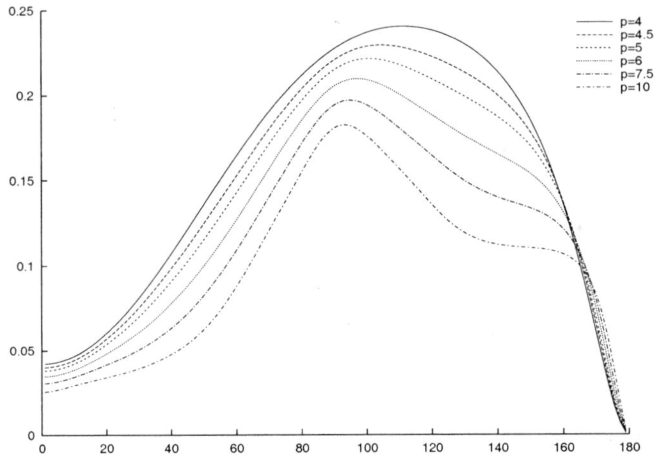

Figure 2   Diffraction coefficients for transversely isotropic material

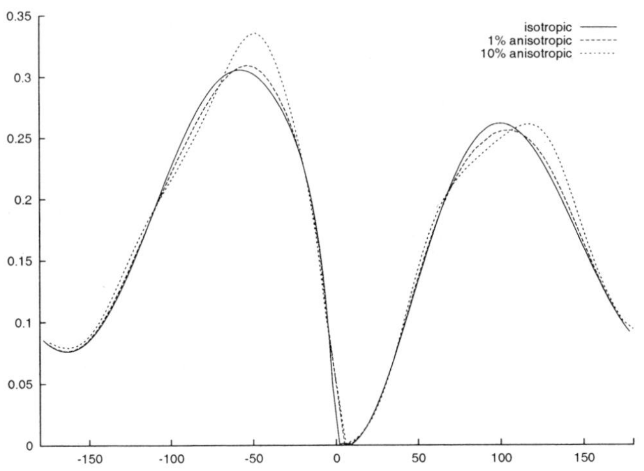

Figure 3   Diffraction coefficients for varying anisotropy

References

[1] Achenbach, J D, Gautesen, A K and McMaken, H (1982) *Ray Methods for Waves in Elastic Solids*, London: Pitman.

[2] Achenbach, J D and Gautesen, A K (1977) *Geometrical theory of diffraction for three-D elastodynamics*, J. Acoust. Soc. Amer. **61**, 413–421.

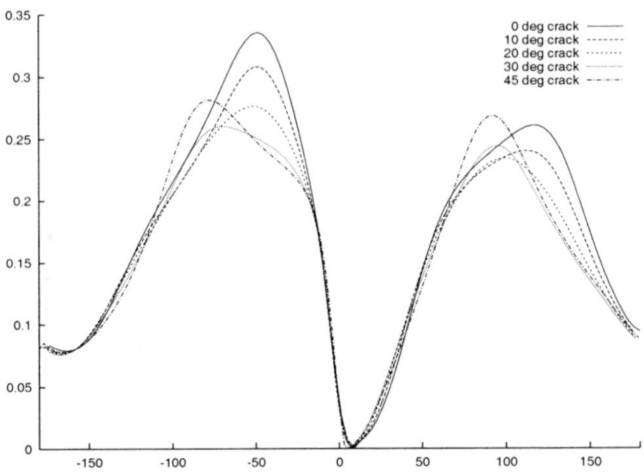

Figure 4   Diffraction coefficients for varying crack angle

[3] Norris, A N and Achenbach, J D (1984) *Elastic wave diffraction by a semi-infinite crack in a transversely isotropic material*, Quart. J. Mech. Appl. Math. **37**, 565–580.

[4] Temple, J A G and White, L (1993) *Numerical calculation of diffraction coefficients in anisotropic media*, Review of Progress in Quantitative Non-Destructive Evaluation (eds. D O Thompson & D E Chimenti), **12**, 49–54, New York: Plenum.

[5] Noble, B (1958) *Methods based on the Wiener–Hopf Technique*, Oxford: Pergamon Press.

[6] Meister, E and Speck, F-O (1989) *Wiener–Hopf factorization of certain non-rational matrix functions in mathematical physics*, The Gohberg anniversary collection, vol. II, 385–394, Basel: Birkhauser.

[7] Lighthill, M J (1960) *Studies of magneto-hydrodynamic waves and other anisotropic wave motions*, Phil. Trans. Roy. Soc. Lond. A **252**, 397–430.

[8] Lewis, P A, Temple, J A G, Walker, E J and Wickham, G R (1998) *Calculation of diffraction coefficients for a semi-infinite crack embedded in an infinite anisotropic linearly elastic solid*, Proc. Roy. Soc. Lond. A **454**, 1781–1803.

[9] Musgrave, M J P (1970) *Crystal Acoustics*, San Francisco: Holden-Day.

[10] Auld, B A (1973) *Acoustic Fields and Waves in Solids*, New York: Wiley.

THEORY OF CRACK FRONT WAVES

J. R. Willis

Department of Applied Mathematics and Theoretical Physics
University of Cambridge, Cambridge CB3 9EW, UK

J.R.Willis@damtp.cam.ac.uk

A. B. Movchan

Department of Mathematical Sciences
University of Liverpool, Liverpool L69 3BX, UK

abm@maths.liv.ac.uk

Abstract   A crack front wave is a disturbance of the edge of a propagating crack, which remains localised about the edge as it propagates. Because it is confined to the vicinity of the edge, a crack front wave propagates without attenuation, unless some local mechanism for dissipation is present. This article presents the theory underlying crack front waves. It is more general than any presented previously. First, the presence of a non-singular term in the stress field near the unperturbed crack edge is shown to introduce dispersion, which becomes negligible as frequency tends to infinity; the previously-published work that neglected this term and predicted that the crack front wave is non-dispersive thus has only asymptotic validity, in the limit of high frequency. In addition, the present analysis is conducted for a crack which propagates through a medium that is viscoelastic rather than elastic; again, the previous elastic result is recovered as frequency tends to infinity. Explicit results are presented in the case that the frequency of the disturbance is high: the leading-order term is the one previously found for elasticity, while the first correction term yields both dispersion and attenuation, proportional to $(\text{frequency})^{-1}$. The virtue of the asymptotic analysis is that it is applicable to any isotropic viscoelastic medium: the properties of the medium enter only through two-term expansions (for high frequency) of the (complex) .phase speeds of longitudinal and shear waves. The analysis reproduces but generalises results recently published elsewhere by the authors, for the case of crack propagation through a Maxwell fluid, with frequency-independent Poisson's ratio.

I.D. Abrahams et al. (eds.),
IUTAM Symposium on Diffraction and Scattering in Fluid Mechanics and Elasticity, 235–250.
© 2002 *Kluwer Academic Publishers. Printed in the Netherlands.*

1.   INTRODUCTION

Crack front waves were discovered during numerical simulation of the perturbation of the leading edge of a propagating crack as it passes through a region which displays a local variation in material toughness [1]. Whereas such a disturbance from straight of the edge of a propagating crack would *a priori* be expected to radiate energy and so attenuate as it spreads along the edge, the distinctive feature of the crack front wave is that it constitutes a disturbance that propagates along the crack edge, without attenuation or dispersion. Thus, the whole disturbance must be localised around the crack edge, radiating nothing into the far field. An analytic solution for the perturbation of the stress near the edge of a crack when its edge deviates slightly from straight was developed by Willis and Movchan [2]. It was used by Ramanathan and Fisher [3] to confirm analytically the existence of the crack front wave. Recent experimental observations [4] suggest that the waves may have the character of solitons, so that nonlinear terms so far neglected in the analysis may also have an influence.

This article presents the theory underlying crack front waves. Implications are then developed explicitly, in greater generality than previously. A non-singular term in the stress that was neglected in the analysis of Ramanathan and Fisher is retained. This term destroys the homogeneity of the dispersion relation and renders the crack front wave dispersive; strictly, the earlier analysis has only asymptotic validity, in the limit of high frequency, when this effect reduces to zero. In addition, the medium through which the crack propagates is allowed to be viscoelastic. This introduces both attenuation and another source of dispersion. The effect of viscoelasticity was studied by Willis and Movchan [5], to the first non-trivial order at high frequency, in the special case of a Maxwell fluid for which an explicit perturbation solution, analogous to that of Willis and Movchan [2], is available, from work of Woolfries, Movchan and Willis [6]. Here, the novelty is that the high-frequency asymptotic calculations are developed explicitly for an arbitrary viscoelastic medium.

The section that follows gives the basic theory, assuming that the required stress analysis has been performed. Then, Section 3 provides this analysis, for a general viscoelastic medium, in the asymptotic limit of high frequencies. Section 4 puts the two together and provides explicit results. A brief concluding Section 5 outlines work in prospect.

2.   CRACK FRONT WAVES: BASIC THEORY

Although the waves observed by Sharon, Cohen and Fineberg [4] show out-of-plane deflection, the simulations of Morrissey and Rice [1] and the theory of Ramanathan and Fisher [3] admit only deflection within the plane of propagation of the crack. Correspondingly, the theory presented here is for the in-plane case. Extension of the reasoning to apply to out-of-plane deflections should be possible, since the basic stress analysis is already available [7], but this is a project for the future.

The basic configuration is that of a semi-infinite crack, propagating with uniform speed V, in the plane $x_3 = 0$. It is loaded symmetrically, that is, under conditions termed Mode I. The basic configuration is slightly perturbed, so that the crack occupies the region

$$S_\varepsilon = \{\mathbf{x} : -\infty < x_1 < Vt + \varepsilon\phi(t, x_2), -\infty < x_2 < \infty, x_3 = 0\}. \quad (26.1)$$

The loading of the crack is assumed to be as follows. If the crack were not present, the loading would generate a stress field $\sigma_{ij}^0(x_1 - Vt, x_3)$. Then, by superposition, the additional stress σ_{ij} induced by the presence of the crack has to satisfy the equations of motion and the boundary conditions

$$\sigma_{i3} = -\sigma_{i3}^0 \quad \text{on} \ \ S_\varepsilon. \quad (26.2)$$

These boundary conditions actually describe the perturbation of any stress field σ_{ij}^0 by the crack. However, in the case of Mode I loading, $\sigma_{13}^0 = \sigma_{23}^0 = 0$ on the plane $x_3 = 0$. The particular dependence, on $x_1 - Vt$, for σ_{ij}^0 is assumed so that it is consistent that the crack could propagate with speed V, with $\varepsilon = 0$. The problem is to determine the subsequent evolution of a perturbation ϕ, once it is introduced by some extraneous means.

It is assumed that, during its motion, the crack maintains the local Griffith energy balance, that the rate of flow of energy into the crack edge is $2\gamma v$ per unit length, where the specific surface energy γ is assumed to be constant, and

$$v = (V + \varepsilon\dot{\phi})/(1 + \varepsilon^2\phi'^2)^{1/2} \quad (26.3)$$

is the local crack speed, where $\dot{\phi}$ denotes $\partial\phi/\partial t$ and ϕ' denotes $\partial\phi/\partial x_2$.

The rate of flow of energy into the crack tip can be found, even for a viscoelastic medium. Stress, strain and strain-rate are all singular at the edge of the crack, and therefore conform locally to the equations of elastodynamics, with moduli and wave speeds corresponding to instantaneous, or high-frequency, response. Thus, from [8] for instance, the rate of flow of energy into the crack edge is $\mathcal{G}v$, where

$$\mathcal{G} = f(v)K_I^2, \quad (26.4)$$

where K_I is the local Mode I stress intensity factor (in the present case, K_{II} and K_{III} are zero), and

$$f(v) = (1 - v^2/a_0^2)^{1/2}v^2/[2\rho b_0^4 R_0(v)]. \quad (26.5)$$

Here, a_0 and b_0 are the high-frequency limits of the speeds a and b of dilatational and shear waves, ρ is the mass density of the material and $R_0(v)$ is the high-frequency limit of the Rayleigh wave function:

$$R(v) = 4(1 - v^2/a^2)^{1/2}(1 - v^2/b^2)^{1/2} - (2 - v^2/b^2)^2, \qquad (26.6)$$

evaluated with a, b replaced by a_0, b_0.

If K_I could be expressed in terms of ϕ, the statement of energy balance

$$\mathcal{G} = 2\gamma \qquad (26.7)$$

would yield a nonlinear functional equation for ϕ. This is too ambitious an undertaking. It is, however, possible to develop an expansion of equation (26.7), in powers of ε. To zeroth order, the stress intensity factor K_I becomes $K_{I,0}$, and v is replaced by V. Since the unperturbed motion is assumed to be possible, it follows that

$$\mathcal{G}_0 \equiv f(V)K_{I,0}^2 = 2\gamma. \qquad (26.8)$$

Thus, $K_{I,0}$ takes a constant value, and the form of the applied loading allows this.

Now to first order in ε, it can be shown [2, 6] that

$$K_I = K_{I,0} + \Delta K_I \sim K_{I,0} + \varepsilon\{Q * (\phi K_{I,0}) + (\pi/2)^{1/2}\phi A_{I,0}\}, \quad (26.9)$$

where the stress component σ_{33} a distance X ahead of the unperturbed crack has the asymptotic form

$$\sigma_{33}(X + Vt, x_2, 0) \sim K_{I,0}/\sqrt{2\pi X} - \sigma_{33}^0(0,0) + A_{I,0}\sqrt{X} \qquad (26.10)$$

as $X \to 0$. The symbol $*$ denotes convolution over x_2 and t. The function $Q(t, x_2)$ is discussed in more detail later. At present, it is noted simply that, in the case of an elastic medium, it is a homogeneous function of degree -1.

Therefore, to first order in ε, the energy balance (26.7) yields

$$\frac{\Delta \mathcal{G}}{\varepsilon \mathcal{G}_0} \equiv 2Q * \phi + \frac{f'(V)}{f(V)}\dot{\phi} + 2m\phi = 0, \qquad (26.11)$$

where $m = (\pi/2)^{1/2}A_{I,0}/K_{I,0}$. The parameter m has the physical dimension $(\text{length})^{-1}$ and thus should be measured relative to some length L characteristic of the pattern of loading. Since equation (26.11) has convolution form, it can be Fourier transformed or, equivalently, a solution can be sought for which ϕ is a constant times $e^{-i(\omega t + kx_2)}$. No solution of such form is possible, unless the dispersion relation

$$2\overline{Q}(\omega, k) - i\omega\frac{f'(V)}{f(V)} + 2m = 0 \qquad (26.12)$$

is satisfied. This is the dispersion relation for crack front waves and is discussed further in the sequel. Here, however, it is noted that, in the limit of high frequencies, it reduces to its elastic form (with $a = a_0$ and $b = b_0$), so $\overline{Q}$ reduces to its elastic form $\overline{Q}_0$, which is a homogeneous function of degree 1. The constant term m becomes negligible and the dispersion relation reduces to

$$2\overline{Q}_0(\omega, k) - \mathrm{i}\omega \frac{f'(V)}{f(V)} = 0. \tag{26.13}$$

Solutions (if any) of (26.13) have the form $\omega/k = u_0$, constant. Ramanathan and Fisher [3] found one real solution $u_0 > 0$, and its negative, $-u_0$, thus confirming that a crack front wave could propagate without attenuation or dispersion, in either direction along the crack front. (In fact, the 'transfer function' that relates $\overline{\Delta \mathcal{G}}$ to $\overline{\phi}$ has a simple zero at $\omega/k = u$, so that its inverse has a simple pole at that point.) The speed u_0 depends on the speed V of the crack, but u_0 exists for all crack speeds V up to the speed of Rayleigh waves. A plot is given later (Fig. 1).

Retention of the term m in (26.12) destroys the homogeneity, even in the case of an elastic medium. The 'crack front wave' becomes dispersive. Viscoelastic material response provides a further source of dispersion and, in addition, induces exponential decay with distance of propagation. These effects are examined, to lowest non-trivial order, in the limit of high frequencies, in the sections that follow.

3.   ANALYSIS

As discussed in [6], the calculation of the function $\overline{Q}$ first requires the Wiener–Hopf factorisation of the function

$$2\tilde{G} = \frac{2\mathrm{i}\omega'^2(\omega'^2/a^2 - \xi_1^2 - \xi_2^2)^{1/2}}{\rho b^4 D(\omega', \xi_1, \xi_2)}, \tag{26.14}$$

where $\omega' = \omega - V\xi_1$, the phase speeds a and b are functions of ω' and

$$D(\omega', \xi_1, \xi_2) = 4|\xi|^2(\omega'^2/a^2 - |\xi|^2)^{1/2}(\omega'^2/b^2 - |\xi|^2)^{1/2} + (\omega'^2/b^2 - 2|\xi|^2)^2.$$

Here we use the notation $|\xi|^2 = \xi_1^2 + \xi_2^2$. The factorisation takes the form

$$2\tilde{G} = U_+(\omega, \xi_1, \xi_2)/\Sigma_-(\omega, \xi_1, \xi_2) \tag{26.15}$$

where, for real ξ_2 and ω lying just above the real axis, the function U_+ is analytic in an upper half of the complex ξ_1-plane, and Σ_- is analytic in a lower half of the same plane. The two half-planes have some overlap, corresponding to the analyticity of the function $\tilde{G}$ in a strip containing the real axis. The function that is required for the present purpose is U_+. It is normalised so that

$$U_+(\omega, z, \xi_2) \sim (2\mathrm{i})^{1/2} z^{-1/2} \quad \text{as} \quad z \to \infty. \tag{26.16}$$

Once U_+ is determined, the function $\overline{Q}$ is defined from its large-z expansion,

$$U_+(\omega, z, \xi_2) \sim \frac{(2\mathrm{i})^{1/2}}{z^{1/2}} \left\{ 1 + \frac{\mathrm{i}\overline{Q}(\omega, \xi_2)}{z} \right\} \quad \text{as} \quad z \to \infty. \tag{26.17}$$

The variable k of the preceding section corresponds to the present ξ_2.

In preparation for performing the factorisation, it is noted that

$$D(\omega', \xi_1, \xi_2) \sim -2(\xi_1^2 + \xi_2^2)\omega'^2 \frac{(a^2 - b^2)}{a^2 b^2} \quad \text{as} \quad \omega' \to 0,$$

while as $\xi_1 \to \infty$,

$$D(\omega - V\xi_1, \xi_1, \xi_2) \sim -\xi_1^4 R_0(V). \tag{26.18}$$

Also, $D(\omega - V\xi_1, \xi_1, \xi_2) = 0$ when

$$(\omega - V\xi_1)^2/c^2 - \xi_1^2 - \xi_2^2 = 0, \tag{26.19}$$

where c is a function of $\omega' = \omega - V\xi_1$ and satisfies the equation

$$R(c) = 0. \tag{26.20}$$

In the absence of explicit formulae for a and b as functions of ω', it is impossible to say exactly how many solutions equation (26.20) may have. However, asymptotically, as $\omega' \to \infty$,

$$a^2 \sim a_0^2[1 - \mathrm{i}f_a/(\omega'\tau)], \quad b^2 \sim b_0^2[1 - \mathrm{i}f_b/(\omega'\tau)] \tag{26.21}$$

for some constants f_a and f_b. Causality requires that f_a and f_b have positive real parts. The parameter τ is a characteristic relaxation time; it is real and positive. The only solutions of equation (26.20) that come close to the real axis are the ones that reduce to $\pm c_0$ as $\omega'\tau \to \infty$, where c_0 is the speed of Rayleigh waves in the case of an elastic medium, with wave speeds a_0 and b_0. Thus, as $\omega' \to \infty$, $c^2 \sim c_0^2[1 - \mathrm{i}f_c/(\omega'\tau)]$, where $R_0(c_0) = 0$ and, by routine expansion of equation (26.20),

$$f_c = \frac{\left(1 - \frac{c_0^2}{a_0^2}\right)^{-1/2}\left(1 - \frac{c_0^2}{b_0^2}\right)^{1/2}\frac{f_a}{a_0^2} + \left(1 - \frac{c_0^2}{a_0^2}\right)^{1/2}\left(1 - \frac{c_0^2}{b_0^2}\right)^{-1/2}\frac{f_b}{b_0^2} - \left(2 - \frac{c_0^2}{b_0^2}\right)\frac{f_b}{b_0^2}}{\left(1 - \frac{c_0^2}{a_0^2}\right)^{-1/2}\left(1 - \frac{c_0^2}{b_0^2}\right)^{1/2}\frac{1}{a_0^2} + \left(1 - \frac{c_0^2}{a_0^2}\right)^{1/2}\left(1 - \frac{c_0^2}{b_0^2}\right)^{-1/2}\frac{1}{b_0^2} - \left(2 - \frac{c_0^2}{b_0^2}\right)\frac{1}{b_0^2}}.$$

In the sequel, c will mean the root that connects to c_0, since roots other than $\pm c$, even if they exist, play no role in the analysis to follow.

It is convenient now to define $T(\omega, \xi_1, \xi_2)$ so that

$$T(\omega, \xi_1, \xi_2) = -\frac{D(\omega - V\xi_1, \xi_1, \xi_2)V^2(1 - V^2/c_0^2)}{(\omega - V\xi_1)^2[(\omega - V\xi_1)^2/c^2 - \xi_1^2 - \xi_2^2]R_0(V)}. \qquad (26.22)$$

When ω has small positive imaginary part, T is analytic and non-zero in a strip containing the real axis in the complex ξ_1-plane. Also, $T \to 1$ as $\xi_1 \to \infty$. Therefore, T can be factorised:

$$T = T_+ T_-. \qquad (26.23)$$

The relevant factor for present purposes is

$$T_+(\omega, z, \xi_2) = \exp\left\{\frac{1}{2\pi i} \int_{-\infty}^{\infty} \ln[T(\omega, \xi_1, \xi_2)]\frac{d\xi_1}{\xi_1 - z}\right\}. \qquad (26.24)$$

This definition applies when z has positive imaginary part; T_+ can be defined elsewhere by analytic continuation. In terms of T,

$$2\tilde{G} = \frac{-2i[(\omega - V\xi_1)^2/a^2 - \xi_1^2 - \xi_2^2]^{1/2}V^2(1 - V^2/c_0^2)}{\rho b^4[(\omega - V\xi_1)^2/c^2 - \xi_1^2 - \xi_2^2]T(\omega, \xi_1, \xi_2)R_0(V)}. \qquad (26.25)$$

The factorisation of $2\tilde{G}$ requires factorisation of $[(\omega - V\xi_1)^2/a^2 - \xi_1^2 - \xi_2^2]^{1/2}$. Asymptotically, as $(\omega - V\xi_1)\tau \to \infty$, this expression reduces to its elastic form, in which the function a is replaced by the constant a_0. The elastic form can be factorised at sight:

$$\left[\frac{(\omega - V\xi_1)^2}{a_0^2} - \xi_1^2 - \xi_2^2\right]^{1/2} = i(1 - V^2/a_0^2)^{1/2}(\xi_1 - \xi_a^{-,0})^{1/2}(\xi_1 - \xi_a^{+,0})^{1/2},$$

where

$$\xi_a^{\pm,0} = \left\{-\frac{\omega V}{a_0^2} \pm \left[\frac{\omega^2}{a_0^2} - \left(1 - \frac{V^2}{a_0^2}\right)\xi_2^2\right]^{1/2}\right\}\left(1 - \frac{V^2}{a_0^2}\right)^{-1}.$$

The required '+' factor of the original function (given as a function of z) is then

$$(z - \xi_a^{-,0})^{1/2}\exp\left\{\frac{1}{4\pi i}\int_{-\infty}^{\infty}\ln\left[\frac{(\omega - V\xi_1)^2/a^2 - \xi_1^2 - \xi_2^2)}{(\omega - V\xi_1)^2/a_0^2 - \xi_1^2 - \xi_2^2)}\right]\frac{d\xi_1}{\xi_1 - z}\right\}. $$
$$ (26.26)$$

The term $[(\omega - V\xi_1)^2/c^2 - \xi_1^2 - \xi_2^2]$ is factorised similarly: a is replaced by c and the square root is not taken (equivalently, the above expression is squared).

The factorisation of $2\tilde{G}$ is now, in effect, complete. The function $b^{-2}(\omega - V\xi_1)$ has to be, for causality, an analytic function in the upper half-plane of its argument and so is a '$-$' function with respect to the variable ξ_1. The task now is to obtain, more explicitly, the high-frequency asymptotic form of $2\tilde{G}$, which is obtained, formally, by letting $\tau \to \infty$.

Considering first the expression (26.26), expansion of the integrand, taking account of (26.21), gives for the exponential,

$$\exp\left\{\frac{-1}{2\pi i}\int_{-\infty}^{\infty}\frac{i f_a \omega(\omega - V\xi_1)\mathrm{d}\xi_1}{2a_0^2(\xi_1 - \xi_a^{+,0})(\xi_1 - \xi_a^{-,0})(1 - V^2/a_0^2)(\xi_1 - z)}\right\} \tag{26.27}$$

which leads, upon closing the contour in the lower half-plane and invoking Cauchy's theorem, to

$$(z - \xi_a^{-,0})^{1/2}\left[1 - \frac{i\xi_a^{-,1}}{2(z - \xi_a^{-,0})\omega\tau}\right], \tag{26.28}$$

where

$$\xi_a^{-,1} = \frac{-f_a}{2a_0^2\left(1 - V^2/a_0^2\right)}\left[\omega V + \omega^2\left[\frac{\omega^2}{a_0^2} - \left(1 - \frac{V^2}{a_0^2}\right)\xi_2^2\right]^{-1/2}\right].$$

This is entirely consistent with the expansion of $(z - \xi_a^-)^{1/2}$, with

$$\xi_a^- \sim \xi_a^{-,0} + i\xi_a^{-,1}/(\omega\tau), \tag{26.29}$$

which can be obtained by direct expansion and factorisation of $[(\omega - Vz)/a^2 - z^2 - \xi_2^2]$. The treatment of the corresponding factor of $[(\omega - V\xi_1)^2/c^2 - \xi_1^2 - \xi_2^2]$ is similar. The expansion of T_+ follows the same pattern but is somewhat more complicated. Specifically,

$$\begin{aligned}
T_+ &= (T_0)_+ \exp\left\{\frac{1}{2\pi i}\int_{-\infty}^{\infty}\ln\left(\frac{T}{T_0}\right)\frac{\mathrm{d}\xi_1}{\xi_1 - z}\right\}\\[2mm]
&= (T_0)_+ \exp\left\{\frac{1}{2\pi i}\int_{-\infty}^{\infty}\ln\left(\frac{D[(\omega - V\xi_1)^2/c_0^2 - |\xi|^2]}{D_0[(\omega - V\xi_1)^2/c^2 - |\xi|^2]}\right)\frac{\mathrm{d}\xi_1}{\xi_1 - z}\right\}\\[2mm]
&\sim (T_0)_+ \exp\left\{\frac{1}{2\pi i}\int_{-\infty}^{\infty}\left(\frac{D_1}{D_0\omega\tau}\right.\right.\\[2mm]
&\qquad\left.\left. + \frac{i f_c \omega(\omega - V\xi_1)}{c_0^2\omega\tau(\xi_1 - \xi_c^{+,0})(\xi_1 - \xi_c^{-,0})(1 - V^2/c_0^2)}\right)\frac{\mathrm{d}\xi_1}{\xi_1 - z}\right\}. \tag{26.30}
\end{aligned}$$

Here, $(T_0)_+$ and D_0 are the corresponding expressions for an elastic medium, with wave speeds a_0, b_0. Also, the expansion of D is written

$$D \sim D_0 + iD_1/(\omega\tau) \tag{26.31}$$

where, by routine expansion,

$$
\begin{aligned}
D_1 &= \left[4|\xi|^2 \left\{ \left[\frac{(\omega - V\xi_1)^2}{a_0^2} - |\xi|^2 \right]^{-1/2} \left[\frac{(\omega - V\xi_1)^2}{b_0^2} - |\xi|^2 \right]^{1/2} \frac{f_a}{2a_0^2} \right. \right. \\
&\quad \left. + \left[\frac{(\omega - V\xi_1)^2}{a_0^2} - |\xi|^2 \right]^{1/2} \left[\frac{(\omega - V\xi_1)^2}{b_0^2} - |\xi|^2 \right]^{-1/2} \frac{f_b}{2b_0^2} \right\} \\
&\quad \left. + 2 \left(\frac{(\omega - V\xi_1)^2}{b_0^2} - 2|\xi|^2 \right) \frac{f_b}{b_0^2} \right] \omega(\omega - V\xi_1).
\end{aligned}
$$

It should be noted that the integrand in the last line of (26.30) apparently has a singularity in the lower half-plane, at $\xi_1 = \xi_c^{-,0}$. It is, however, only apparent because when ξ_1 is close to $\xi_c^{-,0}$,

$$
D_0 \sim (\xi_1 - \xi_c^{-,0}) \frac{\partial D_0}{\partial \xi_1}, \tag{26.32}
$$

the partial derivative being evaluated at $\xi_c^{-,0}$. However,

$$
D(\omega - V\xi_c^-, \xi_c^-, \xi_2) \equiv 0, \tag{26.33}
$$

where $\xi_1 = \xi_c^{\pm}$ give exact zeros of $D(\omega - V\xi_1, \xi_1, \xi_2)$ and correspond to $R(c) = 0$. Also,

$$
D_0(\omega - V\xi_c^{-,0}, \xi_c^{-,0}, \xi_2) = 0. \tag{26.34}
$$

Since, asymptotically, $\xi_c^- \sim \xi_c^{-,0} + i\xi_c^{-,1}/(\omega\tau)$, it follows by expansion of (26.33) that

$$
\frac{i}{\omega\tau} \frac{\partial D}{\partial \xi_1} \xi_c^{-,1} + \frac{iD_1}{\omega\tau} = 0, \tag{26.35}
$$

the functions being evaluated at $\xi_1 = \xi_c^{-,0}$. Since the singularity at $\xi_1 = \xi_c^{-,0}$ is only apparent, the contour of integration can be deformed to one that encloses the branch cut in the lower half-plane, associated with the function $D_0(\omega - V\xi_1, \xi_1, \xi_2)$. This cut connects $\xi_a^{-,0}$ and $\xi_b^{-,0}$ (the latter being defined by replacing a_0 with b_0). The contour was called C_- in [2].

Having closed and deformed the contour in this way, it is now possible finally to obtain an expression for the large-τ expansion of $\overline{Q}$, by taking $z \to \infty$. The result is

$$
\overline{Q} \sim \overline{Q}_0 + i\overline{Q}_1/(\omega\tau), \tag{26.36}
$$

where $\overline{Q}_0$ is the purely elastic term, obtained in [2], and

$$
i\overline{Q}_1 = -\tfrac{1}{2}\xi_a^{-,1} + \xi_c^{-,1} - \frac{1}{2\pi i} \int_{C_-} \frac{D_1}{D_0} \, d\xi_1. \tag{26.37}
$$

The contour C_- is followed anti-clockwise. The term multiplying f_c that rendered the singularity only apparent has no explicit influence in (26.37) because it is continuous across the brach cut and so contributes nothing to the integral around C_-.

The special case of a Maxwell fluid is recovered by setting $f_a = f_b = 1$ so that, also, $f_c = 1$. The results obtained here then agree with those obtained directly for the Maxwell fluid in [5].

4.   RESULTS

The dispersion relation (26.12) will be investigated by regarding it as an equation for k, given ω. When $\omega\tau$ is large, it becomes, asymptotically,

$$2\overline{Q}_0(1, k/\omega) - \mathrm{i}\frac{f'(V)}{f(V)} + \frac{2}{\omega\tau}\left[\mathrm{i}\overline{Q}_1(1, k/\omega) + m\tau\right] \sim 0, \qquad (26.38)$$

having exploited the fact that $\overline{Q}_0$ and $\overline{Q}_1$ are homogeneous functions of degree 1. This equation can be solved by iteration, when the frequency is sufficiently high, to give

$$\frac{k}{\omega} \sim \pm\frac{1}{u_0}\left(1 + \frac{\mathrm{i}k_1}{\omega\tau}\right), \qquad (26.39)$$

where

$$2\overline{Q}_0(1, u_0^{-1}) - \mathrm{i}\frac{f'(V)}{f(V)} = 0 \qquad (26.40)$$

and

$$k_1 = u_0\frac{\mathrm{i}m\tau - \overline{Q}_1(1, u_0^{-1})}{\overline{Q}_0'(1, u_0^{-1})}, \qquad (26.41)$$

with $\overline{Q}_0'(\omega, k) = \partial\overline{Q}_0(\omega, k)/\partial k$.

Equation (26.40) defines the speed u_0 of a high-frequency crack front wave, as a function of crack speed V. In this context, high frequency means that $\omega L/c_0 \gg 1$ and $\omega\tau \gg 1$, where L is a characteristic length associated with the pattern of loading. The medium responds as though it is purely elastic, with longitudinal and shear wave speeds a_0 and b_0. The first striking fact is that $2\overline{Q}_0(1, u^{-1}) - \mathrm{i}f'(V)/f(V)$ is real for all u in the range $[0, (c_0^2 - V^2)^{1/2})$; this can be confirmed by analysis [5]. The next is that (26.40) has exactly one solution in $[0, (c_0^2 - V^2)^{1/2})$ for all V in $[0, c_0)$. This was found by calculation, first by Ramanathan and Fisher [3] and confirmed here in Fig. 1, which shows a plot of u_0/c_0 against V/c_0, where c_0 is the speed of high-frequency Rayleigh waves, for values of V up to c_0, for a medium with $a_0^2 = 3b_0^2$, corresponding to a high-frequency Poisson's ratio of 0.25. The crack front wave travels

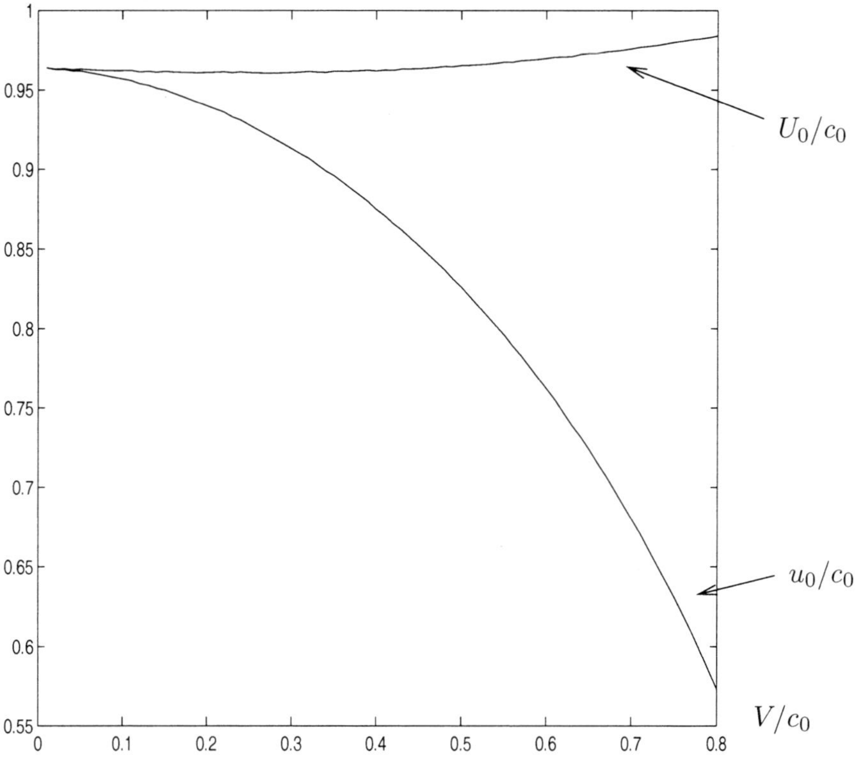

Figure 1   The normalized crack front wave speeds u_0/c_0 and U_0/c_0 versus V/c_0; for this computation we used $c_0 = 1.0, a_0 = 1.857, b_0 = 1.072$.

along the crack at speed u_0; since the crack front itself travels with speed V, the resultant speed of the disturbance relative to the medium is, by Pythagoras' theorem, $U_0 = (u_0^2 + V^2)^{1/2}$. The figure also shows a plot of U_0/c_0. It is not constant, but U_0 is close to c_0 for all V.

Turning now to the perturbed solution (26.39), it is remarked first that $\overline{Q}_0'(1, u^{-1})$ is real and negative for $u \in [0, c_0)$. If the medium is purely elastic, $\overline{Q}_1 = 0$ and k_1 is purely imaginary, giving dispersion of the crack front wave. Viscoelasticity introduces further dispersion and also attenuation.

Fig. 2 shows sample plots of the real and imaginary parts of $\overline{Q}_1(1, u^{-1})$ against u/c_0, for the case $f_a = f_b = 1$. With appropriate choice of τ, this is realised by any viscoelastic material for which a/b is constant, corresponding to a Poisson's ratio independent of frequency. The plot is for the case $a_0^2 = 3b_0^2$ and $V/c_0 = 0.625$. Fig. 3 shows a plot of the

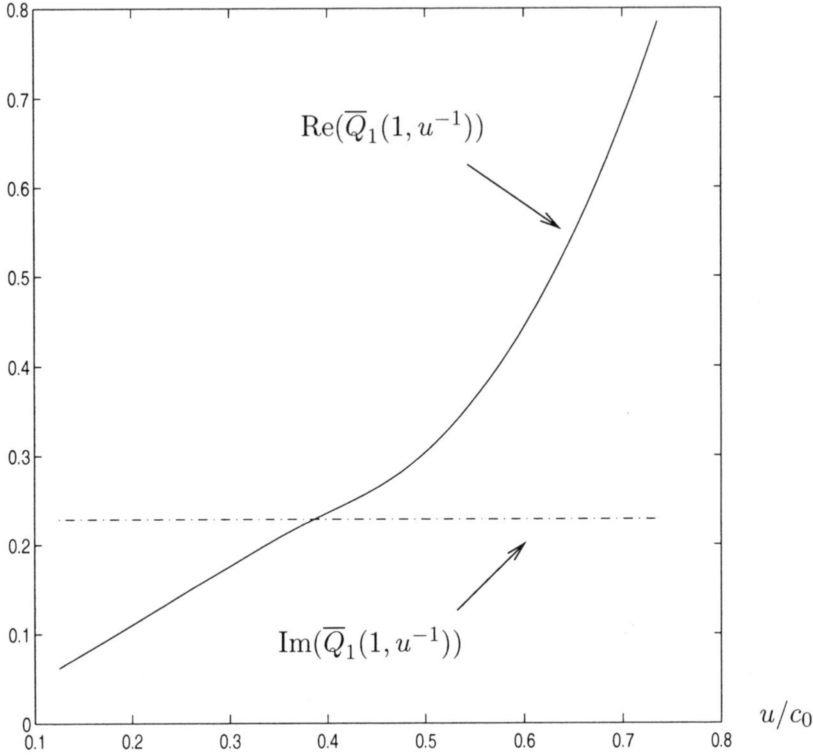

Figure 2   Real and imaginary parts of $\overline{Q}_1(1, u^{-1})$; here we used $V/c_0 = 0.625$.

real part of k_1, the real part of $\overline{Q}_1(1, u_0^{-1})$ and the function $\overline{Q}_0'(1, u_0^{-1})$ against normalised crack speed V/c_0 for the same material.

The interest of this result is that the amplitude of the crack front wave decays in proportion to $\exp[-\text{Re}(k_1)d/u_0\tau]$, where d is distance of propagation, independently of the frequency, except for the 'high-frequency' restriction that $\omega L/c_0 \gg 1$ and $\omega\tau \gg 1$.

The graph in Fig. 3 shows that $\text{Re}(k_1)$ is constant. In fact, it can be verified analytically that the real part of k_1 equals 0.5 when $f_a = f_b = 1$. Since the function $\overline{Q}_0$ is homogeneous, of degree 1, we can write

$$\omega\frac{\partial\overline{Q}_0}{\partial\omega} + k\frac{\partial\overline{Q}_0}{\partial k} = \overline{Q}_0.$$

Next, we know that $\overline{Q}_0 - \mathrm{i}\omega f'(V)/2f(V)$ is real when $0 \le \omega/k < (c_0^2 - V^2)^{1/2}$, so that $\partial\overline{Q}_0/\partial k$ is real in the same range. When $\omega/k = u_0$, we

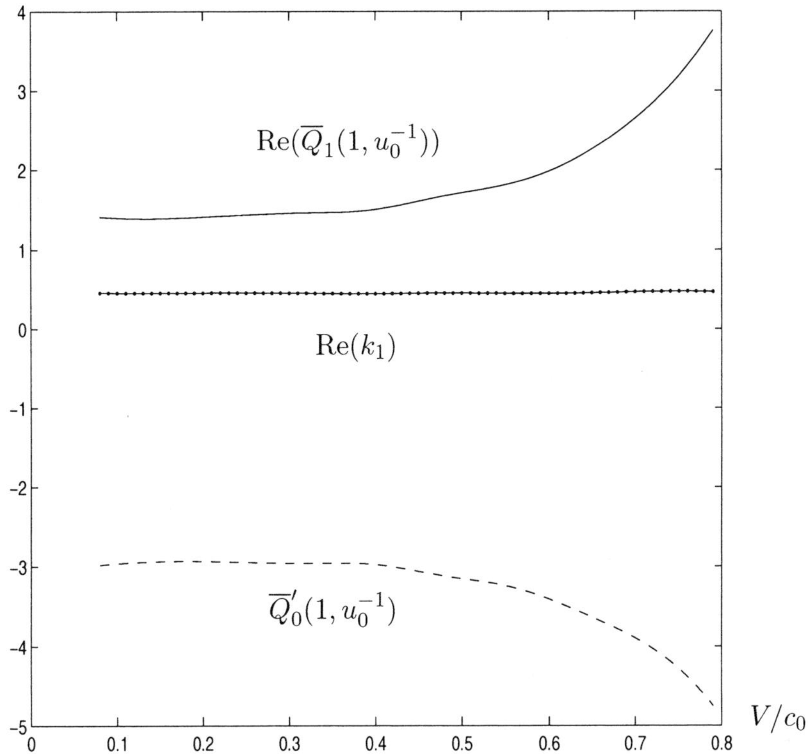

Figure 3   Real part of k_1, real part of $\overline{Q}_1(1, u_0^{-1})$ and the function $\overline{Q}_0^{'}(1, u_0^{-1})$ versus V/c_0; for this computation we used $c_0 = 1.0, a_0 = 1.857, b_0 = 1.072$.

have

$$\overline{Q}_0 - \mathrm{i}\omega f'(V)/2f(V) = 0,$$

and

$$\mathrm{Re}\left\{ \omega \frac{\partial \overline{Q}_0}{\partial \omega} + k \frac{\partial \overline{Q}_0}{\partial k} \right\} = 0.$$

It is also true that, when $f_a = f_b = 1$,

$$\overline{Q}_1 = \tfrac{1}{2}\omega \frac{\partial \overline{Q}_0}{\partial \omega}.$$

Hence,

$$\mathrm{Re}\{\overline{Q}_1\} = -\tfrac{1}{2}k \frac{\partial \overline{Q}_0}{\partial k}$$

when $\omega/k = u_0$, and it yields that $\mathrm{Re}(k_1) = 0.5$.

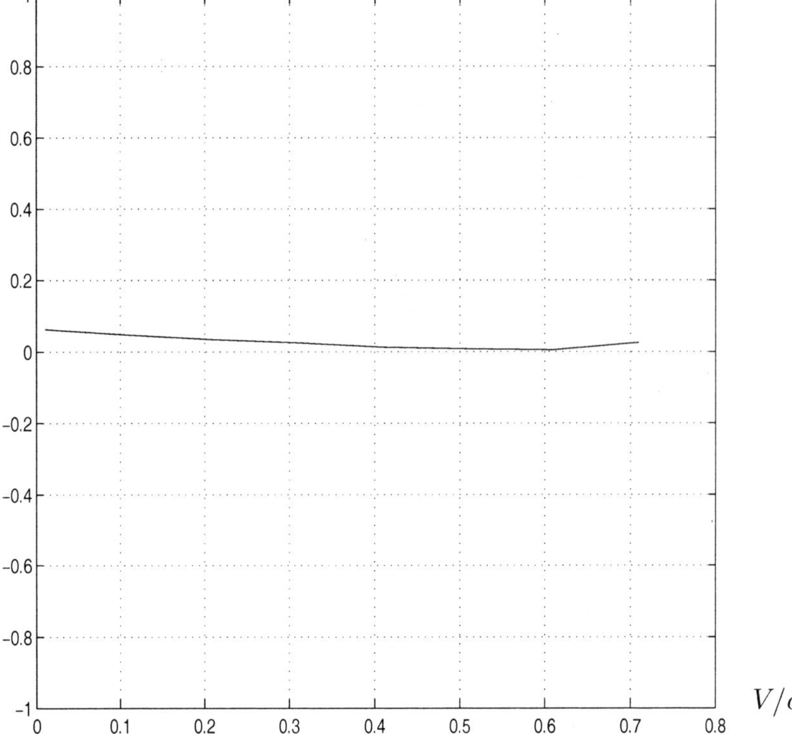

Figure 4   Real part of k_1 versus V/c_0 for the case when $f_a = 1$, $f_b = 0$; we used $c_0 = 1.0, a_0 = 1.857, b_0 = 1.072$.

When $f_a \neq f_b$, the quantity $\mathrm{Re}(k_1)$ is not constant, but its variation is small. The term D_1/D_0 in (26.37) can be written as follows:

$$
\frac{D_1}{D_0} = \frac{\omega(\omega - V\xi_1)}{D_0} \left\{ \frac{f_a[D_0 - \{(\omega - V\xi_1)^2/b_0^2 - 2|\xi|^2\}^2]}{2a_0^2((\omega - V\xi_1)^2/a_0^2 - |\xi|^2)} \right.
$$
$$
+ \frac{f_b[D_0 - \{(\omega - V\xi_1)^2/b_0^2 - 2|\xi|^2\}^2]}{2b_0^2((\omega - V\xi_1)^2/b_0^2 - |\xi|^2)}
$$
$$
\left. + 2[(\omega - V\xi_1)^2/b_0^2 - 2|\xi|^2]f_b/b_0^2 \right\}.
$$

If we consider the particular case $f_b = 0$, the quantity $\mathrm{i}\overline{Q}_1$ reduces to

$$
\mathrm{i}\overline{Q}_1 = -\tfrac{1}{2}\xi_a^{-,1} + \xi_c^{-,1} + f_a(2\pi\mathrm{i})^{-1}I, \tag{26.42}
$$

where

$$I = \int_{C_-} \frac{F(\omega, \xi_1, \xi_2)}{D_0(\omega - V\xi_1', \xi_1', \xi_2)} d\xi_1 \qquad (26.43)$$

and

$$F(\omega, \xi_1, \xi_2) = \frac{\omega(\omega - V\xi_1)}{2a_0^2} \frac{((\omega - V\xi_1)^2/b_0^2 - 2|\xi|^2)^2}{(\omega - V\xi_1)^2/a_0^2 - |\xi|^2}, \qquad (26.44)$$

since only the part of the integrand which contains D_0 in its numerator is discontinuous across the branch cut.

Taking into account that k_1 is a linear function of f_a and f_b, its real part, $k_1^R(f_a, f_b) = \text{Re}(k_1)$, can be expressed as

$$k_1^R(f_a, f_b) = \tfrac{1}{2}f_b + (f_a - f_b)k_1^R(1, 0),$$

since, as already noted, $k_1^R(1, 1) = 0.5$. Fig. 4 shows a plot of $k_1^R(1, 0)$ against crack speed V. The method of evaluating I in (26.42) is outlined in the Appendix.

5.   FURTHER POSSIBILITIES

The perturbation theory that has been exploited here can, in fact, be carried out to higher order. A sample equation, for a 'weakly-nonlinear' wave, was obtained in [5] by developing the perturbation theory to second order in ε. The real need, however, is to allow for deflections out of the plane $x_3 = 0$. The basic perturbation solution has been obtained by Willis and Movchan [7] in the case of elasticity, and the corresponding viscoelastic analysis is under development.

Acknowledgments

Partial support for this research from ONR (Grant reference N00014-1042) is gratefully acknowledged.

Appendix: The integral I

The integral I, defined by (26.43), is around C_-, a contour enclosing the branch cut joining $\xi_a^{-,0}$ and $\xi_b^{-,0}$. We note that the function F, defined by (26.44), takes the same values on both sides of the branch cut. The branch cut can be parametrized by

$$\xi_1' = \xi_a^{-,0} + s(\xi_b^{-,0} - \xi_a^{-,0}), \quad 0 \le s \le 1.$$

The quantity D_0 is equal to $Y - X$ above the branch cut and $Y + X$ below the branch cut, where

$$X = 4\mathrm{i}\alpha\beta(\xi_1' - \xi_a^{+,0})^{1/2}(\xi_1' - \xi_b^{+,0})^{1/2}(\xi_1'^2 + \xi_2^2)(\xi_a^{-,0} - \xi_b^{-,b})\sqrt{s(1-s)},$$

and $Y = \{\xi_1' + \xi_2 + \beta^2(1-s)(\xi_1' - \xi_b^{+,0})(\xi_a^{-,0} - \xi_b^{-,0})\}^2$; the coefficients α and β are defined by $\alpha = (1 - V^2/a_0^2)^{1/2}$ and $\beta = (1 - V^2/b_0^2)^{1/2}$. Finally, the integral I is represented in the form

$$I = (\xi_b^{-,0} - \xi_a^{-,0}) \int_0^1 F(\omega, \xi_1'(s), \xi_2)\frac{2X}{Y^2 - X^2}ds.$$

References

[1] Morrissey, J W and Rice, J R (1998) *Crack front waves*, J. Mech. Phys. Solids **46**, 467–487.

[2] Willis, J R and Movchan, A B (1995) *Dynamic weight functions for a moving crack. I. Mode I loading*, J. Mech. Phys. Solids **43**, 319–341.

[3] Ramanathan, S and Fisher, D S (1997) *Dynamics and instabilities of planar tensile cracks in heterogeneous media*, Phys. Rev. Lett. **79**, 877–880.

[4] Sharon, E, Cohen, G and Fineberg, J (2000) *Crack front waves: localized solitary waves in dynamic fracture*, preprint, Racah Institute of Physics, Hebrew University of Jerusalem.

[5] Willis, J R and Movchan, A B (2001) *The influence of viscoelasticity on crack front waves*, J. Mech. Phys. Solids **49**, 2177–2189.

[6] Woolfries, S, Movchan, A B and Willis, J R (2001) *Perturbation of a dynamic planar crack moving in a model viscoelastic solid*, to be submitted.

[7] Willis, J R and Movchan, A B (1997) *Three-dimensional dynamic perturbation of a propagating crack*, J. Mech. Phys. Solids **45**, 591–610.

[8] Freund, L B (1990) *Dynamic Fracture Mechanics*, Cambridge: University Press.

VI

FLUID STRUCTURAL WAVE INTERACTION

FORMATION OF AN UNKNOWN DISCRETE SPECTRUM IN A GENERAL SPECTRUM FOR STRUCTURES INTERACTING WITH A FLUID

A. K. Abramian

Institute for Problems in Mechanical Engineering, RAS

Bolshoy pr. V.O. 61, St. Petersburg 199178, Russia

abramian@math.ipme.ru, andrei@dv.twi.tudelft.nl

Abstract   During investigation of noise levels from large structures, such as ship hulls, attention is usually drawn to obvious strong vibration mechanisms. However, unexpected noise levels at certain frequencies may be due to radiation from structural details (e.g. joints or walls) when the structure is operating. Three possible reasons causing the existence of such frequencies have been investigated.

1.   LOCALIZED MODES OF ELASTIC STRUCTURES

Infinite length structures with elastic mass inclusions have both continuous and real discrete spectrums [1]. A real discrete spectrum can be found above a cut-off frequency of a system as well as below it. Location of this spectrum depends on the parameters of the structure and its inclusions. The real discrete spectrum corresponds to localized modes containing finite energy (*trapped modes* as they were called in [2]).

It has been found in [1] that the phenomenon of trapped modes explains some cases of noise levels at certain frequencies in finite length structures. The author has sequentially solved the problems of the evaluation of the spectrums for the following hinged structures:

- a finite length beam on the elastic foundation with two similar elastic supports

- a circular cylindrical shell supported by two circular frames

To solve these problems, the solutions for finite length structures were expressed in terms of Green's function for the similar infinite length structures. The solutions found lead to the following results:

253

I.D. Abrahams et al. (eds.),

IUTAM Symposium on Diffraction and Scattering in Fluid Mechanics and Elasticity, 253–260.

© 2002 *Kluwer Academic Publishers. Printed in the Netherlands.*

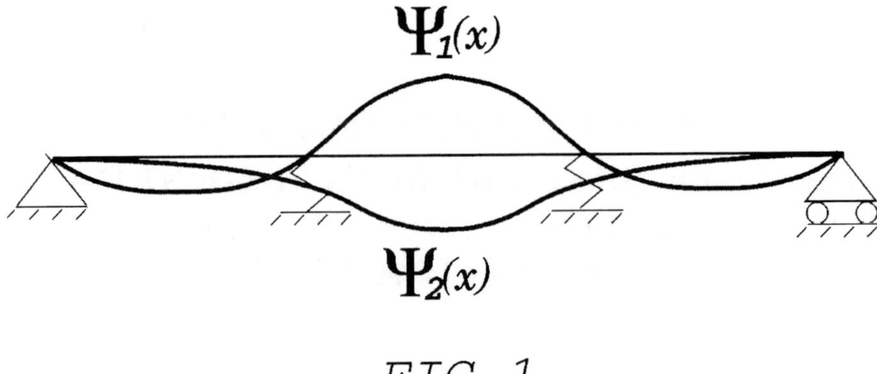

FIG 1

1 There is a real discrete spectrum above the cut-off frequency $\omega^2 = k/m$, where k is the elastic foundation rigidity and m is the mass per unit length of the beam. The spectrum frequencies can be found from the expression

$$\omega_n^2 = D(k + (\pi n/l_1)^4)/m, \quad n = 1, 3, \ldots \qquad (27.1)$$

where D is the beam rigidity and $2l_1$ is the distance between the spring supporters. Obtained values of these frequencies coincide with the trapped frequencies of the infinite length system.

2 For the frequencies from (27.1) to be natural, it is necessary to find the corresponding rigidity spectrum of the supporters.

3 Trapped frequencies become multiple the second order natural frequencies of a beam of length $2l$ under the following condition: $l/l_1 = (2j + 1)/n$ $(j = 1, 2, \ldots)$.

4 Multiple frequency corresponds to two modes: the first mode has nodes at the points of the springs' attachment to the beam, the second mode is localized between the springs. Diagrams of the modes are given in Fig. 1.

A related but more complicated solution has been found for a cylindrical shell. For this case, the spectrum of frequencies corresponding to the trapped modes lies in the range $0 < \omega_{\text{trapped}} < c_0/r$, where c_0 is the speed of sound in the shell material and r is the shell radius.

At the found trapped frequencies, the longitudinal forces cause the vibration in both longitudinal and transverse directions. Modes of such vibration are shown in Fig. 2.

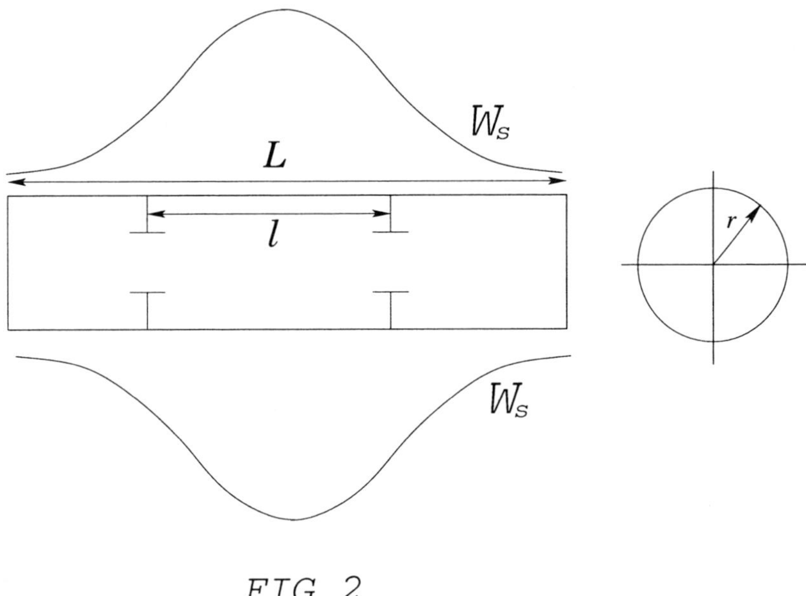

FIG 2

Both above mentioned systems have the following peculiarity in common: under harmonic force loading the main radiating mode becomes a localized one. The reason is that the generalized force corresponding to this mode is much greater than the generalized forces corresponding to the multinode mode. The trapped frequency in the general spectrum is not the lowest and usually it is not taken into account in calculations.

The oscillation mode is *a priori* assumed to be multimode, and the liquid to be incompressible (effect of an added mass). The trapped mode causes considerable radiation for its localized shape.

2. CHOICE OF RHEOLOGICAL MODEL OF THE LIQUID

A choice of a rheological model of the liquid greatly influences structural radiation. Let us consider the axisymmetric oscillation of a system consisting of two coaxial cylindrical shells submerged into a compressible inviscid liquid. Vibration is caused by kinematic excitation of the inner shell: $w(x,t) = w_0 \exp\{i(\alpha x - \omega t)\}$. If H is the distance between the shells, r_1 is the radius of the inner shell and it satisfies the inequality $H/r_1 \ll 1$, the liquid layer between the shells can be considered as a plane one. The analytical solution gives us an expression for the liquid pressure in the space between the shells and outside of the system. Modulus of the system transfer function $|K(i\omega)| = |w_{20}/w_{10}|$ (where ω is the frequency of excitation and w_{10} and w_{20} are the shell's displacements) determines the influence of the outer shell on the radiation pressure. Figs. 3 and 4 show the dependence of $|K(i\omega)|$ on ω for compressible and incompressible liquids, respectively, in the plane layer.

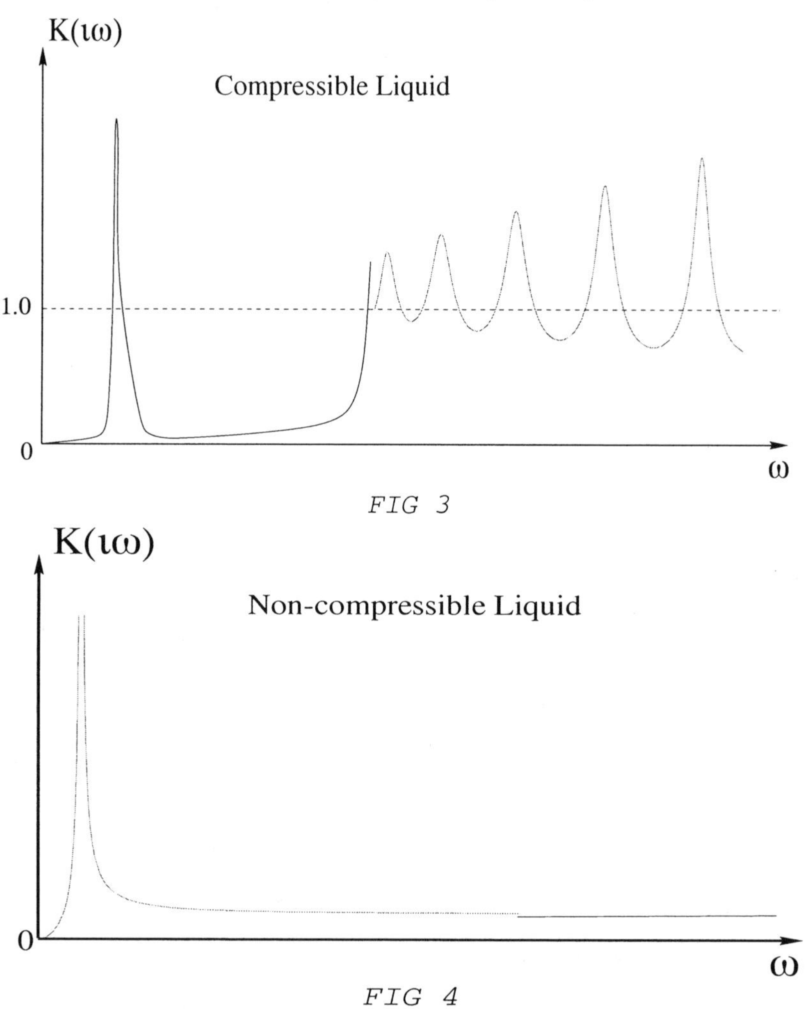

FIG 3

FIG 4

Let us study $|K(i\omega)|$ in three frequency bands. The first band is located between zero and the resonance frequency of the liquid layer. The liquid layer behaves itself as a damper. The same behaviour is seen in the model of the incompressible liquid in this frequency range. The second frequency band lies between the resonance frequency and the cut-off frequency of the system (the cut-off frequency is the frequency with which the outer shell radiation starts).

Provided that the liquid compressibility is taken into account, the outer shell behaves itself as a damper. Application of incompressible liquid model leads to the conclusion that the outer shell is a damper when $\alpha < \omega^2/c_0^2$ (c_0 is the sound speed in liquid), when $\alpha > \omega^2/c_0^2$ both models suggest that the outer shell behaves as a damper. In the third frequency band above the cut-off frequency, application of compressible liquid model leads to describe components in the radiation spectrum, which can not be found in the case of incompressible liquid. The outer shell behaves itself as an oscillation amplifier. Noise radiation of such a structure should be determined in view of a liquid layer compressibility.

3.   GEOMETRIC NONLINEARITY OF A STRUCTURE INTERACTING WITH A LIQUID

Let us study a problem of a nonlinear parametric oscillation of a beam contacting with a semi-infinite channel filled with compressible liquid. In the two-mode Galerkin approximation, equations describing the structure have the form obtained in [3]:

$$\ddot{w}_1 + \Omega_1^2(1 - k_1 \cos\omega t)w_1 + \varepsilon_1 \dot{w}_1 + k_1 w_1^3 + k_2 w_1 w_2^2 = f_1 \cos\omega t,$$
$$\ddot{w}_2 + \Omega_2^2(1 - k_2 \cos\omega t + \gamma)w_2 + \varepsilon_2 \dot{w}_2 + k_3 w_2^3 + k_4 w_2 w_1^2 = f_2 \cos\omega t,$$

where ε_1, ε_2 and γ are magnitudes corresponding to the liquid. This system may be solved only numerically. Computer calculation gives us the possibility to analyze in the paramater space the time evolution of the system and corresponding transition between different regimes: deterministic dynamics $\rightarrow$ quasiperiodic dynamics $\rightarrow$ chaotic dynamics.

At chaotic regimes the structure radiates minimum energy because these regimes contain infinite number of frequencies and each frequency possesses some amount of energy.

4.   ELASTIC NONLINEAR SYSTEMS

Consider the following simplified model (the beam with nonlinear and linear inclusions)

$$Du_{xxxx} + mu_{tt} + Ku + f_\epsilon(x)u + \gamma g_\epsilon u^3 = P(t, x) \qquad (27.2)$$

where $x \in (-\infty, \infty)$. To simplify, firstly let us assume the external force $P = 0$. It is well known that, in the linear case ($g = 0$), for appropriate coefficients f, there exist "localized" and almost localized modes describing special beam oscillations. For example, if we take two inclusions $f = \delta(x) + \delta(x - l)$, then one obtains the solution of the form

$$u = [A\cos(\omega t) + B\sin(\omega t)]\Psi(x, \omega) \qquad (27.3)$$

where

$$\Psi = \exp(-a|x|) + \exp(-a|x - l|) + \lambda(\omega)(\cos(ax) + \cos(a(x - l))),$$

$a(\omega) = (m\omega^2 - K)^{\frac{1}{4}}$ and λ is small, if the "resonance condition" is satisfied: $la \approx \pi(2n + 1)$, $n \in \mathbf{Z}$. This mode is the sum of exponentially damped contribution and the small harmonic tail $\cos(ax) + \cos(a(x - l))$. The pure localized mode corresponds to the case $\lambda = 0$.

The aim of this section is to investigate nonlinear evolution of such modes and show the possibility of the coherent structure formation.

5.   CASE OF A SINGLE MODE

In this section, the case of a single almost localized mode, (27.3), is considered. The natural and classical approach for small nonlinearities (assume that γ is a small parameter) is to suppose that these amplitudes depend on slow time $\tau = \gamma t$. Then one can apply the well known Whitham principle. For the localized modes a simple approach leading to *ordinary* differential equations for slowly oscillating amplitudes $A(\tau)$ and $B(\tau)$ exists.

Let us define the Lagrangian

$$L = \frac{1}{2} \int_0^1 (mu_t^2 - Du_{xx}^2 - Ku^2 - f_\epsilon(x)u^2 - \frac{1}{2}\gamma g_\epsilon u^4)\, dx.$$

Substituting (27.3) into this expression and integrating over "fast" variables x and t, the averaging Whitham Lagrangian $\bar{L}$ is obtained. Up to small corrections, one finds

$$\bar{L} = \frac{1}{2}c \int_0^1 \left[\omega\left(\frac{dA}{d\tau}B - \frac{dB}{d\tau}A\right) + c_1(A^2 + B^2)^2 + c_2(A^2 + B^2)\right] d\tau.$$

The corresponding equations of motion can be easily resolved and they describe slow oscillations of A and B:

$$A = \alpha\cos(\bar{\omega}(\tau - \tau_0)), \quad B = \alpha\sin(\bar{\omega}(\tau - \tau_0)).$$

The frequency $\bar{\omega}$ is proportional to the quantity $A^2 + B^2$ (that does not depend on time τ).

Finally, we see that, in this simplest case, the slow time evolution is defined by an integrable Hamiltonian system with a single degree of freedom.

6.   CASE OF SEVERAL LOCALIZED MODES

If the coexistence of several modes is possible in system (27.2), then the mathematical approach above holds but the results are more interesting. Suppose modes Ψ_1, Ψ_2, ..., Ψ_n coexist, with corresponding frequencies $\omega_1, \omega_2, \ldots$. Let $A_i(\tau)$ and $B_i(\tau)$ be the corresponding slowly oscillating modes. A nontrivial Whitham Lagrangian (with nonlinear contributions that define the mode interactions) can occur if the time resonance condition $\sigma = \omega_{i_1} + \omega_{i_2} + \omega_{i_3} + \omega_{i_4} = 0$ or $\sigma = \omega_{i_1} + \omega_{i_2} - \omega_{i_3} - \omega_{i_4} = 0$ holds, for some ω_{i_k} from the frequency set.

Repeating the calculations from the previous point, we find that the Whitham Lagrangian has the following general form

$$
\bar{L} = \frac{1}{2} \int_0^1 \left\{ \sum_{i=1}^n c \left[\omega \left(\frac{dA_i}{d\tau} B_i - \frac{dB_i}{d\tau} A_i \right) + c_1 (A_i^2 + B_i^2)^2 \right. \right.
$$
$$
\left. + c_2 (A_i^2 + B_i^2) \right] + \lambda^4 \sum_{i_1, i_2, i_3, i_4 : \sigma = 0} c_{i_1, i_2, i_3, i_4} A_{i_1} A_{i_2} A_{i_3} A_{i_4} + \cdots
$$
$$
\left. + d_{i_1, i_2, i_3, i_4} B_{i_1} B_{i_2} B_{i_3} B_{i_4} \right\} d\tau,
$$

where the second sum describes different contributions connected with the fourth-order resonances $\sigma = 0$. The corresponding equations for A_i and B_i, in general, are not integrable. They can be analyzed by the KAM theory which can be used due to the condition $\lambda \ll 1$. Due to classical results, we know that they can describe quasiperiodic motions and moreover, different chaotic occurrences (for example, Arnold's web, homoclinic structures etc.)

To conclude this section, let us note an important point. We can change the form of $\bar{L}$ by a frequency and coefficient f, g choice. Thus, in a sense, these coherent structures are controllable.

7.   NONLINEAR ELASTIC BODIES INTERACTING WITH FLUID

As the simplest model, the beam (27.2) in a fluid flow is considered. For simplicity this flow is taken in acoustic approximations

$$
\Delta \phi(x, y) = c_0^{-2} \phi_{tt}, \quad \phi_y'(x, y)|_{y=0} = u_t,
$$

and the force P in (27.2) has the form $P = \rho_0 \phi_t(x, 0)$. The influence of the fluid can be easily taken into account in the limit of large frequencies $\omega_j \gg 1$. Then for ϕ one has the following asymptotic expansion $\phi = \sum_j \phi_j(x, \omega, y, \tau) \cos(\omega_j t)$, where $\phi_j \approx c \Psi_j(x) \exp(i \omega_j y) + \cdots$. Thus, one can show that, for high frequencies, the term P in (27.2) can replaced by the usual simple dissipative term $c u_t$.

Repeating all the usual procedure of two scale expansion, we can obtain some complicated equations for amplitudes A_i and B_i. They contains some dissipative contributions.

The investigation shows that the coherent structures lead to slowly damped quasiperiodic waves (with slowly oscillating amplitudes) which exist in the beam and the fluid.

The nontrivial structures occur as a result of the interaction of localized modes through their tails.

8.   CONCLUSIONS

1 The reason for radiation at unexpected frequencies of a structure with elastic-mass inclusions could be the localized modes.

2 Localized modes correspond to the trapped frequencies of the infinite length system.

3 The choice of a liquid model essentially influences the estimation of the spectrum of compound structure radiation.

4 Geometric nonlinearity causes quasiperiodic and chaotic regimes. Classification of possible dynamic regimes of a nonlinear systems allows us to predict frequencies which can not be found from linear analysis of the spectrum.

References

[1] Abramian, A K and Indejtchev, D A (1995) *Trapping modes of oscillation in an elastic system*, J. Tech. Ac. Vol. 2, **3**, 3–17.

[2] Ursell, F (1951) *Trapping modes in the theory of surface waves*, Proc. Camb. Phil. Soc. **47**, 347–358.

[3] Abramian, A K, Fedorova, A N and Zeitlin, M G (1995) *From tori to chaos in Galerkin approximations*, Proc. of XXII Summer School on Nonlinear Oscillations in Mechanical Systems, 103–123, St. Petersburg: Russian Academy of Science.

A NEW CLASS OF POINT MODELS IN DIFFRACTION BY THIN ELASTIC PLATES

I. V. Andronov

Department of Mathematical & Computational Physics
University of St. Petersburg
Ulianovskaja 1-1, St. Petersburg 198904, Russia
iva@aa2628.spb.edu

1.   INTRODUCTION

The problems of acoustic diffraction by elastic plates or shells with edges, conjunctions, cracks, inclusions and stiffeners attract wide interest caused by the needs of shipbuilding, noise analysis, acoustics of the ocean and other sciences. Much attention is paid to problems that can be solved in a closed form of integrals or series [1, 2]. Then the analysis of physical effects is the most simple. Such are the problems using point models that are formulated by means of contact conditions fixed in a midpoint of the obstacle. However real joints, cracks, stiffeners, and other inhomogeneities have finite size. Thus, for all the point models the question of applicability is of importance. Partly it is discussed in [3], where the correction due to the height of a protruding stiffener is derived. In [4] the diffraction by a narrow crack is studied. For thin plates the applicability of point model is shown to be justified only for exponentially narrow cracks when $ka \ll \exp\left(-(kh)^{8/5}\right)$, where k is the wavenumber, h is the thickness of plate and a is the width of crack.

Here we present more precise model of cracks and formulate a class of generalized point models. These models are formulated as operator extensions in the form of zero-range potentials [5, 6].

2.   DIFFRACTION BY A NARROW CRACK

Consider an infinite elastic plate bounding a homogeneous acoustic half-space. Let the plate be absent on a segment $[-a, a]$ and let the edges of the two half-plates be free. The field is generated by an incident harmonic plane wave with the time factor $\exp(-i\omega t)$ dropped elsewhere.

261

I.D. Abrahams et al. (eds.),
IUTAM Symposium on Diffraction and Scattering in Fluid Mechanics and Elasticity, 261–268.
© 2002 *Kluwer Academic Publishers. Printed in the Netherlands.*

Mathematically, the problem is formulated as a boundary-contact value problem:

$$\Delta u + k^2 u = 0, \qquad y > 0, \tag{28.1}$$

$$D\frac{d^4\xi}{dx^4} - \rho h\omega^2\xi + u|_{y=0} = 0, \quad \xi = \frac{1}{\rho_0\omega^2}\left.\frac{\partial u}{\partial y}\right|_{y=0}, \qquad |x| > a, \tag{28.2}$$

$$u|_{y=0} = 0 \text{ for } |x| < a, \ \xi''(\pm a) = 0 \text{ and } \xi'''(\pm a) = 0.$$

Besides the Meixner conditions are assumed in a vicinity of the points $(\pm a, 0)$ and the scattered field is subject to radiation condition. In (28.1) and (28.2) the following parameters are introduced: $k = \omega/c$ is the wave number in acoustic media, D is bending stiffness of the plate, ρ and ρ_0 are the densities of the plate and fluid. It is convenient to introduce the wave number k_0 of flexure waves in an isolated plate and the parameter ν as

$$k_0 = \sqrt[4]{\frac{\rho h\omega^2}{D}}, \qquad \nu = \frac{\rho_0\omega^2}{D}.$$

The incident field $u_0 = \exp(ik(x\cos\varphi_0 - y\sin\varphi_0))$ is reflected by the plate, $u_1 = R(\varphi_0)\exp(ik(x\cos\varphi_0 + y\sin\varphi_0))$. Here

$$R(\varphi_0) = \frac{-\overline{L(\varphi_0)}}{L(\varphi_0)}, \qquad L(\varphi_0) = ik\sin\varphi_0\left(k^4\cos^4\varphi_0 - k_0^4\right) + \nu.$$

These two plane waves u_0 and u_1 would solve the problem in the case of a homogeneous plate. Due to the crack some scattered field u_2 is also generated. At large distances from the crack it forms cylindrical wave with a far field amplitude $\Psi(\varphi, \varphi_0)$ and two waves running along the half-plates. By Green function method the problem can be reduced to integral equations on the segment $[-a, a]$. Small-ka asymptotic analysis of these equations allows the far field amplitude to be found [4]

$$\Psi = \frac{i\nu}{\pi}\frac{k^2\sin\varphi\sin\varphi_0}{L(\varphi)L(\varphi_0)}\left\{\frac{k^4}{D_4}\cos^2\varphi\cos^2\varphi_0 - \frac{k^6}{D_6}\cos^3\varphi\cos^3\varphi_0\right.$$
$$\left. - \frac{\pi\nu M(\varphi)M(\varphi_0)}{\ln(ka/4) + \gamma - i\pi/2 + \pi J + \nu B_2^2/D_4} + \dots\right\}. \tag{28.3}$$

Here γ is the Euler constant,

$$M(\varphi) = \frac{1}{\nu}(k^4\cos^4\varphi - k_0^4) + \frac{k^2 B_2}{D_4}\cos^2\varphi,$$

$$D_j = \frac{1}{2\pi}\int e^{+i0}\frac{(i\mu)^j\sqrt{\mu^2 - k^2}}{l(\mu)}d\mu, \tag{28.4}$$

$$B_j = \frac{1}{2\pi} \int \frac{(i\mu)^j}{l(\mu)} d\mu, \qquad J = \frac{1}{2\pi} \int \frac{d\mu}{l(\mu)\sqrt{\mu^2 - k^2}}. \qquad (28.5)$$

The path of integration in (28.4) and (28.5) goes along the real axis avoiding the singularities of $l(\mu)$ according to the limiting absorption principle.

The first two terms in the asymptotics (28.3) coincide with the exact expression for the far field amplitude of scattering by point model [2]. The correction is logarithmically by ka smaller. However the principal order terms vanish at orthogonal direction to the plate, while the correction remains nonzero. Moreover for thin plates ($kh \ll 1$) asymptotic analysis of the integrals D_j, B_j and J involved in the formula (28.3) shows that the last term in (28.3) appears the principal one by the parameter kh. The effective cross-sections Σ, that is the relative energy carried by the scattered field [8], computed for $a = 0, 10^{-20}, 10^{-10}, 10^{-5}, 10^{-3}, 0.01$ and 0.1m are presented on Figure 1. The 1cm steel plate in water is assumed. The noticeable difference can be seen already for very narrow cracks. So, classical point model of crack should be corrected.

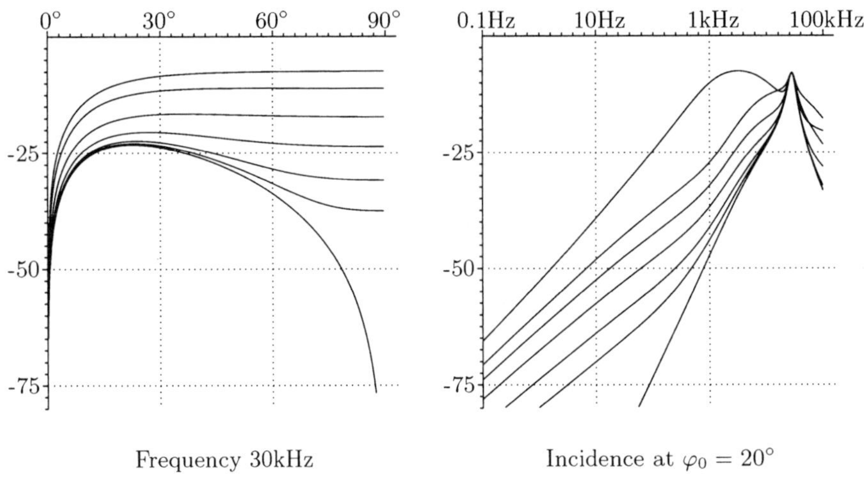

Frequency 30kHz   Incidence at $\varphi_0 = 20°$

Figure 1   Angular and frequency dependencies of Σ (dB) for narrow cracks

3.   GENERALIZED POINT MODELS

In the boundary-contact problems of acoustics the following approach is often used for the construction of the scattered field u_2. First the so-called general solution U is constructed. This solutions satisfies all the equations and boundary conditions of the problem except the contact conditions. It contains a number of parameters that are then determined from a linear algebraic system, when U is substituted into the contact conditions. Generally U satisfies the following boundary-value problem

$$
\begin{cases}
\Delta U + k^2 U = 0, \qquad y > 0 \\
\left(\dfrac{d^4}{dx^4} - k_0^4 \right) \dfrac{\partial U}{\partial y} + \nu U \\
\qquad = c_0 \delta(x) + c_1 \delta'(x) + c_2 \delta''(x) + c_3 \delta'''(x), \qquad y = 0 \,.
\end{cases}
$$

Here $\delta(x)$ is the Dirac delta-function and the constants c_j are arbitrary parameters. That is, physically the solution U is the field generated by a point source on the plate.

By substituting the expression for U into the contact conditions one obtains a linear system of algebraic equations for the coefficients c_j. In order that this system has a unique solution, the boundary conditions should be correctly set [8]. Namely the following boundary form should vanish for the solution

$$
\begin{aligned}
&\left(\frac{\partial^4 u}{\partial x^3 \partial y} \frac{\partial \overline{u}}{\partial y} - \frac{\partial^3 u}{\partial x^2 \partial y} \frac{\partial^2 \overline{u}}{\partial x \partial y} + \frac{\partial^2 u}{\partial x \partial y} \frac{\partial^3 \overline{u}}{\partial x^2 \partial y} \right. \\
&\qquad \left. - \frac{\partial u}{\partial y} \frac{\partial^4 \overline{u}}{\partial x^3 \partial y} \right) \Bigg|_{x=-0}^{x=+0} = 0, \qquad y = 0.
\end{aligned}
\tag{28.6}
$$

To formulate the point model that takes into account the geometry of the obstacle, one needs to add to passive sources in the plate another passive source in the fluid. That is, the solution is sought in the form of a general solution of the following problem

$$
\begin{cases}
\Delta U + k^2 U = C \delta(x) \delta(y), \qquad y > 0 \\
\left(\dfrac{d^4}{dx^4} - k_0^4 \right) \dfrac{\partial U}{\partial y} + \nu U \\
\qquad = c_0 \delta(x) + c_1 \delta'(x) + c_2 \delta''(x) + c_3 \delta'''(x), \qquad y = 0.
\end{cases}
\tag{28.7}
$$

The first equation is understood as a limit. First the position of the passive source is chosen as $(0, y_0)$, then the limit as $y_0 \to 0$ is taken in the solution.

An additional parameter C is presented in the new model. Correspondingly an additional "boundary" condition should be added. This "boundary" condition together with the usual boundary-contact conditions should be taken such that some expression vanishes on the solution. Consider the domain $\Omega_{R,\varepsilon}$, bounded by an arc of large radius R, by the plate and by another arc of small radius ε around the obstacle. Apply Green's formula to u and $\overline{u}$ in this domain and take the limit as

$R \to +\infty$, $\varepsilon \to +0$. If the sum of the integral over a small arc and the substitutes in the points $x = \pm\varepsilon$ tends to zero, this yields the identity called "optical theorem" [8]. Let the class of conditions on the scatter that cause the mentioned above limit to vanish, be studied. Evidently it is sufficient to ask two conditions to be satisfied, namely the condition (28.6) and

$$\lim_{r \to 0} \int_0^\pi \left(\frac{\partial u}{\partial r}\overline{u} - u\frac{\partial \overline{u}}{\partial r} \right) r d\varphi = 0. \qquad (28.8)$$

Consider the asymptotics of the solution U when $r \to 0$

$$U = (C/\pi)\ln(r) + D + \dots , \qquad r \to 0. \qquad (28.9)$$

Here C is the same constant as in (28.7) and the constant D depends on the incident field and on the coefficients C, c_0, c_1, c_2 and c_3. Substituting the asymptotics (28.9) into (28.8) yields the condition

$$C\overline{D} - D\overline{C} = 0.$$

That is, C and D in (28.9) should be proportional with a real coefficient

$$C = \kappa D. \qquad (28.10)$$

If $\kappa = 0$, the generalized model is reduced to a classical one.

It should be noted that mathematically the condition (28.10) fixes some self-adjoint extension of the operator in the form of a zero-range potential [5, 6, 7].

Combining (28.9) with (28.10) and introducing $S = e^{-\pi/\kappa}$, the additional condition can be written as

$$U = (C/\pi)\ln(r/S) + o(1), \qquad r \to 0. \qquad (28.11)$$

Here S is fixed and C is arbitrary.

4.   CHOICE OF THE PARAMETER

In the problem of scattering by a generalized point model, the Meixner condition is replaced by (28.11). The boundary-contact conditions remain. The main question that appears in the formulation of the model of a particular obstacle is the value of the parameter S in (28.11).

The most rigorous way is to find S by comparing the far field asymptotics in the problem of scattering by a generalized point model with the asymptotics (28.3) found for the scattering by a particular obstacle. One can find that logarithmically small corrections in (28.3) correspond to

$$S = a/2.$$

In this case the exact expression for the far field amplitude is

$$\Psi = \frac{i\nu}{\pi} \frac{k^2 \sin\varphi \sin\varphi_0}{L(\varphi)L(\varphi_0)} \left\{ \frac{k^4}{D_4} \cos^2\varphi \cos^2\varphi_0 - \frac{k^6}{D_6} \cos^3\varphi \cos^3\varphi_0 \right.$$
$$\left. - \frac{\pi\nu M(\varphi)M(\varphi_0)}{\ln(ka/4) + \gamma - i\pi/2 + \pi J + \nu B_2^2/D_4} \right\}. \tag{28.12}$$

The difference of (28.12) and (28.3) is of order $o(1/\ln(ka))$.

The value of the parameter S is defined only by the width of the crack and does not depend on the incident field (in particular its frequency) and on the parameters of the plate and fluid. This observation is general for all the generalized models both in two- and three-dimensional problems and allows the following algorithm for the choice of the parameter S to be suggested. Instead of examining the scattering problem for fluid loaded plate and finding the asymptotics of the far field amplitude, one can consider the problem for the absolutely rigid plate and from the far field asymptotics find the value of S.

5.   COMPARISON OF THE MODELS

Combining the condition (28.11) with different boundary-contact conditions yields the following generalized models

1. Point model of crack.   (see [2])

$$C = 0, \qquad \xi''(\pm 0) = 0, \qquad \xi'''(\pm 0) = 0.$$

2. Model of a narrow crack.   (free edges)

$$U \sim (C/\pi)\ln(2r/a) + o(1), \qquad r \to 0,$$
$$\xi''(\pm 0) = 0, \qquad \xi'''(\pm 0) = 0.$$

3. Fixed point.   (see [1])

$$C = 0, \qquad \xi(\pm 0) = 0, \qquad \xi'(\pm 0) = 0.$$

4. Model of a narrow slit with fixed edges.

$$U \sim (C/\pi)\ln(2r/a) + o(1), \qquad r \to 0,$$
$$\xi(\pm 0) = 0, \qquad \xi'(\pm 0) = 0.$$

5. Model of a bubble.

$$U \sim (C/\pi)\ln(2r/a) + o(1), \quad r \to 0, \qquad \xi \in C^3(\mathbb{R}).$$

Note that for $a \to 0$ the models 2 and 4 are close to the corresponding classical models 1 and 3. However, already for very small a/h significant difference can be seen. The bubble (model 5) cannot be described by classical boundary-contact conditions. Figure 2 presents the dependence of effective cross-section on half-width a for all the models formulated above. The incidence of 10kHz plane wave at $\varphi_0 = 10°$ on a 5mm steel plate in water is taken as an example.

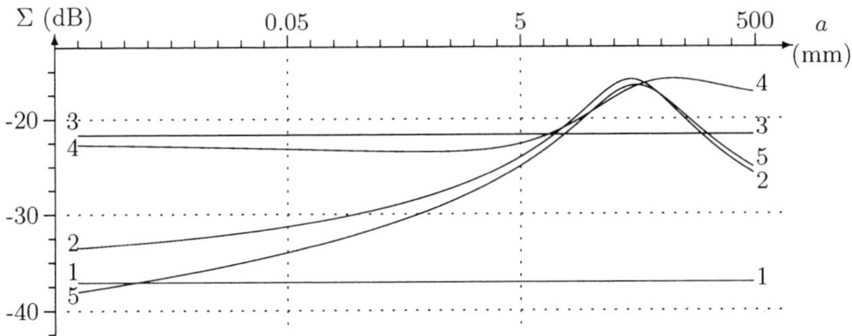

Figure 2   Effective cross-sections Σ as functions of half-width a for the models 1–5

6.   CONCLUSION

The generalized point models were considered for two-dimensional problems of diffraction by fluid loaded plates with small obstacles. In these models additionally to classical boundary-contact conditions an additional condition was specified for the logarithmic derivative of the acoustic field in the central point of the obstacle. The parameter S in this condition was noted independent of the general parameters of the boundary value problem. It is defined only by the width of a gap in the plate. As it was mentioned above this fact allows the parameter S to be chosen from the analysis of a simpler problem for absolutely rigid plate. The other advantage of the parametrization in the form (28.11) is the possibility to extend the model to the case of dynamic problems (nonstationary) and to three-dimensional problems with translational symmetry [9].

The method of zero-range potentials [5, 6] allows three-dimensional defects in fluid loaded plates to be modelled in a similar manner. See e.g. [10], where the model of a short crack is introduced, and [11] for the model of a round hole.

References

[1] Konovalyuk, I P and Krasilnikov, V N (1965) *Influence of a stiffener on plane wave reflection from a thin plate*, Problems of diffraction and wave propagation, **4**, 149–165, Leningrad: University Press.

[2] Kouzov, D P (1963) *Diffraction of plane hydroacoustic waves by a crack in an elastic plate*, Appl. Math. Mech. **27**, 1037–1043.

[3] Belinskiy, B P (1978) *Diffraction of plane waves by a plate supported with a protruding stiffener*, Appl. Math. Mech. **42**, 486–493.

[4] Andronov, I V, Belinskiy, B P and Dauer, J P (1996) *Scattering of acoustic waves by a narrow crack in elastic plate*, Wave Motion **24**, 101–115.

[5] Pavlov, B S (1987) *Theory of extensions and explicitly solvable models*, Uspekhi Matem. Nauk **42**(6), 99–131.

[6] Albeverio, S and Kurasov, P (2000) *Singular perturbations of differential operators and solvable Schroedinger type operators*, London Mathematical Society Lecture Notes Series **221**, Cambridge: University Press.

[7] Andronov, I V (1995) *Application of zero-range potentials in the problems of diffraction by small inhomogeneities in elastic plates*, Appl. Math. Mech. **59**, 451–463.

[8] Belinskiy, B P and Kouzov, D P (1980) *An optical theorem for plate–fluid systems*, Akust. Zh. **26**(1), 13–19.

[9] Andronov, I V (2000) *Waves propagating along a narrow crack in an elastic plate*, Mathematical and Numerical Aspects of Wave Propagation (eds. Bermudez et al), 103–107, Philadelphia: SIAM Publications.

[10] Andronov, I V (1990) *Modelling of scattering processes on a crack in an elastic plate by means of zero-range potentials*, Zapiski LOMI **186**, 14–19, (see J. Math. Sci. **73**, 304–307).

[11] Andronov, I V (1993) *Application of zero-range models to the problem of diffraction by a small hole in an elastic plate*, J. Math. Phys. **34**, 2226–2241.

A 2-D MODEL OF THE INTERACTION BETWEEN A PRESSURE WAVE AND A SUBMERGED SANDWICH PANEL

S. Kadyrov

Department of Mathematics

State Marine Technical University

Lotsmanskaya 3, St. Petersburg 190008, Russia

kadyrov@westpost.ru

1.   INTRODUCTION

The interaction between an underwater shock wave and homogeneous plates and shells has been investigated by many authors and therefore is elaborated in details in most respects. However it is not possible to assert the same about extending the developed approaches to the interaction of a pressure wave and submerged sandwich plates [1, 2]. In this case it is necessary to point out two main features of the problem which make its solution rather complicated.

At first, although a number of theories to solve static problems of sandwich plates and beams are now developed, it is not at all clear whether it is possible to apply them for solving transient problems in dynamics, where the effect of transverse squashing is essential, so that a use of 'classical' sandwich plates theories may be dangerous.

Secondly, a negative pressure generated by a deformable surface of a fast moving panel surface leads to appearance of cavitation zones in the fluid which may drastically change the process of a fluid-structure interaction. If the process of a shock wave reflection from an elastic panel is considered, then it is found that the pressure in a reflected wave is decreasing with a decrease in a panel thickness and at last gaps in water appear. The 'tail part' of an incident shock wave reflects from these gaps as a wave of rarefaction, and may generate new gaps at a certain distance from them and so on. The amount of such gaps depends on the shape and the amplitude of an incident shock wave. Thus, details of the pressure field at each time instant become important even if simplified sandwich plate theory in general may correctly predict the deformation of plate

I.D. Abrahams et al. (eds.),

IUTAM Symposium on Diffraction and Scattering in Fluid Mechanics and Elasticity, 269–277.

surface with cavitation not taken into account. Apparently, numerical modelling is a more realistic way to describe all possible physical effects. Some of such numerical modelling is done in this paper.

2.   2-D PROBLEM FORMULATION

An infinitely long sandwich strip of unit width which may be treated as a beam is considered (Fig. 1a). The strip rests on a set of immobile equally spaced supports, so the deformation of a single span may be examined independently of the others. An upper half-space is occupied by water, another part beyond the strip is occupied by air. At time $t = 0$ the front of an incident shock wave coincides with the panel's surface.

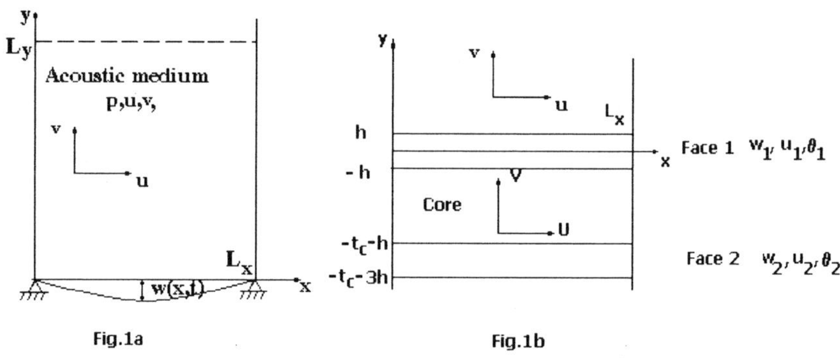

3.   MODELLING OF SANDWICH BEAM

Simplified theory of sandwich beam.   The generalisation of equations suggested in [3] to dynamical problems is used. As was shown in [4] dispersive properties of these equations in a low-wave part of spectra are very close to exact theory, based on the theory of elasticity

Hereafter all the parameters of faces (supposed to be identical) have subscript f, t_f, E_f, ρ_f are the thickness, Young's modulus and density accordingly, the same parameters for internal core have subscript c. Poisson' ratio ν for the faces and the core materials is the same. The dimensionless parameters are induced to describe an internal structure of a sandwich plate: $\epsilon = t_f/t_c$ as a thickness parameter, $\delta = \rho_c/\rho_f$ as a density parameter, $\gamma = E_c/E_f$ as a longitudinal stiffness parameter and $\gamma_G = G_c/G_f$ as a shear stiffness parameter. The deformation of a sandwich beam element is governed by two independent variables: a displacement of the mid-surface of the whole element w (which is the same for all plies), and a shear angle θ between mid-surfaces of skin.

The moments and forces are related to w and θ by

$$M_1 = -D_1 \frac{\partial^2 w}{\partial x^2}, \quad M_2 = D_2 \frac{\partial \theta}{\partial x}, \quad Q = \Gamma \left(\theta + \frac{\partial \theta}{\partial x} \right)$$

Here M_1 is a bending moment of the 'first kind'; it is composed as a sum of bending moments acting in each ply. It is produced by normal stresses linearly distributed in each ply in accordance with classic Bernoulli–Euler model applied to each ply individually. These stresses are related to curvature $\partial^2 w/\partial x^2$, which is the same for all plies. M_2 is a bending moment of the 'second kind'; it is generated by a uniform part of normal stresses acting in skin plies. These stresses are produced by a deformation $\partial \theta/\partial x$. Shear force in a core ply is uniformly distributed and it is proportional to the shear angle in a core ply $\theta + (\partial w/\partial x)$. The elastic and inertial parameters are

$$D_1 = \frac{E_f t_f^3 (2 + \gamma \epsilon^{-3})}{12(1 - \gamma^2)}, \quad D_2 = \frac{E_f t_f^3 (1 + \epsilon^{-3})}{2(1 - \gamma^2)}, \quad \Gamma = \frac{E_f t_f \gamma_G}{2(1 + \gamma)} (1 + \epsilon),$$

$$m = \rho_f t_f \left(2 + \frac{\delta}{\epsilon} \right), \quad I_1 = \frac{\rho_f t_f^3}{12} \left(2 + \frac{\delta}{\epsilon^3} \right), \quad I_2 = \frac{\rho_f t_f^3}{2} \left(1 + \frac{1}{\epsilon} \right).$$

Thus, dynamic equations are

$$D_1 \frac{\partial^4 w}{\partial x^4} - \Gamma \left(\frac{\partial \theta}{\partial x} + \frac{\partial^2 w}{\partial x^2} \right) + m \frac{\partial^2 w}{\partial t^2} - I_1 \frac{\partial^4 w}{\partial x^2 \partial t^2} = p(x, 0, t),$$

$$-D_1 \frac{\partial^2 \theta}{\partial x^2} + \Gamma \left(\theta + \frac{\partial \theta}{\partial x} \right) + I_2 \frac{\partial^2 \theta}{\partial t^2} = 0,$$

where $p(x, 0, t)$ is a hydrodynamic pressure at the surface of a plate. Boundary conditions at edges are

$$w(0, t) = \frac{\partial w(0, t)}{\partial x} = \theta(0, t) = 0, \quad w(L_x, t) = \frac{\partial w(L_x, t)}{\partial x} = \theta(L_x, t) = 0.$$

Refined theory of sandwich beam.   In this theory a sandwich strip is considered as a system of two identical thin plates (Fig. 1b) with elastic characteristics E_f, G_f, ν and an elastic layer between them. Layer's elastic characteristics are E_c, G_c, ν. The face thickness is $t_f = 2h$. If displacements of plates along axes Oy and Ox are denoted as w_1, w_2, u_1, u_2 and shear angles as θ_1, θ_2, then we get equations of Timoshenko type of dynamics of faces 1 and 2 (see Fig. 1b).
Face 1:

$$\frac{4G_f h}{1 - \nu^2} \frac{\partial^2 u_1}{\partial x^2} - \tau(x, -h, t) = 2\rho_f h \frac{\partial^2 u_1}{\partial t^2},$$

$$2k^2 G_f h \left(\frac{\partial^2 w_1}{\partial x^2} - \frac{\partial \theta_1}{\partial x} \right) - \sigma(x, -h, t) - p(x, 0, t) = 2\rho_f h \frac{\partial^2 w_1}{\partial t^2},$$

$$\frac{4 G_f h}{3(1 - \nu^2)} \frac{\partial^2 \theta_1}{\partial x^2} + 2k^2 G_f \left(\frac{\partial w_1}{\partial x} - \theta_1 \right) - \tau(x, -h, t) = \frac{2}{3} \rho_f h^2 \frac{\partial^2 \theta_1}{\partial t^2}.$$

Face 2:

$$\frac{4 G_f h}{1 - \nu^2} \frac{\partial^2 u_2}{\partial x^2} - \tau(x, -t_c - h, t) = 2\rho_f h \frac{\partial^2 u_2}{\partial t^2},$$

$$2k^2 G_f h \left(\frac{\partial^2 w_2}{\partial x^2} - \frac{\partial \theta_2}{\partial x} \right) - \sigma(x, -t_c - h, t) = 2\rho_f h \frac{\partial^2 w_2}{\partial t^2},$$

$$\frac{4 G_f h}{3(1 - \nu^2)} \frac{\partial^2 \theta_2}{\partial x^2} + 2k^2 G_f \left(\frac{\partial w_2}{\partial x} - \theta_2 \right) - \tau(x, -t_c - h, t) = \frac{2}{3} \rho_f h^2 \frac{\partial^2 \theta_2}{\partial t^2}.$$

Here k is a stiffness coefficient (typically $k = 5/6$), τ and σ are tangential and transversal stress components respectively.

Equations for an internal core are (Lamé equations of elastic medium):

$$(\lambda + \mu) \frac{\partial s}{\partial x} + \mu \Delta U = \rho_c \frac{\partial^2 U}{\partial t^2}, \quad (\lambda + \mu) \frac{\partial s}{\partial y} + \mu \Delta V = \rho_c \frac{\partial^2 V}{\partial t^2},$$

where $s = (\partial U / \partial x) + (\partial V / \partial y)$ and $\Delta = \partial^2 / \partial x^2 + \partial^2 / \partial y^2$. Here, λ, μ are Lamé constants for the core material which may be expressed via 'technical' elastic parameters as $\lambda = E_c / [(1 + \nu)(1 - 2\nu)]$ and $\mu = G_c = E_c / [2(1 + \nu)]$. From Hooke's law, the dynamic boundary conditions are

$$\sigma(x, y, t) = \lambda s + \mu \left(\frac{\partial U}{\partial y} + \frac{\partial V}{\partial x} \right) \quad \text{at } y = -h \text{ and at } y = -t_c - h$$

$$\tau(x, y, t) = \mu \left(\frac{\partial U}{\partial y} + \frac{\partial V}{\partial x} \right) \quad \text{at } y = -h \text{ and at } y = -t_c - h.$$

Kinematic boundary conditions on the compatibility of the layers are $u_1(x, t) = U(x, -h, t) + h\theta_1(x, t)$, $w_1(x, t) = V(x, -h, t)$, $u_2(x, t) = U(x, -t_c - h, t) - h\theta_2(x, t)$ and $w_2(x, t) = V(x, -t_c - h, t)$. At the edges of structure we have $U(0, t) = V(0, t) = U(L_x, t) = V(L_x, t) = 0$, $u_1(0, t) = u_2(0, t) = u_1(L_x, t) = u_2(L_x, t) = 0$, $w_1(0, t) = w_2(0, t) = w_1(L_x, t) = w_2(L_x, t) = 0$ and $\theta_1(0, t) = \theta_2(0, t) = \theta_1(L_x, t) = \theta_2(L_x, t) = 0$.

4.   MODELLING OF A CAVITATING FLUID

The system of equations of 2-D acoustics in Cartesian coordinates is

$$\frac{\partial u}{\partial t} + \frac{1}{\rho_0} \frac{\partial p}{\partial x} = 0, \quad \frac{\partial v}{\partial t} + \frac{1}{\rho_0} \frac{\partial p}{\partial y} = 0, \quad \frac{\partial p}{\partial t} + \frac{1}{\rho_0 c_0^2} \left(\frac{\partial u}{\partial x} + \frac{\partial v}{\partial y} \right) = 0.$$

Here $u(x, y, t)$, $v(x, y, t)$ are the velocity components, ρ_0, c_0 are the density and sound velocity of the fluid. Then a free-field pressure and a velocity are used as initial values ($t = 0$) in the numerical modelling. The initial shock wave is assumed to have a peak pressure P_m and decay constant λ and it reaches the front face of a sandwich plate at $t = 0$. The initial conditions for system are $p_i = P_m \exp\{-y/(c_0\lambda)\}$, $v_i = -p_i/(\rho_0 c_0)$. Boundary conditions are formulated for the surface of a plate (conditions of compatibility): $v(x, 0, t) = w_1(x, t)$ and zero for all values on some fictitious line $y = L_y$ posed sufficiently far from the plate to ensure that a signal radiated from this boundary cannot reach the areas near the plate surface. The condition of the appearance of cavitation is formulated as: $p + p_h \leq p_c$, where p_h is a hydrostatic pressure (taken to be zero), p_c is some threshold value of the pressure relevant to the appearance of developed cavitation zones. Experiments showed [5] that for sea water $-0.2 < p_c < -0.35\text{MPa}$. The pressure within cavitation zones varies quite weakly and we assume it to be constant and equal to a vapour pressure in water, somewhere in the range of $-100 \sim -200\text{Pa}$. Therefore the equation of motion in cavitation is [5] $du/dt = 0$ and $dv/dt = 0$, which means that water particles in cavitation zones move without acceleration and interaction with each other.

5.   NUMERICAL METHOD

For numerical solution a CFD technique is used as the combination of an explicit scheme for a 2-D sandwich theory and a non-explicit scheme for 1-D sandwich theory with a 2-D version of so-called Godunov's scheme [6] for fluid. The magnitudes of steps in space and time domains were selected experimentally to satisfy requirements of an accuracy and a stability of the CFD scheme. It is important to recognize that they also have a certain physical meaning, namely they characterise space and time scales of 'elementary' cavity. Zones with cavitation smaller than the spatial step in such an approach cannot exist.

6.   RESULTS OF NUMERICAL EXPERIMENTS AND DISCUSSION

The calculations were done for realistic parameteres of a sandwich beam typical for naval structures [2], namely, beam span 2m, core thickness 0.09m, core E-modulus 140MPa, core density 60kg/m^3, face thickness 0.007m, face E-modulus 15GPa, face density 1800kg/m^3, $\nu = 0.3$. It is assumed that cavitation arises if a pressure in a fluid is less than -0.3MPa and that a pressure inside cavitation zones is equal to -100Pa. To verify the simplified sandwich beam model, a time history of maxi-

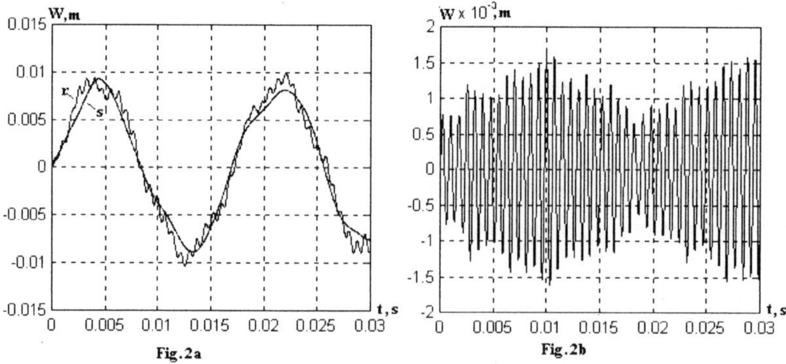

Fig.2a   Fig.2b

mum of the displacement (in the centre point of a beam) was calculated in the case when there is no fluid in front of the beam and loading does not vary in time and is uniform along the y-axis: $p(t) = P_m = 30\mathrm{MPa}$. Calculations have proved (Fig. 2a) closeness of results relevant to the refined (curve 'r' presents the displacement of the outer face) and the simplified (curve 's') theories. Nevertheless, beam's faces are moving by no means synchronously. To illustrate this fact (which is important in further calculations within the framework of the coupled formulation) the difference between displacements of faces at the centre of a beam is plotted in Fig. 2b It appears that this difference may reach 20% of the displacement's maxima. Thus, it is clear that, if the duration of an incident pressure wave is close to a time of travelling of an elastic wave in the core from one face to another, then the simplified theory should give inadequate results. In the numerical example explored here, such a characteristic time scale is about $2 \cdot 10^{-4}$s. The presence of a fluid produces a damping effect and even for $\lambda = 1 \cdot 10^{-4}$s the simplified theory predicts believable values of displacements (see Fig. 3a and Fig. 3b).

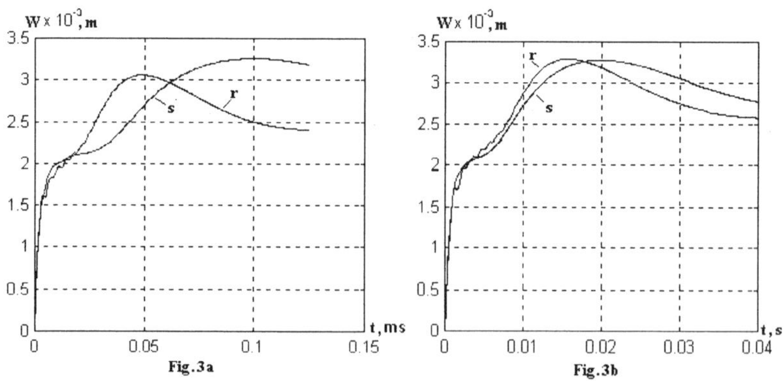

Fig.3a   Fig.3b

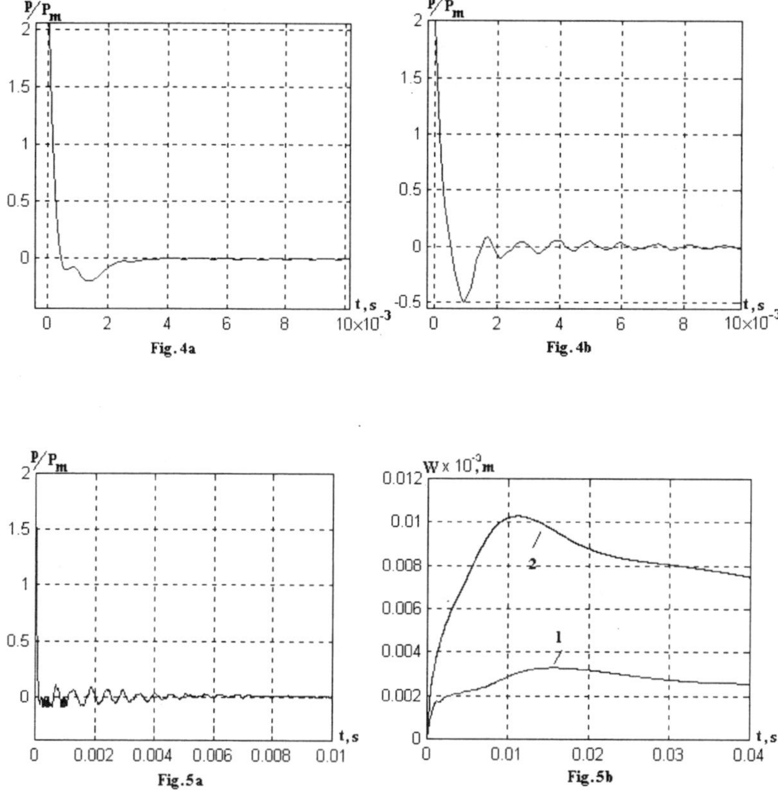

Fig. 4a

Fig. 4b

Fig.5a

Fig.5b

All the process of sandwich plate - pressure wave interaction may be divided into three stages. At the first stage, the flexural wave propagates from the edge to the centre and the main part of a plate moves as a rigid body, as the core material is rather soft, velocity of a flexural wave is controlled by the stiffness of an outer face. At the second stage, the form of an outer face is of sinusoidal type and the pressure near a plate is small and almost uniform. At the last stage, a plate begins to move backward and the pressure increases again. The time history of a pressure in the centre point of an outer face ($P_m = 30\text{MPa}$, $\lambda = 4 \cdot 10^{-4}\text{s}$) is shown in Fig. 4a (simplified theory) and in Fig. 4b (refined theory). There are two important features of these graphs.

First, the negative pressure appears very quickly and its actual magnitude is larger than that predicted by a simplified theory. This result should be expected, since the stiffness of an outer face is much less, then the fictitious one, calculated via formulas of simplified theory. Consequently, in reality generation of the cavitation zones is more likely than it is promised by the simplified theory.

Second, according to the simplified theory, the process of an interaction almost finishes at $t = L_x/c_0$ and then as it was shown in [5] fluid may be treated as incompressible. In reality elastic waves of compression-extension travel inside the core, periodically come to the outer face surface and essentially delay the process of the interaction. So far cavitation effects were not included into consideration. As calculations according to the simplified theory demonstrated an influence of cavitation zones to be in fact rather small and these zones exist for a short time not so essential to change pressure field. At the same time, the refined theory predicts noticeable changes in all the processes (see Figs. 5–7). Comparing Figs. 4b and 5a one may see how cavitation modifies a pressure at the centre point of a plate. In fact, as soon as cavitation occurs, a plate moves freely in a medium with an internal pressure $p_v = -100$Pa. As result, a negative pressure on the surface of a plate almost vanishes and maximum of a plate's displacement calculated for fluid with cavitation (curve 2 in Fig. 5b) is much larger than for the fluid without cavitation effects (curve 1 in Fig. 5b). The process of creation and disappearance of cavitation bubbles is displayed in Figs. 6a–6d where boundaries of cavitation zones at several times are shown. Cavitation arises at some distance from the plate surface (Fig. 6a) than spreading quickly reaches the surface of a plate and steadily exists during the first and the second stages of the process mentioned above. It begins to vanish only at the last stage of deformation when a plate begins to move backward. While cavitation effects are not included in the model, the problem stays linear and results for different amplitudes of a pressure shock wave P_m and the same incident wave duration λ are proportional (curve 1 in Fig. 7a). In reality such a dependence has to be nonlinear (curve 2 in Fig. 7a). The influence of the duration of the incident wave on the final maximum of the plate displacement for the fluid without and with cavitation is weaker (Fig. 7b, curves 1 and 2).

References

[1] Hayman, B (1995) *Underwater explosion loading on foam-cored sandwich panels*, Sandwich constructions **2**(3), EMAS.

[2] Makinen, K and Kadyrov, S (1998) *A 1-D fluid-sandwich interaction model for time underwater shock loading*, Proc. of 4th Int. Conf. on Sandwich Constructions, 781–793, Stockholm, Sweden.

[3] Skvortsov, V R (1993) *Symmetrically inhomogeneous through thickness plate as a sandwich plate having a soft core*, Trans. Acad. Sci. USSR. Mech. Rigid body, **1**, 162–168 (in Russian).

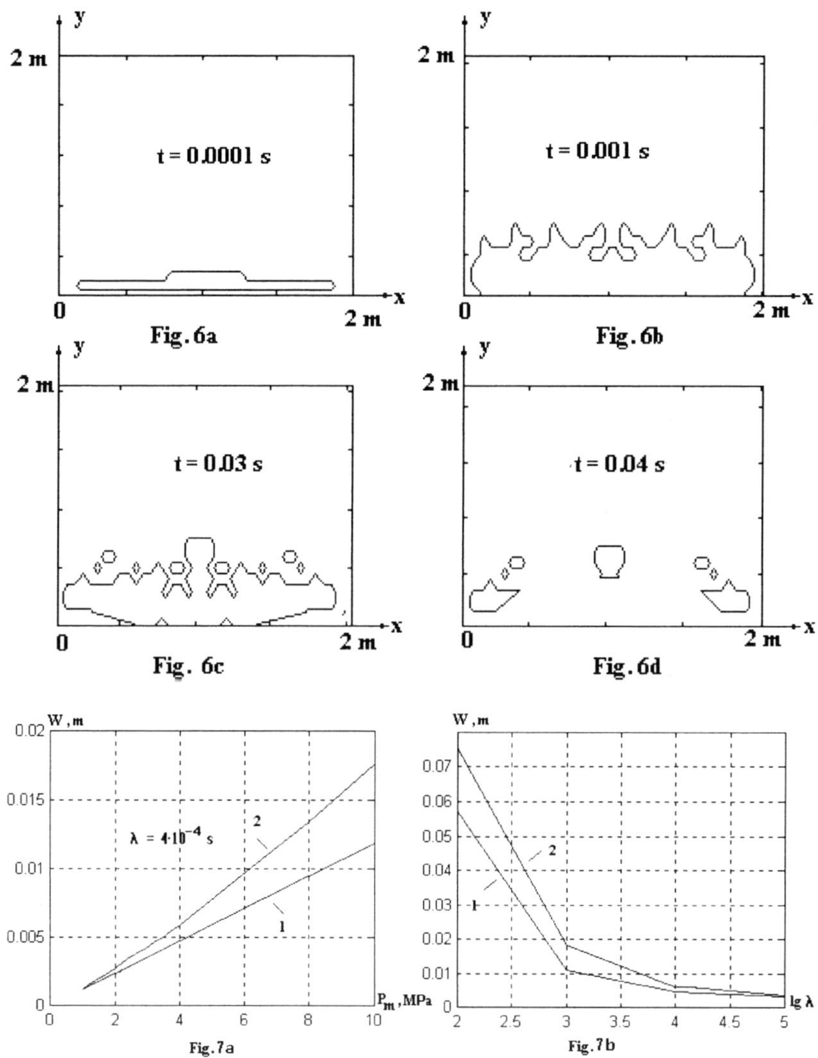

[4] Sorokin, S V (2000) *Vibration of and sound radiation from sandwich plates in heavy fluid conditions*, Composite Structures **48**, 219–230.

[5] Galiev, S U (1981) *The dynamic of hydro-elastic - plastic systems*, Kiev: Naukova Dumka (in Russian).

[6] Godunov, S K et al. (1978) *Numerical solution of multi-dimensional problems of gas dynamics*, Moscow: Nauka (in Russian).

ON THE PROPAGATION AND SCATTERING OF FLUID-STRUCTURAL WAVES IN A THREE-DIMENSIONAL DUCT BOUNDED BY THIN ELASTIC WALLS

J. B. Lawrie

Department of Mathematical Sciences
Brunel University, Uxbridge UB8 3PH, UK
jane.lawrie@brunel.ac.uk

I. D. Abrahams*

Department of Mathematics
University of Manchester, Manchester M13 9PL, UK
i.d.abrahams@ma.man.ac.uk

1.   INTRODUCTION

The propagation of acoustic waves along ducts or pipes has long been of interest to scientists and engineers. Acoustic scattering is a feature that becomes relevant whenever there is an abrupt change in duct geometry or material property. Most of the analytic work relating to scattering in waveguides concerns ducts with two-dimensional or circular cylindrical geometries. The Wiener–Hopf technique has proved a powerful tool in those cases where the geometry is uniform but the material properties change discontinuously. Alternatively, where there is more than one change in the material properties or the geometry undergoes abrupt change (e.g. in duct height), eigenfunction expansions and their associated orthogonality relations are often an effective means of reducing the problem to a system of linear algebraic equations that can be truncated and solved numerically. Problems involving wave propagation and mode-conversion/scattering are much more difficult when the duct

*This research was partially supported by a *Grant in Aid of Research* from the Leverhulme Trust, UK.

I.D. Abrahams et al. (eds.),
IUTAM Symposium on Diffraction and Scattering in Fluid Mechanics and Elasticity, 279–288.
© 2002 *Kluwer Academic Publishers. Printed in the Netherlands.*

walls are compliant, i.e. when fluid-structural interactions need to be taken into account. In particular, the authors are interested in problems involving compressible fluids contained within ducts which have wave-bearing surfaces, such as membranes or elastic plates. It is a feature of such problems that, due to the presence of high order derivatives in the boundary conditions, the relevant eigen-sub-problems are not Sturm-Liouville in type. Nevertheless, appropriate orthogonality relations can be derived and these enable the successful solution of many such models.

A broad class of orthogonality relations is discussed by Lawrie and Abrahams [1]; their derivation hinges on the separability of solutions of Helmholtz' equation for problems involving two-dimensional or axisymmetric circular cylindrical geometries. In many engineering applications, however, the model problems are intrinsically three-dimensional and, for those cases involving wave-bearing boundaries, whether or not the eigenmodes are separable in form depends on the conditions imposed at the duct corners. In Section 2 of this article these eigenfunctions, and the corresponding wavenumbers for propagating modes, are determined for a duct with one elastic wall clamped along its edges. (The analysis is easily extended to other corner conditions as will be shown in a forthcoming article by the authors.) It is shown that duct wavenumber, s say, satisfies the unusual dispersion relation $L(s) = \sum_{n=0}^{\infty} L_n(s, \gamma_n) = 0$ where the function L_n is an expression related to the eigenmodes, γ_n, for the two-dimensional y-z cross-sectional cavity (bounded by three rigid walls and an elastic plate). Despite its complicated form, it is a relatively straightforward procedure to numerically solve the dispersion relation and thereby determine the duct modes. The axial wavenumbers are the roots of $L(s)$ and once these are known, the wave-field in any region of the duct can be specified in terms of an infinite sum of eigenmodes of unknown amplitudes. For any scattering problem the aim is to determine these latter coefficients and to do this a fully three-dimensional orthogonality relation would be of great assistance. To this end a 'partial' orthogonality relation is presented and its use in the solution of a typical boundary-value problem involving scattering of fluid-structural waves in a three-dimensional semi-infinite duct is discussed in Section 3. Finally, some numerical results are presented and extensions of the method discussed.

2.   TRAVELLING WAVE SOLUTIONS

Before attempting to solve problems involving the scattering of fluid-structural waves in a three dimensional duct with one compliant wall, it is necessary to construct the explicit form of the travelling waves and, of course, the admissible wavenumbers. In this section a non-separable ansatz for the duct modes is posed; the validity of this expression is verified and the dispersion relation for the eigenvalues obtained. The duct occupies the region $-\infty < x < \infty$, $0 \leq y \leq a$, $-b \leq z \leq b$ where (x, y, z) are Cartesian coordinates that have been non-dimensionalised with respect to k^{-1}, where k is the fluid wavenumber. An elastic plate of infinite length and width $2b$ bounds the duct at $y = a$, whilst the remaining three sides are rigid (Fig. 2). A compressible fluid of density ρ and sound speed c occupies the interior region of the duct, but the exterior region is *in vacuo*. The travelling waves are assumed to have harmonic time dependence, $\exp(-it)$, where t has been non-dimensionalised with respect to ω^{-1}, with $\omega = ck$. Thus, the non-dimensional time-independent velocity potential, $\phi(x, y, z)$, satisfies Helmholtz' equation, that is

$$\left\{ \frac{\partial^2}{\partial x^2} + \frac{\partial^2}{\partial y^2} + \frac{\partial^2}{\partial z^2} + 1 \right\} \phi(x, y, z) = 0, \tag{30.1}$$

and the normal component of fluid velocity vanishes at the three rigid walls which, on assuming now even eigenmodes, $\phi(x, y, -z) = \phi(x, y, z)$, can by expressed as

$$\frac{\partial \phi}{\partial y} = 0, \ y = 0, \ 0 \leq z \leq b; \qquad \frac{\partial \phi}{\partial z} = 0, \ z = 0 \text{ and } b, \ 0 \leq y \leq a.$$

The case of antisymmetric eigenmodes can be dealt with in an identical fashion, but for brevity is not discussed here. The boundary condition that describes the deflections of a thin elastic plate bounding the top of the duct is given by

$$\left\{ \left(\frac{\partial^2}{\partial z^2} + \frac{\partial^2}{\partial x^2} \right)^2 - \mu^4 \right\} \phi_y - \alpha\phi = 0, \quad y = a, 0 \leq z \leq b, -\infty < x < \infty,$$

where μ is the *in vacuo* plate wavenumber and α the fluid loading parameter (see [2] for definitions of these physical quantities). Note that the plate vertical displacement is proportional to $\phi_y(x, a, z)$, where the subscript y here and henceforth denotes differentiation with respect to that variable. We further specify that the elastic plate is clamped to the adjacent rigid sides of the duct, hence the displacement and gradient are zero:

$$\frac{\partial \phi}{\partial y} = 0 \ \text{ and } \ \frac{\partial^2 \phi}{\partial y \partial z} = 0, \quad y = a, \ z = b, \ -\infty < x < \infty. \tag{30.2}$$

These conditions will henceforth be referred to as the *corner conditions*. Note that alternative conditions, such as free plate edges, could equally well have been taken.

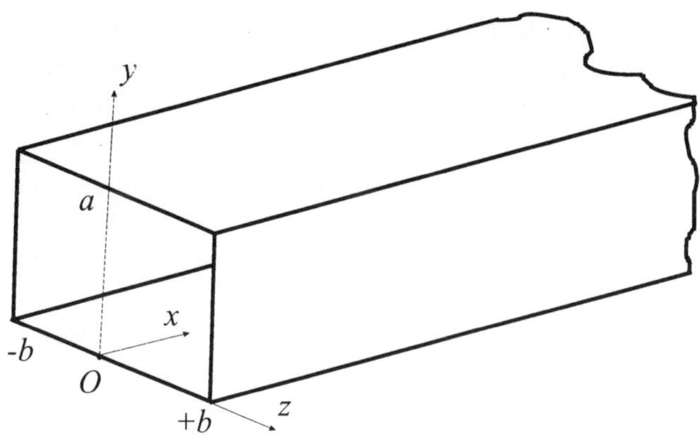

Figure 1   The physical configuration.

The aim of this section is to determine the form of the waves travelling in the x-direction along the duct. Without loss of generality it can be assumed that they propagate in the positive sense, so that ϕ assumes the form

$$\phi(x, y, z) = e^{isx}\psi_s(y, z) \tag{30.3}$$

where s is the, as yet unknown, axial wavenumber (with real or imaginary parts positive) and the nonseparable reduced potential ψ_s is to be determined. In terms of ψ_s, the governing equation is

$$\left\{ \frac{\partial^2}{\partial y^2} + \frac{\partial^2}{\partial z^2} + 1 - s^2 \right\} \psi_s(y, z) = 0, \tag{30.4}$$

whilst the boundary conditions are recast as

$$\frac{\partial \psi_s}{\partial y} = 0, \qquad y = 0, \ 0 \leq z \leq b, \tag{30.5}$$

$$\frac{\partial \psi_s}{\partial z} = 0, \qquad z = 0, \ 0 \leq y \leq a, \tag{30.6}$$

$$\frac{\partial \psi_s}{\partial z} = 0, \qquad z = b, \ 0 \leq y \leq a, \tag{30.7}$$

$$\left\{ \left(\frac{\partial^2}{\partial z^2} - s^2 \right)^2 - \mu^4 \right\} \psi_{sy} - \alpha \psi_s = 0, \qquad y = a, \ 0 \leq z \leq b \tag{30.8}$$

and the corner conditions become

$$\frac{\partial \psi_s}{\partial y} = 0, \qquad y = a, \ z = b, \tag{30.9}$$

$$\frac{\partial^2 \psi_s}{\partial y \partial z} = 0, \qquad y = a, \ z = b. \tag{30.10}$$

Let us suppose that an appropriate ansatz for ψ_s, which is composed of discrete eigenfunctions, is

$$\psi_s(y, z) = \sum_{m=0}^{\infty} E_{ms} Y_m(y) \cosh(\tau_{ms} z) \qquad (30.11)$$

which respects the fact that ψ_s is chosen to be even in z (30.6). The functions $Y_m(y)$, $m = 0, 1, 2, \ldots$, from (30.4) and (30.5), satisfy the following simple ordinary differential equation and boundary condition,

$$Y_m''(y) - \gamma_m^2 Y_m(y) = 0 \quad \text{and} \quad Y_m'(0) = 0, \qquad (30.12)$$

which yields $Y_m(y) = \cosh(\gamma_m y)$, where γ_m is related to τ_{ms} through $\tau_{ms}^2 + \gamma_m^2 + 1 - s^2 = 0$. A further condition for $Y_m(y)$ can be obtained by ensuring that ψ_s satisfies (30.8). On substituting (30.11) into (30.8) it is found that the quantities γ_m, $m = 0, 1, 2, \ldots$ are defined by

$$K(\gamma_m) = \{(\gamma_m^2 + 1)^2 - \mu^4\} Y_m'(a) - \alpha Y_m(a) = 0, \qquad (30.13)$$

where the $'$ denotes differentiation with respect to y.

The eigen-system specified by (30.12), (30.13) is *non-Sturm-Liouville* so that the eigenfunctions are not orthogonal in the usual sense. However, orthogonality relations do exist for systems of this class. Andronov and Belinskii [3] and Warren and Lawrie [4] considered specific examples of such systems whilst Lawrie and Abrahams [1] derived an orthogonality relation for a broad class of such systems. It can be shown, although the proof is omitted here, that the appropriate orthogonality relation for the system specified above is

$$\alpha \int_0^a Y_m(y) Y_j(y) \, dy = C_j \delta_{jm} - (\gamma_m^2 + \gamma_j^2 + 2) Y_j'(a) Y_m'(a) \qquad (30.14)$$

where δ_{jm} is the usual Kronecker delta and

$$C_j = 2(\gamma_j^2 + 1)[Y_j'(a)]^2 + \tfrac{1}{2}\alpha\gamma_j^{-2} Y_j'(a) Y_j(a) + \tfrac{1}{2}\alpha a. \qquad (30.15)$$

Expression (30.14) can now be employed to impose the condition of no normal velocity on the rigid face at $z = b$, that is expression (30.7). On multiplying (30.14) by $E_{js}\tau_{js}\sinh(\tau_{js}b)$ and summing over j, it is found that

$$\alpha \int_0^a \left\{ \sum_{j=0}^{\infty} E_{js}\tau_{js}\sinh(\tau_{js}b) Y_j(y) \right\} Y_m(y) \, dy = \qquad (30.16)$$

$$\sum_{j=0}^{\infty} E_{js}\tau_{js}\sinh(\tau_{js}b) C_j \delta_{jm} - Y_m'(a) \sum_{j=0}^{\infty} E_{js}\tau_{js}\sinh(\tau_{js}b)\gamma_j^2 Y_j'(a)$$

$$- \; Y_m'(a)(\gamma_m^2 + 2)\left\{\sum_{j=0}^{\infty} E_{js}\tau_{js}\sinh(\tau_{js}b)Y_j'(a)\right\}.$$

The terms in parentheses are zero as a consequence of the boundary condition (30.7) and corner constraint (30.10), and if we write

$$P_s = \sum_{j=0}^{\infty} E_{js}\tau_{js}\sinh(\tau_{js}b)\gamma_j^2 Y_j'(a) \tag{30.17}$$

then (30.16) yields

$$E_{ms} = \frac{Y_m'(a)P_s}{C_m\tau_{ms}\sinh(\tau_{ms}b)}. \tag{30.18}$$

The ansatz for $\psi_s(y,z)$, (30.11), can be multiplied by any arbitrary constant (or function of s) and so without loss of generality we can set $P_s = 1$. It follows that the duct modes have the form

$$\psi_s(y,z) = \sum_{m=0}^{\infty} \frac{Y_m'(a)Y_m(y)\cosh(\tau_{ms}z)}{C_m\tau_{ms}\sinh(\tau_{ms}b)}, \tag{30.19}$$

and as a consequence of (30.17) and (30.18) we have the identity

$$\sum_{j=0}^{\infty} \gamma_j^2 \left[Y_j'(a)\right]^2 C_j^{-1} = \sum_{j=0}^{\infty} \gamma_j^4 \sinh^2(\gamma_j a)C_j^{-1} \equiv 1. \tag{30.20}$$

Equation (30.19) contains the quantity s which corresponds to the axial wavenumber of the mode and is, as yet, unspecified. However, condition (30.9) has not been imposed and it is this that gives rise to the dispersion relation, viz.

$$L(s) = \sum_{m=0}^{\infty} \frac{[Y_m'(a)]^2 \cosh\{(s^2 - \gamma_m^2 - 1)^{1/2}b\}}{C_m(s^2 - \gamma_m^2 - 1)^{1/2}\sinh\{(s^2 - \gamma_m^2 - 1)^{1/2}b\}} = 0. \tag{30.21}$$

Despite its complicated form, it is a straightforward procedure to numerically solve (30.21) and the roots s_n, $n = 0,1,2,\ldots$ are the admissible wavenumbers for the duct modes. The notation ψ_n will henceforth be used to indicate that s has been replaced by s_n in the function ψ_s. Likewise, τ_{ms} becomes τ_{mn} etc. For an aluminium plate of thickness 5mm, and forcing frequency of 2000Hz, typical values of the coefficients are $\alpha = 0.002736$, $\mu = 1.05$. For these values, it transpires that $L(s)$ has two real roots together with an infinite number of pure imaginary roots

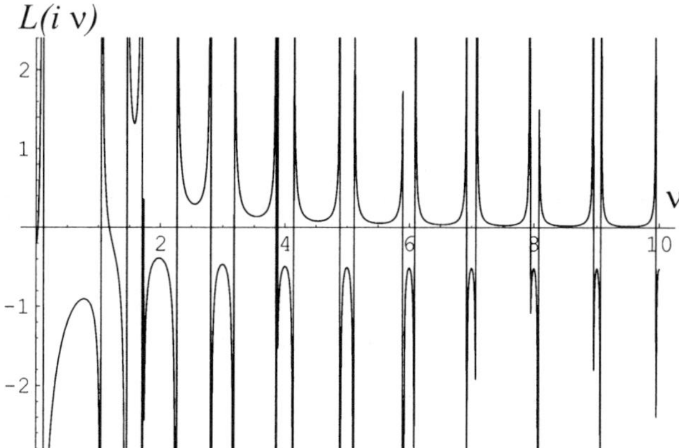

Figure 2   Plot of dispersion function $L(i\nu)$ for real ν.

at $s_n = i\nu_n$, where ν_n are shown as the zeros of the curve in Fig. 1. Although difficult to discern the zeros from this figure, as their locations are irregularly spaced, it is straightforward to show that the first few lie at the points $s = 0.829, 1.001, 0.0288i, 1.174i, 1.732i, 2.828i$.

A useful alternative form for $\psi_{ny}(a, z)$ is obtained by using the Fourier cosine series representation for $\cosh(\tau_{mn}z)$. It follows that

$$\psi_{ny}(a, z) = \frac{2}{b} \sum_{m=0}^{\infty} \frac{[Y_m'(a)]^2}{C_m} \sum_{q=0}^{\infty} \frac{(-1)^q \cos(q\pi z/b)}{\epsilon_q(\tau_{mn}^2 + (q\pi/b)^2)} \qquad (30.22)$$

where ϵ_q takes the value 2 for $q = 0$ and is 1 for $q = 1, 2, 3, \ldots$. Similar forms are available for $\psi_{nyyy}(a, z)$ etc. and these are needed in Section 3.

It is not sufficient simply to know the explicit form of the travelling waves in the duct. Any scattering problem will involve the propagation of all such modes and usually it will be necessary to determine the amplitudes. Analogous two-dimensional problems can be solved using orthogonality relations. Whilst it has not proved possible to determine a full orthogonality relation for the set of functions $\psi_n(y, z)$, $n = 0, 1, 2, \ldots$ the following is a useful 'partial' orthogonality relation (the derivation will be provided in a forthcoming article by the authors)

$$\alpha \int_0^b \int_0^a \psi_n(y, z)\psi_\ell(y, z)\, dy\, dz = D_\ell \delta_{\ell n} - \int_0^b \Big\{ \psi_{nyyy}(a, z)\psi_{\ell y}(a, z)$$

$$+ \psi_{ny}(a, z)\psi_{\ell yyy}(a, z) + 2\psi_{ny}(a, z)\psi_{\ell y}(a, z) \Big\}\, dz \qquad (30.23)$$

where $D_n = -L'(s_n)/(2s_n)$. Equation (30.23) it not in itself sufficient to perform mode-matching or, indeed, solve even simple scattering problems. However, supplementary identities can be derived which, when used in conjunction with (30.23), enable the solution of scattering problems in three-dimensional ducts of the type discussed in this article. One such identity is

$$\sum_{m=0}^{\infty} \frac{\psi_{ny}(a,z)\psi_{ny}(a,w)}{s_n D_n} = \frac{2}{b}\sum_{m=0}^{\infty}\frac{[Y_m'(a)]^2}{C_m}\sum_{n=0}^{\infty}\frac{\cos(n\pi z/b)\cos(n\pi w/b)}{\epsilon_n(\gamma_m^2 - (n\pi/b)^2 + 1)^{1/2}}$$
$$- \frac{2}{\pi i}\sum_{m=0}^{\infty}\sum_{j=0}^{\infty}\frac{[Y_m'(a)]^2}{C_m}\frac{[Y_j'(a)]^2}{C_j}I_{jm}(z,w) \quad (30.24)$$

for all $0 \le z, w \le b$ where

$$I_{jm}(z,w) = \int_0^{\infty} \frac{\cosh(\tau_{ms}z)\cosh(\tau_{js}w)}{\tau_{ms}\sinh(\tau_{ms}b)\tau_{js}\sinh(\tau_{js}b)L(s)} \, ds. \quad (30.25)$$

Note, for reasons of brevity, the derivation of (30.23) and (30.24) cannot be given here, but their effective use will be demonstrated below.

3.   A SEMI-INFINITE SCATTERING PROBLEM

In this section the use of (30.23) and (30.24) in solving a typical scattering problem is demonstrated. A fluid-structural wave is incident in the negative x-direction towards $x = 0$ along the elastic surface of a semi-infinite duct which occupies the space $x > 0$, $0 \le y \le a$, $-b \le z \le b$. The duct is closed at $x = 0$ by a rigid rectangular surface but in all other respects the duct is identical to that described at the beginning of Section 2. The fluid velocity potential can be expressed in terms of the eigenmodes $\psi_n(y,z)$, which are given in (30.19), as

$$\phi(x,y,z) = e^{-is_0 x}\psi_0(y,z) + \sum_{n=0}^{\infty}B_n e^{is_n x}\psi_n(y,z) \quad (30.26)$$

and the boundary condition at $x = 0$, $0 \le y \le a$, $-b \le z \le b$ is

$$\phi_x(0,y,z) = -is_0\psi_0(y,z) + i\sum_{n=0}^{\infty}B_n s_n\psi_n(y,z) = 0. \quad (30.27)$$

In addition to (30.27) two edge conditions must be applied along the line segment $x = 0$, $y = a$, $0 \le z \le b$ to specify the way in which the plate is 'attached' to the end barrier. Arbitrarily, it is assumed that the top elastic surface is clamped along this edge and so it follows that

$$\phi_y(0,a,z) = \psi_{0y}(a,z) + \sum_{n=0}^{\infty}B_n\psi_{ny}(a,z) = 0, \quad (30.28)$$

$$\phi_{yx}(0, a, z) \;=\; -is_0\psi_{0y}(a, z) + i\sum_{n=0}^{\infty} B_n s_n \psi_{ny}(a, z) = 0. \quad (30.29)$$

On multiplying (30.23) by $B_{\ell}s_{\ell}$, summing across ℓ and using (30.27) and (30.29), it is found that

$$\bar{B}_n = \frac{1}{s_n D_n} \sum_{\ell=0}^{\infty} \bar{B}_{\ell}s_{\ell} \int_0^b \psi_{ny}(a, w)\psi_{\ell yyy}(a, w)\, dw \quad (30.30)$$

where $\bar{B}_n = B_n - \delta_{n0}$ and the dummy variable z has been replaced by w. Now, using the alternative form for $\psi_{ny}(a, z)$ given by (30.22), expression (30.30) can, after significant rearrangement, be recast in the form

$$\bar{B}_n = \frac{2}{bs_n D_n} \sum_{p=0}^{\infty} \frac{1}{\epsilon_p} \sum_{m=0}^{\infty} \frac{[Y_m'(a)]^2 C_m^{-1}}{(\tau_{mn}^2 + (p\pi/b)^2)} \sum_{\ell=0}^{\infty} \bar{B}_{\ell}s_{\ell} \sum_{j=0}^{\infty} \frac{\gamma_j^2[Y_j'(a)]^2 C_j^{-1}}{(\tau_{jl}^2 + (p\pi/b)^2)}. \quad (30.31)$$

There remains one condition that has not yet been utilised, that is (30.28). On multiplying (30.30) by $\psi_{ny}(a, z)$, summing over n and using this last constraint, it is found that

$$-2\psi_{0y}(a, z) = \sum_{\ell=0}^{\infty} \bar{B}_{\ell}s_{\ell} \int_0^b f(z, w)\psi_{\ell yyy}(a, w)\, dw \quad (30.32)$$

where $f(z, w)$ is the right-hand side of (30.24). On multiplying both sides of (30.32) by $\cos(p\pi z/b)$ and integrating with respect to z, $0 \le z \le b$, it is found, after extensive algebraic manipulation, that

$$\sum_{\ell=0}^{\infty} \bar{B}_{\ell}s_{\ell} \sum_{j=0}^{\infty} \frac{\gamma_j^2[Y_j'(a)]^2}{C_j(\tau_{jl}^2 + (p\pi/b)^2)} \;=\; -\frac{2}{S(p)} \int_0^b \psi_{0y}(a, z)\cos(p\pi z/b)\, dz$$

$$+ \frac{2i}{\pi S(p)} \sum_{\ell=0}^{\infty} \bar{B}_{\ell}s_{\ell}F(\ell, p), \quad (30.33)$$

where

$$S(p) = (-1)^p \sum_{m=0}^{\infty} \frac{[Y_m'(a)]^2}{C_m(\gamma_m^2 - (p\pi/b)^2 + 1)^{1/2}} \quad (30.34)$$

and

$$F(\ell, p) = \sum_{m,j=0}^{\infty} \frac{[Y_m'(a)Y_j'(a)]^2}{C_m C_j} \int_0^b \int_0^b \psi_{\ell yyy}(a, w)\cos\frac{p\pi z}{b} I_{jm}(z, w)dzdw.$$

$$(30.35)$$

It now follows directly from (30.33) and (30.31) that the scattered wave amplitudes, B_n, $n = 0, 1, 2, \ldots$, are given by

$$B_n - \delta_{n0} = \frac{4i}{\pi b s_n D_n} \sum_{p=0}^{\infty} \frac{1}{\epsilon_p S(p)} \sum_{m=0}^{\infty} \frac{[Y'_m(a)]^2}{C_m(\tau_{mn}^2 + (p\pi/b)^2)} \times \qquad (30.36)$$

$$\left(i\pi \int_0^b \psi_{0y}(a, z) \cos(p\pi z/b) \, dz + \sum_{\ell=0}^{\infty} (B_\ell - \delta_{0\ell}) s_\ell F(\ell, p) \right).$$

Thus, by using the partial orthogonality relation (30.23) and the identity (30.24), the boundary value problem posed in (30.26)–(30.29) has been reduced to a (Fredholm) system of linear algebraic equations of the second kind for the unknown amplitudes B_n. Other identities, similar to (30.24), are available and may be employed if edge constraints different from the clamped conditions specified in (30.28) and (30.29) are to be applied.

In concludion, this article has been concerned with the propagation of coupled fluid structural waves along a three dimensional duct whose sides are a combination of rigid and elastic walls. It was found that the eigenfrequencies for propagating or decaying modes are determined from the roots of an unusual dispersion relation consisting of an infinite sum of terms. Although a complete representation for the acoustic/structural field in the duct can be constructed by summing over all eigenmodes, the absence of an orthogonality relation precludes explicit determination of the coefficients in this sum for any given scattering problem. A 'partial' orthogonality relation can be posed for the infinite duct, and from this it is shown, for a specific model problem, that the coefficients of the eigenmodes may be obtained by numerical solution of an infinite algebraic system of equations. In a forthcoming article the authors will show that this system of equations is well behaved for a variety of duct configurations (corner conditions etc.) and scattering conditions.

References

[1] Lawrie, J B and Abrahams, I D (1999) Quart. J. Mech. Appl. Math. **52**, 161–181.

[2] Abrahams, I D (1999) J. Acoust. Soc. Amer. **105**, 3009–3020.

[3] Andronov, I V and Belinskii, B P (1990) PMM J. Appl. Math. Mech. **54**, 366–371.

[4] Warren, D P and Lawrie, J B (2001) Tech. Rept TR/17/99, Dept. Math. Sci., Brunel University, UK (to be published in Wave Motion).

ULTRASONIC BACKSCATTERING ENHANCEMENTS FOR TRUNCATED OBJECTS IN WATER: QUANTITATIVE MODELS, TESTS AND SPECIAL CASES

P. L. Marston, F. J. Blonigen, B. T. Hefner
Department of Physics, Washington State University
Pullman, WA 99164-2814, USA
marston@wsu.edu

K. Gipson
Department of Physics, Grand Valley State University
Allendale, MI 49401, USA

S. F. Morse
Physical Acoustics Branch, Code 7136, Naval Research Laboratory
Washington, DC 20375, USA

1. INTRODUCTION

This survey is concerned with the scattering of sound in water by objects whose size significantly exceeds the acoustic wavelength. Diffraction from the edges of the object can be important for regions where reflected contributions are weak. Elastic contributions to the scattering, however, can be much larger in magnitude for certain viewing angles than simple edge diffraction. Such elastic contributions can greatly enhance the visibility of truncated cylinders imaged with a sonar [1]. For situations where the dominant elastic contributions are the result of leaky waves excited on truncated objects, an approximation method has been developed [2, 3]. Leaky waves have phase velocities that exceed the speed of sound in water. Examples include Rayleigh waves on metallic objects, the symmetric generalization of Lamb waves on plates and shells, and (at sufficiently high frequencies) the lowest antisymmet-

I.D. Abrahams et al. (eds.),
IUTAM Symposium on Diffraction and Scattering in Fluid Mechanics and Elasticity, 289–292.
© 2002 *Kluwer Academic Publishers. Printed in the Netherlands.*

ric generalization of a Lamb wave. This type of scattering enhancement has been measured by us for metallic tilted bluntly truncated cylinders [4, 5, 6] and cubes [7] and glass disks. For most of these examples, the leaky wave is significantly attenuated upon traveling the length of the scatterer. As a consequence of that damping, the quality factor Q is small for the global modes associated with the reverberations of the leaky waves between truncations on the object. Consequently resonances associated with the global modes of the scatterer can be unimportant. In the case of shells, the leaky Lamb waves are sufficiently dispersive that frequency-dependent synchronization conditions must be met to facilitate strong coupling between the leaky waves and the incident acoustic field [4].

Targets fabricated from materials where either the shear wave velocity or both the shear and the longitudinal wave velocities are less than the speed of sound in water involve special considerations. Examples of such materials include rubbers and plastics in contrast to the metallic targets noted earlier. Significant backscattering enhancements associated with refracted waves and caustics have been verified and modeled for tilted truncated plastic and rubber cylinders [8]. Rayleigh waves on plastics have subsonic velocities relative to water and must be modeled differently than for the corresponding metallic target [9].

2.   THEORETICAL FORMULATION

Various methods had been previously investigated for approximating the acoustic radiation and scattering contributions for leaky waves for a wide variety of objects. Existing results were not directly applicable to the problems of interest here either because the assumptions of thin shell mechanics are no longer applicable or because of geometric complications. For the high frequencies under consideration, the physically relevant leaky wave paths were sufficiently close to specific rays of interest that for the purpose of approximating the outgoing amplitude from any specific ray, the propagation anisotropy introduced by surface curvature could be neglected. The formulation [2, 3] first approximates the pressure amplitude p_l radiated by the leaky wave at the surface of the scatterer by convolving the local amplitude p_i of the incident wave on the illuminated portion of the surface with a two-dimensional response function. The resulting two-dimensional integral for p_l at a surface point S may be expressed as

$$p_l(S) \approx \int_{\mathcal{D}} p_i(S') \left[\kappa H_0^{(1)}(k_p s) \right] d\mathcal{A}',$$

where $H_0^{(1)}$ is the Hankel function having an argument proportional to the geodesic distance s between S and an illuminated point at S' where the incident wave's complex amplitude is $p_i(S')$ and $d\mathcal{A}'$ is the differential area of the contributing surface patch. The domain $\mathcal{D}$ is restricted depending on the directional properties of the waves of interest. The leaky wavenumber for the lth class of leaky wave is $k_p = k_l + i\alpha$ where $k_l = kc/c_l$ and k is the wavenumber in water, $\kappa \approx -\alpha a k_l \exp(i\phi_{bl})$, and ϕ_{bl} is a background phase. The farfield amplitude is calculated through the evaluation of a Rayleigh propagation integral. Previous results for a circular cylinder at zero tilt were recovered [2]. The benchmark problem of meridional rays on an infinite circular cylinder was analyzed [3] leading to predictions for the scattering to observers in the meridional plane as the tilt is varied near the optimum coupling angle. These ray theoretic predictions were confirmed by comparison with exact results for large enhancements for tilted solid [3] and empty shell [10] cylinders. Scattering [4, 5, 6, 7, 11, 12] and holographic measurements [13] for truncated targets also support this formulation provided the partial leaky wave reflection at truncations is included in the analysis. In related research, a component of the complex farfield scattering amplitude for objects having inversion symmetry has been expressed for an arbitrary scattering angle in terms of an integral involving scattering amplitudes at all angles. This is a generalized optical theorem for acoustical scattering in the absence of dissipation [14]. Many of the aforementioned targets that we have studied have the required inversion symmetry.

Acknowledgments

This research was supported by the Office of Naval Research.

References

[1] Kaduchak, G, Wassmuth, C M and Loeffler, C M (1996) *Elastic wave contributions in high resolution acoustic images of fluid-filled finite cylindrical shells in water*, J. Acoust. Soc. Amer. **100**, 64–71.

[2] Marston, P L (1997) *Spatial approximation of leaky wave surface amplitudes for three-dimensional high-frequency scattering: Fresnel patches and application to edge-excited and regular helical waves on cylinders*, J. Acoust. Soc. Amer. **102**, 1628–1638.

[3] Marston, P L (1997) *Approximate meridional leaky ray amplitudes for tilted cylinders: end-backscattering enhancements and comparisons with exact theory for infinite solid cylinders*, J. Acoust. Soc. Amer. **102**, 358–369; **103** (1998) 2236.

[4] Morse, S F, Marston, P L and Kaduchak, G (1998) *High-frequency backscattering enhancements by thick finite cylindrical shells in water at oblique incidence: experiments, interpretation, and calculations*, J. Acoust. Soc. Amer. **103**, 785–794.

[5] Gipson, K and Marston, P L (1999) *Backscattering enhancements due to reflection of meridional leaky Rayleigh waves at the blunt truncation of a tilted solid cylinder in water: observations and theory*, J. Acoust. Soc. Amer. **106**, 1673–1680.

[6] Gipson, K and Marston, P L (2000) *Backscattering enhancements from Rayleigh waves on the flat face of a tilted solid cylinder in water*, J. Acoust. Soc. Amer. **107**, 112–117.

[7] Gipson, K and Marston, P L (1999) *Backscattering enhancements due to retroreflection of ultrasonic leaky Rayleigh waves at corners of solid elastic cubes in water*, J. Acoust. Soc. Amer. **105**, 700–710.

[8] Blonigen, F J and Marston, P L (2000) *Backscattering enhancements for tilted solid plastic cylinders in water due to the caustic merging transition: observations and theory*, J. Acoust. Soc. Amer. **107**, 689–698.

[9] Hefner, B T and Marston, P L (2000) *Backscattering enhancements associated with subsonic Rayleigh waves on polymer spheres in water: observation and modeling for acrylic spheres*, J. Acoust. Soc. Amer. **107**, 1930–1936.

[10] Morse, S F and Marston, P L (1999) *Meridional ray contributions to scattering by tilted cylindrical shells above the coincidence frequency: ray theory and computations*, J. Acoust. Soc. Amer. **106**, 2595–2600.

[11] Morse, S F (1998) *High frequency acoustic backscattering enhancements for finite cylindrical shells in water at oblique incidence*, Ph.D. Thesis, Washington State University.

[12] Morse, S F and Marston, P L (2001) *Degradation of meridional ray backscattering enhancements for tilted cylinders by mode conversion: wide-band observations using a chirped PVDF sheet source*, IEEE J. Ocean. Eng. **26**, 152–155.

[13] Hefner, B T and Marston, P L (2001) *Backscattering enhancements associated with the excitation of symmetric Lamb waves on a circular plate: direct and holographic observations*, ARLO, Acoust. Res. Lett. Online **2**, to appear.

[14] Marston, P L (2001) *Generalized optical theorem for scatterers having inversion symmetry: applications to acoustic backscattering*, J. Acoust. Soc. Amer. **109**, 1291–1295.

VII

ELASTIC WAVE PROPAGATION IN LAYERED AND ANISOTROPIC MATERIALS

KISS SINGULARITIES OF GREEN'S FUNCTIONS FOR NON-STRICTLY HYPERBOLIC EQUATIONS

V. A. Borovikov

Benemerita Universidad Autonoma de Puebla, Puebla Pue. 72570, Mexico

borovik@fcfm.buap.mx

D. Gridin

Centre for Waves and Fields, School of EEIE, South Bank University

103 Borough Road, London SE1 0AA, UK

gridind@sbu.ac.uk

1.   INTRODUCTION

Let us consider the equation

$$L\left(\frac{\partial}{\partial t}, \frac{\partial}{\partial x_1}, \ldots, \frac{\partial}{\partial x_n}\right) u(t, \mathbf{x}) = f(t, \mathbf{x}), \tag{32.1}$$

where $L(v, \boldsymbol{\xi}) = v^m + \sum_{k=1}^{m} v^{m-k} L_k(\boldsymbol{\xi})$ is a homogeneous polynomial of degree m and $\mathbf{x} = (x_1, \ldots, x_n)$. It is called strictly hyperbolic if, for any $\boldsymbol{\xi} \neq 0$, the equation $L(v, \boldsymbol{\xi}) = 0$ has m distinct real roots $v_1, \ldots, v_m$. Surface $S : L(1, \boldsymbol{\xi}) = 0$ is known as the slowness surface. It consists of $[m/2]$ non-overlapping ovals which enclose the origin. When m is odd, one more component, an unpaired part, is added [1]. A set Γ of points $(t, \mathbf{x})$, such that the plane $t + (\mathbf{x}, \boldsymbol{\xi}) = 0$ is tangential to S, is called a characteristic cone and consists of $[(m+1)/2]$ sheets. The section of the cone by the plane $t = 1$ is known as the wave surface.

When $f(t, \mathbf{x}) = \delta(t, \mathbf{x})$, the solution $G(t, \mathbf{x})$, such that $G(t, \mathbf{x}) = 0$ for $t < 0$, is called the Green's function of (32.1). It is analytic outside the characteristic cone Γ and has a singularity on the cone. At regular points of the characteristic cone the singularities of Green's function of strictly hyperbolic equations have been described previously in [2]. The above definitions can be extended to the case of systems of equations.

However, the condition of strict hyperbolicity is not always satisfied, and there are equations (and systems) such that, for some values of $\boldsymbol{\xi}$,

295

I.D. Abrahams et al. (eds.),
IUTAM Symposium on Diffraction and Scattering in Fluid Mechanics and Elasticity, 295–302.
© 2002 *Kluwer Academic Publishers. Printed in the Netherlands.*

the equation $L(v, \boldsymbol{\xi}) = 0$ has double roots v. For instance, elastic media of cubic symmetry have conical points of the slowness surface [3], and transversely isotropic media possess points of tangential contact of the SH and qSV sheets of the slowness surface (see, e.g., [4]). The equations of crystal optics are also non-strictly hyperbolic since the corresponding slowness surfaces contain conical points for biaxial crystals and tangential points for uniaxial crystals (see, e.g., [5]).

It is well-known that if a slowness surface has singular points, then the corresponding Green's function may have additional singular surfaces. For instance, a conical point maps onto the so-called plane lid of the wave surface (see, e.g., [5]). For elastic media of cubic symmetry the singularities of Green's function on a plane lid far from the lid edge have been described in [3]. These results have been generalised for arbitrary non-strictly hyperbolic equations with conical points in [6].

The case of the tangential contact of the SH and qSV sheets of the slowness surface for a transversely isotropic solid has been treated by two different techniques in [4] and [7]. A special description has been obtained for the so-called kiss singularity, that is the singularity of the SH and qSV parts of Green's tensor near the tangential point of the characteristic cone.

In this paper, we consider the case of tangential contact of two sheets of slowness surface for an arbitrary hyperbolic equation. A method is developed to describe the kiss singularity in terms of a sum of three terms, one of which is an integral of special type. A fuller presentation of the results, including several examples, is given in [8].

2.   SINGULARITY AT A REGULAR POINT

Following [6], in this section we consider a singularity of Green's function at a regular point of the characteristic cone. For $t > 0$, Green's function can be expressed (up to a polynomial in t for even $n < m - 1$) as a sum of integrals over the unit sphere (see, e.g., [9]):

$$G(t, \mathbf{x}) = \frac{i^{n+1}}{(2\pi)^{n-1}} \int\limits_{|\boldsymbol{\xi}|=1} \sum_{j=1}^{m} c_j f_p^s(\mathbf{x} \cdot \boldsymbol{\xi} + v_j t) \, d\Omega, \qquad (32.2)$$

where $p = m - n - 1$, $s = (-1)^{n+1}$ and $d\Omega$ is a surface element of the unit sphere. Here, coefficients c_j are given by

$$c_j = [(v_1 - v_j)(v_2 - v_j) \cdots (v_{j-1} - v_j)(v_{j+1} - v_j) \cdots (v_m - v_j)]^{-1}.$$

Functions $f_p^1(\tau)$ are expressed as follows:

$$f_p^1(\tau) = \begin{cases} [2\Gamma(p+1)]^{-1}\tau^p \mathrm{sgn}\,\tau, & p \geq 0, \\ \delta^{-p-1}(\tau), & p < 0. \end{cases} \tag{32.3}$$

Functions $f_p^{-1}(\tau)$ are Hilbert transforms of $f_p^1(\tau)$ and are given by

$$f_p^{-1}(\tau) = \begin{cases} -(\mathrm{i}/\pi)[\Gamma(p+1)]^{-1}\tau^p \log|\tau|, & p \geq 0, \\ (\mathrm{i}/\pi)(-1)^p\Gamma(-p)\,\tau^p, & p < 0. \end{cases} \tag{32.4}$$

An alternative representation of Green's function as an integral over the slowness surface is furnished by Herglotz–Petrovski formulae (e.g., [9]).

The singularities of $G(t, \mathbf{x})$ are determined by the asymptotics as $\omega \to \pm\infty$ of its Fourier transform $F(\omega, \mathbf{x})$. Performing the Fourier transform of the jth-term in (32.2), applying the multidimensional stationary phase method and taking the inverse Fourier transform, we obtain, up to an infinitely differentiable function of t [2],

$$\begin{aligned} G_j(t, \mathbf{x}) &\equiv \frac{\mathrm{i}^{n+1}}{(2\pi)^{n-1}} \int\limits_{|\boldsymbol{\xi}|=1} c_j f_p^s(\mathbf{x} \cdot \boldsymbol{\xi} + v_j t)\, \mathrm{d}\Omega \tag{32.5} \\ &\simeq -\frac{\mathrm{i}^{\sigma^+}}{(2\pi)^{(n-1)/2}} g_j(\boldsymbol{\xi}^0) f_{m-(n+3)/2}^{s'}[\mathbf{x} \cdot \boldsymbol{\xi}^0 + v_j(\boldsymbol{\xi}^0)t]. \end{aligned}$$

In (32.5), $\boldsymbol{\xi}^0$ is a stationary point of the phase function $\varphi(\mathbf{x}, \boldsymbol{\xi}) = -\mathbf{x} \cdot \boldsymbol{\xi}/v_j(\boldsymbol{\xi})$, H_0 is the Hessian matrix of φ at $\boldsymbol{\xi}^0$, $\sigma = \sigma^+ - \sigma^-$ is its signature ($\sigma^\pm$ is the number of positive/negative eigenvalues of H_0, respectively), $s' = (-1)^{\sigma^+}$ and we have

$$g_j(\boldsymbol{\xi}^0) = |\det H_0|^{-1/2} c_j(\boldsymbol{\xi}^0) v_j^{(1-n)/2}(\boldsymbol{\xi}^0)(\xi_n^0)^{-1}. \tag{32.6}$$

For integer values of $m - (n+3)/2$, that is for odd n, the functions $f_{m-(n+3)/2}^s(\tau)$ are given by (32.3) and (32.4), and otherwise we have

$$f_{k+1/2}^1(\tau) = \begin{cases} \tau^{k+1/2}/\Gamma(k+3/2), & \tau > 0, \\ 0, & \tau < 0, \end{cases} \tag{32.7}$$

and

$$f_{k+1/2}^{-1}(\tau) = \begin{cases} 0, & \tau > 0, \\ -\mathrm{i}(-1)^k|\tau|^{k+1/2}/\Gamma(k+3/2), & \tau < 0. \end{cases} \tag{32.8}$$

Equation (32.5) describes the singularity of Green's function at a regular point of the characteristic cone.

3.   KISS SINGULARITY

Now let us consider the situation when the jth and $(j+1)$th sheets of the slowness surface S (and wave surface) touch tangentially over a closed manifold of dimension q. Let us denote the corresponding manifold of the unit sphere as $\mathcal{K}$. In general, the Gaussian curvatures of the two sheets of the slowness surface at the tangential manifold do not coincide; as $\mathbf{x}$ approaches the tangential manifold of the wave surface, the sum of the jth and $(j+1)$th terms of type (32.5) tends to infinity, and therefore the case needs a special treatment.

It is sufficient to consider only the sum of the corresponding terms. Their Fourier transform is given by

$$F(\omega, \mathbf{x}) \simeq \frac{(\mathrm{i}\,\mathrm{sgn}\omega)^{n+1}}{(2\pi)^{n-1}} \left(\frac{\mathrm{i}}{\omega}\right)^{p+1} \times \tag{32.9}$$

$$\int_{|\boldsymbol{\xi}|=1} \left[c_j v_j^p e^{-\mathrm{i}\omega\mathbf{x}\cdot\boldsymbol{\xi}/v_j} + c_{j+1} v_{j+1}^p e^{-\mathrm{i}\omega\mathbf{x}\cdot\boldsymbol{\xi}/v_{j+1}} \right] \nu(\boldsymbol{\xi})\, \mathrm{d}\Omega,$$

where $\nu(\boldsymbol{\xi})$ is a neutraliser singling out a small vicinity of point $\boldsymbol{\xi}^*$ belonging to $\mathcal{K}$. Each of the terms of the integrand has poles at $\boldsymbol{\xi} \in \mathcal{K}$ due to the $(v_{j+1}-v_j)^{-1}$ and $(v_j-v_{j+1})^{-1}$ factors in c_j and c_{j+1} respectively.

NO STATIONARY POINTS CLOSE TO $\mathcal{K}$

First, let us show that the tangential manifold does not lead to a new singular surface of the Green's function, that is, the non-uniform contribution of the tangential manifold to the Fourier transform of Green's function is asymptotically negligible.

It is convenient to introduce slowness $s_j(\boldsymbol{\xi}) = 1/v_j(\boldsymbol{\xi})$. The integral in (32.9) can be re-written as follows:

$$J = \int_{|\boldsymbol{\xi}|=1} a(\boldsymbol{\xi}) e^{-\mathrm{i}\omega\mathbf{x}\cdot\boldsymbol{\xi} s_j} \frac{1 - b(\boldsymbol{\xi}) e^{-\mathrm{i}\omega(s_{j+1}-s_j)\mathbf{x}\cdot\boldsymbol{\xi}}}{s_{j+1} - s_j} \nu(\boldsymbol{\xi})\, \mathrm{d}\Omega. \tag{32.10}$$

Since $b(\boldsymbol{\xi}) = 1$, $\boldsymbol{\xi} \in \mathcal{K}$, we can decompose the integral J as follows

$$J = J_a + J_b = \int_{|\boldsymbol{\xi}|=1} a(\boldsymbol{\xi}) \frac{1 - b(\boldsymbol{\xi})}{s_{j+1} - s_j} e^{-\mathrm{i}\omega\mathbf{x}\cdot\boldsymbol{\xi} s_{j+1}} \nu(\boldsymbol{\xi})\, \mathrm{d}\Omega$$

$$+ \int_{|\boldsymbol{\xi}|=1} a(\boldsymbol{\xi}) e^{-\mathrm{i}\omega\mathbf{x}\cdot\boldsymbol{\xi} s_j} \frac{1 - e^{-\mathrm{i}\omega(s_{j+1}-s_j)\mathbf{x}\cdot\boldsymbol{\xi}}}{s_{j+1} - s_j} \nu(\boldsymbol{\xi})\, \mathrm{d}\Omega.$$

Using the identity

$$\frac{1 - e^{-i\omega\alpha}}{\alpha} = i\omega \int_0^1 e^{-i\omega\alpha y}\,dy, \tag{32.11}$$

the integral J_b may be re-written as

$$J_b = i\omega \int_0^1 K(\mathbf{x},\boldsymbol{\xi},y)\,dy, \tag{32.12}$$

where we use the notations

$$K(\mathbf{x},\boldsymbol{\xi},y) = \int_{|\boldsymbol{\xi}|=1} a(\boldsymbol{\xi})\mathbf{x}\cdot\boldsymbol{\xi}\,e^{i\omega\varphi(\mathbf{x},\boldsymbol{\xi},y)}\nu(\boldsymbol{\xi})\,d\Omega, \tag{32.13}$$

and

$$\varphi(\mathbf{x},\boldsymbol{\xi},y) = -\mathbf{x}\cdot\boldsymbol{\xi}/\tilde{v}(\boldsymbol{\xi},y), \quad \tilde{v}(\boldsymbol{\xi},y) = \{s_j(\boldsymbol{\xi}) + [s_{j+1}(\boldsymbol{\xi}) - s_j(\boldsymbol{\xi})]y\}^{-1}.$$

To evaluate the non-uniform contribution of the tangential manifold we assume that the phase functions do not have any stationary points within the support of $\nu(\boldsymbol{\xi})$. Then the integral J_a does not have any critical points, and thus decays more rapidly than any power of $1/\omega$. The phase function of the integral K does not have any stationary points, since near the tangential manifold $|s_{j+1} - s_j|$ is much smaller than s_j. Therefore, the integral K does not have any critical points, and tends to zero more rapidly than any power of $1/\omega$. This means that J_b is also asymptotically negligible, and we have $F(\omega,\mathbf{x}) \simeq O(\omega^{-\infty})$. The above argument is valid for any $\boldsymbol{\xi}^* \in \mathcal{K}$, and therefore the tangential manifold $\mathcal{K}$ does not lead to any singular surface of Green's function.

STATIONARY POINTS CLOSE TO $\mathcal{K}$

Now let us consider the case when there are two isolated stationary points, one for each sheet, close to some point $\boldsymbol{\xi}^* \in \mathcal{K}$ in the domain where $\nu(\boldsymbol{\xi}) = 1$. We represent Green's function as the sum

$$G(t,\mathbf{x}) = G_a(t,\mathbf{x}) + G_b(t,\mathbf{x}) + G_c(t,\mathbf{x}). \tag{32.14}$$

Here, $G_a(t,\mathbf{x}) + G_b(t,\mathbf{x})$ is the inverse Fourier transform of the function $F(\omega,\mathbf{x})$ (see (32.9)); $G_a(t,\mathbf{x})$ corresponds to J_a, and $G_b(t,\mathbf{x})$, to J_b. The inverse Fourier transform of $F(\omega,\mathbf{x})$ is determined up to a function G_c, which is infinitely differentiable with respect to t but can have singularities as a function of $\mathbf{x}$. These singularities, which are required for the description of all singularities of Green's function, may be found by using the analytical properties of $G(t,\mathbf{x})$ outside the characteristic cone.

The integral J_a may be evaluated by the stationary phase method as when dealing with a regular point. Applying the inverse Fourier transform, we obtain

$$G_a(t,\mathbf{x}) \simeq -\frac{\mathrm{i}^{\sigma_1^+}}{(2\pi)^{(n-1)/2}} g_1(\boldsymbol{\xi}^1) f_{m-(n+3)/2}^{k_1}[\mathbf{x}\cdot\boldsymbol{\xi}^1 + v_{j+1}(\boldsymbol{\xi}^1)t], \quad (32.15)$$

where we use the following notations:

$$g_1(\boldsymbol{\xi}^1) = |\det H_1|^{-1/2} a(\boldsymbol{\xi}^1) \frac{1-b(\boldsymbol{\xi}^1)}{s_{j+1}(\boldsymbol{\xi}^1) - s_j(\boldsymbol{\xi}^1)} (\xi_n^1)^{-1} s_{j+1}^{m-(n+3)/2}(\boldsymbol{\xi}^1),$$

$$\varphi_1(\mathbf{x},\boldsymbol{\xi}) = \varphi(\mathbf{x},\boldsymbol{\xi},1) = -\mathbf{x}\cdot\boldsymbol{\xi}\, s_{j+1}(\boldsymbol{\xi}), \quad k_1 = (-1)^{\sigma_1^+}.$$

Here and below, $\boldsymbol{\xi}^y$ is the stationary point of $\varphi_y = \varphi(\mathbf{x},\boldsymbol{\xi},y)$, H_y is the Hessian matrix of φ_y at $\boldsymbol{\xi}^y$, and σ_y^+ is the number of positive eigenvalues for H_y. Obviously, $\boldsymbol{\xi}^y$ depends on both $\mathbf{x}$ and y.

Let us consider the integral K. Its phase function involves the slowness

$$\tilde{s}(\boldsymbol{\xi},y) = \tilde{v}^{-1}(\boldsymbol{\xi},y) = s_j(\boldsymbol{\xi}) + [s_{j+1}(\boldsymbol{\xi}) - s_j(\boldsymbol{\xi})]y, \quad (32.16)$$

which is a function of y. Now, we need to distinguish two cases, a simpler particular case and a more complicated general case.

Case I.   For $y \in [0,1]$ the slowness surface for $\tilde{s}(\boldsymbol{\xi},y)$ does not have locally any parabolic points, that is, the Gaussian curvature of the surface and the Hessian $\Delta(y) = \det H_y$ are both non-zero, and $\sigma_y = \sigma_1$. This means that we can apply the standard stationary phase method to the integral K as in the case of a regular point. Then, applying the inverse Fourier transform we obtain

$$G_b(t,\mathbf{x}) \simeq \frac{\mathrm{i}^{\sigma_1^+}}{(2\pi)^{(n-1)/2}} \int_0^1 g_2(\boldsymbol{\xi}^y,y) f_{m-(n+5)/2}^{k_1}[\mathbf{x}\cdot\boldsymbol{\xi}^y + \tilde{v}(\boldsymbol{\xi}^y,y)t]\,\mathrm{d}y,$$

$$(32.17)$$

where we use the notation

$$g_2(\boldsymbol{\xi}^y,y) = |\Delta(y)|^{-1/2} a(\boldsymbol{\xi}^y)\mathbf{x}\cdot\boldsymbol{\xi}^y(\xi_n^y)^{-1}\tilde{v}^{-m+(n+5)/2}(\boldsymbol{\xi}^y,y). \quad (32.18)$$

Case II.   There are $y \in (0,1)$ such that some points of the tangential manifold are parabolic points of the slowness surface for $\tilde{s}(\boldsymbol{\xi},y)$, and thus the Hessian $\Delta(y)$ has zeros on the same interval. (For example, let us consider a two-dimensional case when two curves of the slowness surface kiss at a point and have curvatures of the opposite sign at it. When

$y = 0$, $\tilde{s}(\boldsymbol{\xi}, y)$ coincides with one of the slownesses, and when $y = 1$, with the other. Thus the curvature of the corresponding curve is of the opposite sign for $y = 0$ and $y = 1$. It follows that there must be some $y \in (0, 1)$ such that the curvature of the slowness curve for $\tilde{s}(\boldsymbol{\xi}, y)$ is zero.)

The application of the standard stationary phase method to the integral K leads to (32.17) with branch points due to the factor $[\Delta(y)]^{-1/2}$. However, this difficulty may be overcome if we shift the contour of integration in the integral J_b into the complex y-plane, that is, away from zeros of the Hessian. This should be done in such a way that the integrand remains bounded. Let us assume that $s_{j+1} \geq s_j$. Then we should shift the integration contour to the upper half-plane for $\omega > 0$, and to the lower half-plane for $\omega < 0$. Applying the saddle-point method and then shifting the integration contour back just above or below the real y-axis we obtain

$$F_b(\omega, \mathbf{x}) \equiv \frac{(\mathrm{i}\,\mathrm{sgn}\omega)^{n+1}}{(2\pi)^{n-1}} \left(\frac{\mathrm{i}}{\omega}\right)^{p+1} J_b = \int\limits_{0+\mathrm{i}\,\mathrm{sgn}\omega\,0}^{1+\mathrm{i}\,\mathrm{sgn}\omega\,0} H(\omega, \mathbf{x}, y)\, dy,$$

where $H(\omega, \mathbf{x}, y)$ is the asymptotics as $\omega \to \pm\infty$ of the integral (32.13):

$$H(\omega, \mathbf{x}, y) \simeq \frac{(\mathrm{i}\,\mathrm{sgn}\omega)^{\sigma_y^+}}{(2\pi)^{(n-1)/2}} \left(\frac{\mathrm{i}}{\omega}\right)^{m-(n+3)/2} g_2(\boldsymbol{\xi}^y, y)\tilde{v}^{m-(n+5)/2} e^{-\mathrm{i}\omega\mathbf{x}\boldsymbol{\xi}^y/\tilde{v}}.$$

Now, when applying the inverse Fourier transform, the integration should be performed separately for $\omega > 0$ and $\omega < 0$. Taking it, we obtain

$$G_b(t, \mathbf{x}) \simeq \frac{(-1)^m}{(2\pi)^{(n+1)/2}} \left[\int\limits_{0+\mathrm{i}0}^{1+\mathrm{i}0} + \int\limits_{0-\mathrm{i}0}^{1-\mathrm{i}0} (-1)^{\sigma_y^+ - n}\right] \mathrm{i}^{\sigma_y^+ - n} g_2(\boldsymbol{\xi}^y, y)$$

$$\times \mu_{m-(n+3)/2}(\mathbf{x} \cdot \boldsymbol{\xi}^y + \tilde{v}t)\, dy, \tag{32.19}$$

where the functions $\mu_\nu(\tau)$ are defined for integer $k > 0$ by

$$\mu_k(\tau) = \frac{(-1)^k}{\Gamma(k)} \tau^{k-1} \log \tau, \qquad \tau > 0, \tag{32.20}$$

and otherwise by

$$\mu_\nu(\tau) = \Gamma(1 - \nu)\tau^{\nu-1}, \qquad \tau > 0. \tag{32.21}$$

For $\tau < 0$ the functions $\mu_\nu(\tau)$ are defined by analytic continuation over the lower half-plane $\mathrm{Im}\,\tau < 0$.

4.   CONCLUSIONS

It has been shown that a manifold of tangential contact of two sheets of the slowness surface does not lead to a new singular surface of Green's function. However, near the tangential manifold, a special description of the singularity of Green's function is required. The description has been given in the form of a sum of three terms: (i) a term similar to the regular case ((32.15)); (ii) an integral term ((32.17) in Case I and (32.19) in Case II); and (iii) an additional term which is required sometimes for the analyticity of Green's function outside the characteristic cone. The integral term may often be evaluated analytically (see [8] for examples). The method developed may be useful in problems of anisotropic wave propagation.

Acknowledgments

The work has been carried out under the EPSRC Grant GR/M31552. We are grateful to Dr Larissa Fradkin for her encouragement of the project.

References

[1] Petrovski, I G (1945) *On the diffusion of waves and lacunas for hyperbolic equations*, Mat. Sbornik **17**, 289–370.

[2] Borovikov, V A (1959) *Fundamental solutions of linear partial differential equations with constants coefficients*, Trans. Moscow Math. Soc. **8**, 199–257.

[3] Burridge, R (1967) *The singularity on the plane lids of the wave surface of elastic media with cubic symmetry*, Quart. J. Mech. Appl. Math. **20**, 41–56.

[4] Vavryčuk, V (1999) *Properties of S waves near a kiss singularity*, Geophys. J. Int. **138**, 581–589.

[5] Courant, R and Hilbert, D (1962) *Methods of Mathematical Physics*, vol. II, New York: Interscience.

[6] Borovikov, V A (2000) *Singularities of the Green function for non-strictly hyperbolic operators*, Russ. J. Math. Phys. **7**, 261–278.

[7] Gridin, D (2000) *Far-field asymptotics of the Green's tensor for a transversely isotropic solid*, Proc. Roy. Soc. Lond. A **456**, 571–591.

[8] Borovikov, V A and Gridin, D (2001) *Kiss singularities of Green's functions of non-strictly hyperbolic equations*, Proc. Roy. Soc. Lond. A **457**, 1059–1078.

[9] Gelfand, I M and Shilov, G E (1964) *Generalized Functions*, vol. I, New York: Academic Press.

DEFLECTION OF A PARTIALLY CLAMPED ELASTIC PLATE

I. D. Abrahams*
Department of Mathematics
University of Manchester, Manchester M13 9PL, UK
i.d.abrahams@ma.man.ac.uk

A. M. J. Davis
Department of Mathematics
The University of Alabama, Tuscaloosa, AL 35487, USA
adavis@euler.math.ua.edu

Abstract   This article is concerned with a long-standing problem in bending plate theory, which is the static or dynamic deflection of a thin elastic strip clamped along one (infinite) edge. The other edge has mixed boundary conditions; clamped on a semi-infinite part of the edge and free on the remaining half. Of primary interest is the scattering of incident plate flexural waves by the discontinuity in the edge condition, but for brevity only a uniform static loading is considered here. The problem is reduced to a matrix Wiener-Hopf equation which is solved by an approximate factorization scheme introduced recently to tackle systems of this class.

1.   INTRODUCTION

The transverse oscillations of a thin elastic plate, which may be loaded, are governed by a differential equation, fourth order in space and second order in time, which is a standard example in texts devoted to separation of variables and Fourier transform techniques, e.g. [1], [2]. The general equilibrium state, including the boundary conditions at a free edge, is discussed by [3]. However, a classic unsolved problem in bending plate theory is the static or dynamic displacement of a forced infinite strip having one edge clamped and the other clamped or free on two semi-

---

*This research was partially supported by a *Grant in Aid of Research* from the Leverhulme Trust, UK.

I.D. Abrahams et al. (eds.),
IUTAM Symposium on Diffraction and Scattering in Fluid Mechanics and Elasticity, 303–312.
© 2002 *Kluwer Academic Publishers. Printed in the Netherlands.*

infinite intervals; henceforth referred to as a partially clamped plate. Forcing may be a uniform loading over the plate, an incident flexural wave from infinity etc. The 'parallel lines' geometry suggests the use of the Wiener-Hopf technique; however, the advantage of the parallel edges in creating constant bending profiles far away in either direction is off-set by the appearance of a matrix Wiener-Hopf system. Such vectorial equations are not, in general, amenable to exact solutions. Thus, the following analysis of the deflection of a partially clamped plate is both an application of a new Padé approximant procedure [5] and the first solution of this classic problem by the Wiener-Hopf technique. For ease of exposition only brief details of the static case are given here; a fuller account of both static and dynamic motions is discussed elsewhere [6].

2.   THE WIENER-HOPF PROBLEM

In terms of Cartesian coordinates (x, y), the out-of-plane displacement $w(x, y)$ of a plate, situated at $-\infty < x < \infty$, $0 \le y \le 1$, is governed by the fourth order equation

$$\nabla^4 w = \frac{q}{D}, \tag{33.1}$$

where q is a uniform load and D is the constant flexural rigidity. Possible boundary conditions, given by [3], are

$$w = 0 = \frac{\partial w}{\partial y} \text{ at a clamped edge} \tag{33.2}$$

and

$$\nu\frac{\partial^2 w}{\partial x^2} + \frac{\partial^2 w}{\partial y^2} = 0, \qquad (2 - \nu)\frac{\partial^3 w}{\partial x^2 \partial y} + \frac{\partial^3 w}{\partial y^3} = 0 \text{ at a free edge.} \tag{33.3}$$

Thus, for a plate clamped at $y = 0, 1$,

$$w = \frac{q}{24D}y^2(1 - y)^2 \tag{33.4}$$

while, for a plate clamped at $y = 0$ but free at $y = 1$,

$$w = \frac{q}{24D}y^2(6 - 4y + y^2). \tag{33.5}$$

Consider a plate *clamped* at $y = 0$ and $y = 1$, $x < 0$ but *free* at $y = 1$, $x > 0$. Then, in terms of two sets of Papkovitch-Fadle eigenfunctions whose details are not needed here, (33.4,33.5) yield the solution form

$$w = \frac{q}{24D}y^2(1 - y)^2 + \sum_{n\neq0} A_n e^{\lambda_n x}\phi_n(y) \qquad x < 0,$$

$$w = \frac{q}{24D}y^2(6 - 4y + y^2) + \sum_{n\neq0} B_n e^{-\mu_n x}\psi_n(y) \qquad x > 0. \tag{33.6}$$

Now, in view of (33.4), it is advantageous to write the total displacement as

$$w = \frac{q}{24D}y^2(1-y)^2 + \frac{q}{D}W, \qquad (33.7)$$

where W satisfies (33.2) at $y = 0$ and $y = 1$, $x < 0$ and, from (33.1) and (33.3), W is biharmonic and, at $y = 1$, $x > 0$:

$$\nu\frac{\partial^2 W}{\partial x^2} + \frac{\partial^2 W}{\partial y^2} = -\frac{1}{12}e^{-\epsilon x}, \qquad (2-\nu)\frac{\partial^3 W}{\partial x^2\partial y} + \frac{\partial^3 W}{\partial y^3} = -\frac{1}{2}e^{-\epsilon x}. \quad (33.8)$$

Here the exponential factors have been introduced for mathematical convenience, with ϵ a small positive number that will revert to zero after application of the Wiener-Hopf technique.

In terms of the Fourier transform

$$\Phi(k,y) = \int_{-\infty}^{\infty} W(x,y)e^{ikx}\,dx, \qquad (33.9)$$

the boundary conditions (33.2), (33.8) yield

$$\Phi(k,0) = 0 = \Phi_y(k,0), \qquad (33.10)$$

$$\int_{-\infty}^{0} W(x,1)e^{ikx}\,dx = \Phi^-(k,1) = 0 = \Phi_y^-(k,1) = \int_{-\infty}^{0} W_y(x,1)e^{ikx}\,dx,$$
$$(33.11)$$

$$\int_0^{\infty} [W_{yy} - \nu k^2 W]e^{ikx}\,dx = \Phi_{yy}^+(k,1) - \nu k^2\Phi^+(k,1) = \frac{-i}{12(k+i\epsilon)},$$
$$(33.12)$$

$$\int_0^{\infty} [W_{yyy} - (2-\nu)k^2 W_y]e^{ikx}\,dx = \Phi_{yyy}^+(k,1) - (2-\nu)k^2\Phi_y^+(k,1) = \frac{-i}{2(k+i\epsilon)}.$$
$$(33.13)$$

Convergence of the above Fourier full and half-range transforms is ensured if k lies in an infinite strip containing the real line, here and henceforth referred to as $\mathcal{D}$, with its width limited from below by the singularity at $k = -i\epsilon$. Evidently (see Noble [7]) the unknown pairs of (half-range transform) functions $\Phi^+(k,1)$, $\Phi_y^+(k,1)$ and $\Phi_{yy}^-(k,1)$, $\Phi_{yyy}^-(k,1)$ are regular in the region above and including $\mathcal{D}$, denoted $\mathcal{D}^+$, and the region below and including $\mathcal{D}$, denoted $\mathcal{D}^-$, respectively. Thus, $\mathcal{D}^+ \cap \mathcal{D}^- \equiv \mathcal{D}$.

In view of the behaviour of W at $x = \pm\infty$, the biharmonic equation can be Fourier transformed (33.9) to give

$$\left(\frac{d^2}{dy^2} - k^2\right)^2 \Phi = 0 \qquad (33.14)$$

and hence a general solution which satisfies (33.10) is

$$\Phi(k, y) = A(k) ky \sinh ky + B(k)(ky \cosh ky - \sinh ky). \qquad (33.15)$$

Application of the conditions (33.11) now yields:

$$\begin{pmatrix} A(k) \\ B(k) \end{pmatrix} = \frac{1}{\sinh^2 k - k^2} \times \qquad (33.16)$$

$$\begin{pmatrix} k \sinh k & -(k \cosh k - \sinh k) \\ -(k \cosh k + \sinh k) & k \sinh k \end{pmatrix} \begin{pmatrix} \Phi^+(k, 1) \\ k^{-1}\Phi_y^+(k, 1) \end{pmatrix}.$$

Then the use of the conditions (33.12,33.13) facilitates the deduction of the following **matrix** Wiener-hopf equation

$$\begin{pmatrix} -\Phi_{yyy}^-(k, 1) \\ \Phi_{yy}^-(k, 1) \end{pmatrix} - \frac{i}{2(k + i\epsilon)} \begin{pmatrix} -1 \\ 1/6 \end{pmatrix} = \mathbf{K}(k) \begin{pmatrix} \Phi^+(k, 1) \\ \Phi_y^+(k, 1) \end{pmatrix}, \quad (33.17)$$

where

$$\mathbf{K}(k) = \begin{pmatrix} 0 & -k \\ 1 & 0 \end{pmatrix} \begin{pmatrix} -e(k) & g(k) - f(k) \\ -g(k) - f(k) & e(k) \end{pmatrix} \begin{pmatrix} k & 0 \\ 0 & 1 \end{pmatrix}, \qquad (33.18)$$

$$e(k) = k \left(1 + \nu + \frac{2k^2}{\sinh^2 k - k^2} \right), \qquad (33.19)$$

$$f(k) = \frac{2k^2}{\sinh^2 k - k^2}, \quad \text{and} \quad g(k) = \frac{k \sinh 2k}{\sinh^2 k - k^2}. \qquad (33.20)$$

The determinant of the kernel is

$$|\mathbf{K}(k)| = \frac{k^4[4 + (3 + \nu)(1 - \nu)\sinh^2 k + (1 - \nu)^2 k^2]}{\sinh^2 k - k^2}, \qquad (33.21)$$

and the complex numbers $\{\mu_n, \lambda_n; n \geq 1\}$ appearing in (33.6) are the zeros in the first quadrant of the numerator and denominator respectively. Negative values of n denote complex conjugates.

It is required to solve the Wiener-Hopf system (33.17), but for a unique solution information is required on the growth of the unknown column vectors appearing on both sides of this equation. This can be obtained by a local analysis of the total deflection about the boundary jump point at $(0, 1)$, by which means it can be shown that the total deflection $w = \mathcal{O}(r^{3/2})$ as $r = [x^2 + (1 - y)^2]^{1/2} \to 0$.

3.   FACTORIZATION OF THE KERNEL

The aim of this section is to summarise the method of factorization of $\mathbf{K}(k)$ (33.18) into a product of two matrices

$$\mathbf{K}(k) = \mathbf{K}^-(k)\mathbf{K}^+(k), \qquad (33.22)$$

one containing those singularities of $\mathbf{K}(k)$ lying in the lower half-plane, $\mathbf{K}^+(k)$ say, and the other $\mathbf{K}^-(k)$ whose singularities lie above the strip $\mathcal{D}$. Further, it is necessary for successful completion of the Wiener-Hopf procedure that $\mathbf{K}^\pm(k)$ are at worst of algebraic growth (see Noble [7]). Unfortunately, although matrix kernel factorization with the requisite growth behaviour has been proven to be possible for a wide class of kernels (Gohberg & Krein [8]), of which the kernel (33.18) belongs, no constructive method has been found to complete this in general. There are classes of matrices for which product factorization can be achieved explicitly, the most important of which are those amenable to Hurd's method [9] and Khrapkov commutative matrices [10]. Details of these, and an extensive bibliography on matrix kernel factorization can be found in [11, 5]. The present problem yields a kernel which falls outside of the classes permitting an exact factorization and so an approximate decomposition will be performed here. The approach follows that developed recently by one of the authors and has been successfully applied to problems in elasticity [11] and acoustics [5]. Essentially, the procedure is to rearrange the kernel into an appropriate form, namely to resemble a Khrapkov (commutative) matrix, and then to replace a scalar component of it by a function which approximates it accurately in the strip of analyticity $\mathcal{D}$. The new approximate kernel is able to be factorized exactly (into an explicit non-commutative decomposition) and, in the previous cases cited above, strong numerical evidence was offered for convergence of the resulting approximate factors to the exact ones as the scalar approximator is increased in accuracy. Further, the convergence to the solution has been validated for one particular matrix kernel [12] where an **exact** non-commutative factorization can be derived by an alternative procedure.

The matrix $\mathbf{K}(k)$ is characterized by its elements $e(k)$, $f(k)$, $g(k)$, given in (33.19)-(33.20), and in particular by their behavior for large and small k; i.e.

$$e(k) \sim k(1+\nu), \quad f(k) \sim 0, \quad g(k) \sim 2|k| \text{ as } |k| \to \infty, k \in \mathcal{D} \quad (33.23)$$

and

$$f(k) \sim \frac{6}{k^2}, \quad g(k) - f(k) \sim \frac{2}{3}k^2 f(k), \quad e(k) \sim k f(k) \text{ as } k \to 0. \quad (33.24)$$

It is appropriate to arrange the kernel to be diagonally dominant as $|k| \to \infty$ in $\mathcal{D}$ by writing (33.18) as

$$\mathbf{K}(k) = \frac{1}{2} \begin{pmatrix} 0 & -k \\ 1 & 0 \end{pmatrix} \begin{pmatrix} 1 & -1 \\ i & i \end{pmatrix} \mathbf{L}(k) \begin{pmatrix} i & 1 \\ i & -1 \end{pmatrix} \begin{pmatrix} k & 0 \\ 0 & 1 \end{pmatrix}, \quad (33.25)$$

where $\mathbf{L}(k)$ may be expressed as

$$\mathbf{L}(k) = g(k)\mathbf{I} + \sqrt{\frac{f^2(k) + e^2(k)}{1 + k^2}}\,\mathbf{J}(k), \quad (33.26)$$

with $\mathbf{I}$ the identity,

$$\mathbf{J}(k) = \begin{pmatrix} 0 & d(k)(1 + ik) \\ d^{-1}(k)(1 - ik) & 0 \end{pmatrix}, \quad (33.27)$$

in which

$$d(k) = \sqrt{\left(\frac{f(k) + ie(k)}{f(k) - ie(k)}\right)\left(\frac{1 - ik}{1 + ik}\right)} \quad (33.28)$$

and

$$\mathbf{J}^2(k) = \Delta^2(k)\mathbf{I}, \qquad \Delta^2(k) = 1 + k^2. \quad (33.29)$$

Evidently, a branch of $d(k)$ can be chosen that is regular in $\mathcal{D}$, equal to unity at $k = 0$ and, because of (33.23), tends to unity at infinity in the strip. The matrix $\mathbf{L}(k)$ now appears to be in Khrapkov form [10] except for the infinite sequences of finite branch-cuts associated with $d(k)$, lying at symmetric locations in the upper and lower half-planes. These will have to be considered once the partial Khrapkov decomposition is complete.

For brevity, all details of the derivation of the partial factorization of $\mathbf{L}(k)$ is omitted here. It is merely stated that

$$\mathbf{K}(k) = \mathbf{Q}^-(k)\mathbf{Q}^+(k), \quad (33.30)$$

with factors

$$\mathbf{Q}^-(k) = \begin{pmatrix} -ik & -ik \\ 1 & -1 \end{pmatrix} \mathbf{R}^-(k)\mathbf{T}^-(k), \quad (33.31)$$

$$\mathbf{Q}^+(k) = \mathbf{T}^+(k)\mathbf{R}^+(k) \begin{pmatrix} ik & 1 \\ ik & -1 \end{pmatrix}, \quad (33.32)$$

where

$$\mathbf{R}^\pm(k) = \frac{3}{4k(k \pm i\sqrt{3}/2)}\left[(1 \mp ik/\sqrt{3})\mathbf{I} + \mathbf{J}(k)\right], \quad (33.33)$$

and

$$\mathbf{T}^{\pm}(k) = r^{\pm}(k)\left(\cosh[\Delta(k)\theta^{\pm}(k)]\mathbf{I} + \frac{1}{\Delta(k)}\sinh[\Delta(k)\theta^{\pm}(k)]\mathbf{J}(k)\right),$$
(33.34)

in which $r^{\pm}(k)$, $\theta^{\pm}(k)$ are **scalar** functions of k given by factorizing

$$[r^{+}(k)r^{-}(k)]^{2} = \frac{1}{3}k^{2}(1 + \frac{4}{3}k^{2})(g^{2} - f^{2} - e^{2})$$
(33.35)

and

$$\tanh[\Delta(k)(\theta^{+}(k) + \theta^{-}(k))] = \frac{(1 + \frac{2}{3}k^{2})\sqrt{f^{2} + e^{2}} - \Delta g}{(1 + \frac{2}{3}k^{2})g - \Delta\sqrt{f^{2} + e^{2}}}$$
(33.36)

via the usual product- and sum-split formulae [7]. Note that $\mathbf{Q}^{\pm}(k)$ are without poles at $k = 0$ even though $\mathbf{R}^{\pm}(k)$ contain this singularity. All that remains is to remove the residual singularities appearing in $\mathbf{J}(k)$.

There is no exact procedure known for eliminating the finite branch-cuts in $d(k)$ present in the upper (lower) half-planes of the matrix factor $\mathbf{Q}^{+}(k)$ $(\mathbf{Q}^{-}(k))$. Instead an approximate factorization is sought where the original matrix $\mathbf{K}(k)$ is replaced by a new one, $\mathbf{K}_{N}(k)$ say, in which all the functions are as given above except for $d(k)$. This latter scalar is replaced by its $[N/N]$ two-point Padé approximant (see [6, 13]), henceforth called $d_{N}(k)$, which is a ratio of two polynomials of degree N and which is derived by the Taylor series expansion of $d(k)$ about both the origin and the point at infinity. The definition (33.28) displays a symmetry which must be reflected in the approximant behaviour:

$$d_{N}(-k) = 1/d_{N}(k),$$
(33.37)

and from this it can be deduced, for example, that

$$d_{4}(k) = \frac{\frac{6}{1+\nu} + \frac{1}{4}k^{2} + k^{3}(\frac{i}{2} + k)}{\frac{6}{1+\nu} + \frac{1}{4}k^{2} + k^{3}(-\frac{i}{2} + k)}.$$
(33.38)

Thus, the factors of $\mathbf{K}_{N}(k)$, $\mathbf{Q}_{N}^{\pm}(k)$, are analytic in their respective half-planes $\mathcal{D}^{\pm}$ apart from sequences of poles, arising from the zeros and poles of $d_{N}(k)$. The removal of these simple singularities will then achieve an explicit exact factorization of $\mathbf{K}_{N}(k)$ which approximates the actual factors $\mathbf{K}^{\pm}(k)$ in their regions of analyticity. The exact factorization of $\mathbf{K}_{N}(k)$ may be written as

$$\mathbf{K}_{N}(k) = \mathbf{K}_{N}^{-}(k)\mathbf{K}_{N}^{+}(k),$$
(33.39)

$$\mathbf{K}_N^-(k) = \mathbf{Q}_N^-(k)\mathbf{M}(k), \qquad \mathbf{K}_N^+(k) = \mathbf{M}^{-1}(k)\mathbf{Q}_N^+(k) \qquad (33.40)$$

in which $\mathbf{M}(k)$ is a meromorphic matrix chosen to eliminate the poles of $\mathbf{Q}_N^-(k)$ in the lower half-plane and the poles of $\mathbf{Q}_N^+(k)$ in $\mathcal{D}^+$. It is straightforward to derive $\mathbf{M}(k)$ by solving certain systems of algebraic equations [5], and a suitable form can be shown to have the properties:

$$|\mathbf{M}(k)| \equiv 1, \qquad (33.41)$$

and

$$\mathbf{M}(k) = \frac{1}{\sqrt{2}} \begin{pmatrix} 1 & -1 \\ 1 & 1 \end{pmatrix} \qquad (33.42)$$

as $k \to \infty$. By estimating the large $|k|$ form of $\mathbf{Q}_N^\pm(k)$ in (33.31), (33.32), the asymptotic growth of $\mathbf{K}_N^\pm(k)$ in (33.40) is found to be

$$\mathbf{K}_N^-(k) \sim k^{1/2} \begin{pmatrix} \mathcal{O}(k) & \mathcal{O}(k) \\ \mathcal{O}(1) & \mathcal{O}(1) \end{pmatrix}, \quad \mathbf{K}_N^+(k) \sim k^{1/2} \begin{pmatrix} \mathcal{O}(k) & \mathcal{O}(1) \\ \mathcal{O}(k) & \mathcal{O}(1) \end{pmatrix}.$$
$$(33.43)$$

Hence

$$[\mathbf{K}_N^-(k)]^{-1} \sim k^{-3/2} \begin{pmatrix} \mathcal{O}(1) & \mathcal{O}(k) \\ \mathcal{O}(1) & \mathcal{O}(k) \end{pmatrix} \qquad (33.44)$$

and the kernel decomposition is now complete.

4.   SOLUTION OF THE WIENER-HOPF EQUATION

Having obtained an approximate factorization of $\mathbf{K}(k)$ it is now a straightforward matter to complete the solution of the Wiener-Hopf equation (33.17). This can be recast into the form (dropping the suffix N on $\mathbf{K}_N^\pm(k)$ henceforth for brevity)

$$[\mathbf{K}^-(k)]^{-1} \begin{pmatrix} -\Phi_{yyy}^-(k,1) \\ \Phi_{yy}^-(k,1) \end{pmatrix} - \qquad (33.45)$$

$$\frac{i}{2(k+i\epsilon)} \left\{ [\mathbf{K}^-(k)]^{-1} - [\mathbf{K}^-(-i\epsilon)]^{-1} \right\} \begin{pmatrix} -1 \\ 1/6 \end{pmatrix} = \mathbf{E}(k)$$

$$= \mathbf{K}^+(k) \begin{pmatrix} \Phi^+(k,1) \\ \Phi_y^+(k,1) \end{pmatrix} + \frac{i}{2(k+i\epsilon)}[\mathbf{K}^-(-i\epsilon)]^{-1} \begin{pmatrix} -1 \\ 1/6 \end{pmatrix}$$

where $k \in \mathcal{D}$. The left hand side is analytic in $\mathcal{D}^-$, whereas the right hand side is regular in $\mathcal{D}^+$. The two sides offer analytic continuation into the whole complex k-plane which must therefore be equal to an entire function, say $\mathbf{E}(k)$, which is determined by examining the growth at infinity of both sides of (33.45) in their respective half-planes of analyticity. This requires the large k behaviour of $\Phi^+(k,1)$, $\Phi_y^+(k,1)$, $\Phi_{yy}^-(k,1)$, $\Phi_{yyy}^-(k,1)$, which relate directly to the values of the untransformed physical variables near to $(0,1)$, where the edge condition changes. As mentioned earlier, $w = \mathcal{O}(r^{3/2})$ as $r = [x^2 + (1-y)^2]^{1/2} \to 0$, from which it can be deduced that

$$\Phi^+(k,1) = \mathcal{O}(k^{-5/2}), \quad \Phi_y^+(k,1) = \mathcal{O}(k^{-3/2}) \tag{33.46}$$

as $|k| \to \infty$, $k \in \mathcal{D}^+$ and

$$\Phi_{yy}^-(k,1) = \mathcal{O}(k^{-1/2}), \quad \Phi_{yyy}^-(k,1) = \mathcal{O}(k^{1/2}), \quad |k| \to \infty, k \in \mathcal{D}^-. \tag{33.47}$$

These are used, together with the asymptotic forms (33.43), (33.44) to reveal that both elements of the left hand side of (33.45) decay as $\mathcal{O}(k^{-1})$ in the lower half plane and similarly the right hand side has the form $o(1)$ as $|k| \to \infty$ in the upper half plane. Hence, $\mathbf{E}(k)$ is an entire function which decays to zero at infinity, and so, by Liouville's theorem,

$$\mathbf{E}(k) \equiv \mathbf{0}. \tag{33.48}$$

Thus, the solution of the Wiener-Hopf equation is

$$\begin{pmatrix} -\Phi_{yyy}^-(k,1) \\ \Phi_{yy}^-(k,1) \end{pmatrix} = \frac{i}{2(k+i\epsilon)} \left\{ \mathbf{I} - \mathbf{K}^-(k)[\mathbf{K}^-(-i\epsilon)]^{-1} \right\} \begin{pmatrix} -1 \\ 1/6 \end{pmatrix} \tag{33.49}$$

or, equivalently,

$$\begin{pmatrix} \Phi^+(k,1) \\ \Phi_y^+(k,1) \end{pmatrix} = -\frac{i}{2(k+i\epsilon)} [\mathbf{K}^+(k)]^{-1} [\mathbf{K}^-(-i\epsilon)]^{-1} \begin{pmatrix} -1 \\ 1/6 \end{pmatrix}. \tag{33.50}$$

From this, the coefficients $A(k), B(k)$ are readily deduced from (33.16) and hence $\Phi(k,y)$ is established for all y, $0 < y < 1$, from (33.15). Finally, on setting the convergence factor ϵ equal to zero in (33.49), the additional displacement due to the semi-infinite free edge is

$$\frac{q}{D}W = \frac{q}{2\pi D} \int_{-\infty}^{\infty} \Phi(k,y)e^{-ikx}dk, \tag{33.51}$$

where the integral runs along the real line indented above the origin and

$$\Phi(k,y) = \frac{-i}{2(\sinh^2 k - k^2)} \begin{pmatrix} y\sinh ky \\ y\cosh ky - k^{-1}\sinh ky \end{pmatrix}^T \times \tag{33.52}$$

$$\begin{pmatrix} k\sinh k & -(\cosh k - k^{-1}\sinh k) \\ -(k\cosh k + \sinh k) & \sinh k \end{pmatrix} [\mathbf{K}^+(k)]^{-1} [\mathbf{K}^-(0)]^{-1} \begin{pmatrix} -1 \\ 1/6 \end{pmatrix}.$$

It is a straightforward matter to verify that, when this solution is substituted into (33.7), the total displacement satisfies the governing equation (33.1) and the boundary conditions (33.2), (33.3). Moreover, it evidently has the structure predicted by (33.6); for $x < 0$, the contour in (33.51) is completed in $\mathcal{D}^+$ and residues at the zeros of $\sinh^2 k - k^2$ are obtained while, for $x > 0$, completion in $\mathcal{D}^-$ yields $qy^2(5 - 2y)/24D$ from the pole at the origin and residues at the zeros of the numerator of (33.21).

References

[1] Lebedev, N N, Skalskaya, I P and Uflyand, Y S (1965) *Worked Problems in Applied Mathematics*, New York: Dover.

[2] Sneddon, I N (1951) *Fourier Transforms*, New York: McGraw-Hill.

[3] Landau, L D and Lifshitz, E M (1986) *Theory of Elasticity*, Third ed., Oxford: Pergamon Press.

[4] Moore, A M, Buchwald, V T and Brewster, M E (1990) *A Stokesian entry flow*, Quart. J. Mech. Appl. Math. **43**, 107–133.

[5] Abrahams, I D (1997) *On the solution of Wiener-Hopf problems involving noncommutative matrix kernel decompositions*, SIAM J. Appl. Math. **57**, 541–567.

[6] Abrahams, I D and Davis, A M J (2002) *Matrix Wiener-Hopf approximation for a partially clamped plate*, to be submitted.

[7] Noble, B (1988) *Methods based on the Wiener-Hopf technique*, New York: Chelsea Press.

[8] Gohberg, I C and Krein, M G (1960) *Systems of integral equations on a half-line with kernels depending on the difference of arguments*, Amer. Math. Soc. Transl. Ser. 2, **14**, 217–287.

[9] Hurd, R A (1976) *The Wiener-Hopf Hilbert method for diffraction problems*, Canad. J. Phys. **54**, 775–780.

[10] Khrapkov, A A (1971) *Certain cases of the elastic equilibrium of an infinite wedge with a non-symmetric notch at the vertex, subjected to concentrated forces*, PMM Appl. Math. Mech. **35**, 625–637.

[11] Abrahams, I D (1996) *Radiation and scattering of waves on an elastic half-space; a noncommutative matrix Wiener-Hopf problem*, J. Mech. Phys. Solids **44**, 2125–2154.

[12] Abrahams, I D (1998) *On the non-commutative factorization of Wiener-Hopf kernels of Khrapkov type*, Proc. Roy. Soc. Lond. A **454**, 1719–1743.

[13] Baker Jr, G A and Graves-Morris, P (1996) *Padé Approximants*, Second ed., Cambridge: University Press.

SEISMIC WAVE DYNAMICS IN REGULAR AND SINGULAR POINTS OF THE RAY

A. A. Duchkov*

Institute of Geophysics

pr. ac. Koptyuga 3

Novosibirsk 630090, Russia

dooch@uiggm.nsc.ru

S. V. Goldin†

Novosibirsk State University

st. Pirogova 2

Novosibirsk 630090, Russia

goldin@uiggm.nsc.ru

Abstract   Wave fields near caustics may be expressed very simply when one calculates them in the time domain rather than in the more usual frequency domain. The technique is described allowing one to calculate the form of the seismic signal that propagates along the ray passing a caustic. Practical use of the technique is illustrated for the case of a caustic cusp.

1.   INTRODUCTION

In this paper we will consider seismic wave propagation described by the ray series approximation. Although ray theory is valid in many circumstances, it breaks down in common situations, e.g. it is not valid near caustics. A lot of work was done (starting from [1]) in order to find appropriate extensions of the ray method. These efforts have resulted in the development of formalized techniques allowing us to treat different types of singularities, e.g. Gaussian beams and the Maslov method.

*Partial funding provided by RFBR (99-05-64425) and INTAS (YSF99-211).

†Partial funding provided by Russian Ministry of Education (in the field of natural sciences).

313

I.D. Abrahams et al. (eds.),

IUTAM Symposium on Diffraction and Scattering in Fluid Mechanics and Elasticity, 313–320.

© 2002 *Kluwer Academic Publishers. Printed in the Netherlands.*

In most cases the wave field $\mathbf{u}(\mathbf{x}, \omega)$ is represented in the frequency domain by the oscillatory integral:

$$\mathbf{u}(\mathbf{x}, \omega) \sim \int a(\mathbf{q}, \mathbf{x}) e^{i\omega \Phi(\mathbf{q}, \mathbf{x})} d\mathbf{q}, \tag{34.1}$$

where ω is the frequency, $a(\mathbf{q}, \mathbf{x})$ denotes the smooth amplitude function, and the phase function $\Phi(\mathbf{q}, \mathbf{x})$ takes a limited number of standard forms depending on the type of caustic (see Table 1).

It should be noted that for practical modelling of wave propagation two problems are to be solved:

- derivation of the oscillatory integral (34.1), i.e. for the particular problem (equations of motion and initial data) we have to find $a(\mathbf{q}, \mathbf{x})$ and coefficients of $\Phi(\mathbf{q}, \mathbf{x})$;

- technical problem of evaluating the integral (34.1).

The second problem alone gives a general insight into the behaviour of the wave field near caustics. Such analysis is interesting but for numerical modelling both steps are crucial.

The first problem was considered in [2] for seismic wave field modelling near simple caustics and cusps. Results of standard dynamic ray tracing were used for derivation of $a(\mathbf{q}, \mathbf{x})$ and $\Phi(\mathbf{q}, \mathbf{x})$. The other approaches to this problem are described in [3]. We use one of them in this paper.

Evaluation of the integral (34.1) is a well-studied problem. The asymptotic approximation of (34.1) gives the uniformly valid description for different caustics in terms of the Airy, Pearcey functions, etc. Alternatively the ray theory can be developed in the time domain where it describes sharp changes of the signal that happen at the wave front. It was found that sometimes it is more effective to calculate the time-domain integrals directly rather then deal with (34.1). In [4] it was proposed to apply the stationary phase method (at large λ) to the integral of arbitrary periodic function p with mean zero:

$$\int a(q) p(\lambda \Phi(q)) \, dq, \tag{34.2}$$

This formalism is still close to the analysis of the harmonic fields. Then, in [5] the theory was extended to treat the time-domain waveforms associated with double integrals of rapidly-varying isolated pulses $F(t)$:

$$\iint a(\mathbf{q}) F(\lambda[t - \Phi(\mathbf{q})]) \, d\mathbf{q}, \tag{34.3}$$

Isolated short function $F(t)$ is closer to objects considered in the time-domain ray theory. Integrals (34.2) and (34.3) are valid for the asymptotic analysis due to the large parameter λ. It is possible to get rid of

λ if we consider the Dirac δ-function in (34.3) instead of $F(t)$ (see [6]). We derive such integrals considering propagation of elastic waves.

2.   METHOD OF DISCONTINUITIES

According to [7] the time-domain ray theory describes propagation of discontinuities situated at the wave front. In our research we neglect a smooth part of the wave field and use the "method of discontinuities" (see [8]) when the displacement vector $\mathbf{u}(\mathbf{x}, t)$ is described by the series

$$\mathbf{u}(\mathbf{x}, t) \sim \sum_{r=0}^{\infty} \mathbf{U}^{(r)}(\mathbf{x}) R_{q+r,\nu}^{(+)}(t - \tau(\mathbf{x})), \qquad (34.4)$$

where $\tau(\mathbf{x})$ denotes the wave eikonal, $\mathbf{U}^{(r)}(\mathbf{x})$ are the amplitude vectors and $R_{q+r,\nu}^{(+)}(t)$ are discontinuous functions ("discontinuities") defined as

$$R_q^{(+)}(t) = \begin{cases} t_+^q / \Gamma(q+1), & q \neq -1, -2, \ldots \\ \delta^{(-q+1)}(t), & q = -1, -2, \ldots \end{cases} \qquad (34.5)$$
$$R_{q,\nu}^{(+)}(t) = \mathbf{H}^\nu[R_q^{(+)}(t)],$$

where the fractional Hilbert transform $\mathbf{H}^\nu \equiv \cos(\pi\nu/2)\mathbf{E} + \sin(\pi\nu/2)\mathbf{H}$, $\mathbf{E}$ and $\mathbf{H}$ being, respectively, the identity and the Hilbert operators. Function $R_{q,\nu}^{(+)}(t)$ is discontinuous at $t = 0$ and is characterized by the order q and the index ν (both can be real). The lower the order q, the sharper the discontinuity. The value of the index ν correspond to the value of the signal phase shift. It is introduced for treating caustics as it is known that after passing a caustic the original form of the signal changes into its Hilbert transform. A set of functions (34.5) is complete for discontinuous part of the wave field: it describes all perturbations of the signal that may happen near caustics.

3.   PROBLEM FORMULATION

Let us consider P-wave propagation in a homogeneous isotropic elastic three-dimensional medium. We perform the local analysis of the wave dynamics along the chosen seismic ray. Initial wave is given in the ray-series approximation in the vicinity of the point $\mathbf{x}_0 = (0, 0, h)$ of this ray

$$\mathbf{u}^{(in)}(x, y, h, t) = A(x, y) R_{q,0}^{(+)}(t - \tau(x, y)) \nabla \tau(x, y), \qquad (34.6)$$

where $A(x, y) \equiv A(x, y, h)$ and $\tau(x, y) \equiv \tau(x, y, h)$ are the amplitude and the eikonal of the P-wave given on the plane $z = h$.

The ray is parallel to the z-axis and is directed downwards. The displacement vector $\mathbf{u}$ is to be found at the point $\mathbf{x}_1 = (0, 0, 0)$ in the form (34.4). Point $\mathbf{x}_1$ is on the same ray.

4.   SOLUTION

We give a brief outline of the proposed technique (details are in [9]).

Integral derivation.   To solve the problem it is proposed to apply the Green's integral formula:

$$\mathbf{u}(\mathbf{x}_1, t_1) \;=\; \int_0^{t_1} \iint_S \{ \mathbf{G}(\mathbf{x}_1; \mathbf{x}, t_1 - t) \mathbf{T_n}[\mathbf{u}^{(in)}(\mathbf{x}, t)]$$
$$- \, \mathbf{T_n}[\mathbf{G}(\mathbf{x}_1; \mathbf{x}, t_1 - t)] \mathbf{u}^{(in)}(\mathbf{x}, t) \} dS(\mathbf{x}) \, dt, \quad (34.7)$$

where $\mathbf{u}(\mathbf{x}_1, t_1)$ is the displacement vector to be calculated; $\mathbf{T_n}$ is the differential operator assigning the Cauchy stress applied to the surface element $dS(\mathbf{x})$ with normal $\mathbf{n}$, the Green's tensor $\mathbf{G}(\mathbf{x}_1; \mathbf{x}, t_1 - t)$ for the infinite three-dimensional space and the incident wave $\mathbf{u}^{(in)}(\mathbf{x}, t)$ are to be known at the surface of integration S. In our analysis we take surface S as a plane $z = h$ perpendicular to the ray of interest.

All integrands are substituted by their discontinuous representations. The incident field $\mathbf{u}^{(in)}$ is given in (34.6) and the Green's tensor for homogeneous medium is known. After some simplification of the integral formula (34.7) we come to a series of integrals,

$$u_k(0,0,0,t) \;\sim\; \sum_{r=0}^{\infty} I_r^{(k)}(t),$$

$$I_r^{(k)}(t) \;\equiv\; \iint_{z=h} L_r^{(k)}(x, y) R_{q-1+r,0}^{(+)}(t - \tilde{\tau}(x, y)) \, dx \, dy, (34.8)$$

where u_k are the components of $\mathbf{u}$; $\tilde{\tau}(x, y) = \tau(x, y) + T(x, y)$ is a sum of the incident wave eikonal $\tau(x, y)$ and the Green's tensor eikonal $T(x, y)$ given on the plane $z = h$, $L_r^{(k)}(x, y)$ are smooth functions derived in [9].

As was mentioned already we do not take into account the smooth part of the wave field. For the discontinuity differentiation followed by integration will give exactly the original discontinuity. Thus, the identity operator may be represented as $\mathbf{E} \sim (\mathbf{D}_t^q)^{-1} \mathbf{D}_t^q$, where $\mathbf{D}_t^q$ is the time differentiation operator (for negative q it corresponds to integration). Applying the first part of $\mathbf{E}$ to (34.8) we get

$$I_r^{(k)}(t) \sim \mathbf{D}_t^{-q-r} \iint_{z=h} L_r^{(k)}(x, y) \delta(t - \tilde{\tau}(x, y)) \, dx \, dy, \quad (34.9)$$

Integration.   Now we have to perform the integration in (34.9). It is a structure of the argument $\tilde{\tau}(x, y)$ that is completely responsible for regularity or singularity of the wave field in the point of interest.

Table 1   A list of the typical caustics for wave propagation in 3D space.

Type of caustic	$\Phi(\mathbf{q})$ *from* (34.1) *or from* (34.10)
Regular case	$\pm q_1^2 \pm q_2^2$
Simple caustic	$\pm q_1^2 + q_2^3 + \alpha q_2$
Caustic cusp	$\pm q_1^2 \pm q_2^4 + \beta q_2^2 + \alpha q_2$
Swallow tail	$\pm q_1^2 + q_2^5 + \gamma q_2^3 + \beta q_2^2 + \alpha q_2$
Umbilics	$q_1^2 q_2 + \gamma q_1^2 + \beta q_1 \pm q_2^3 + \alpha q_2$

In principle, function $\tilde{\tau}(x, y)$ may be arbitrary but according to the catastrophe theory during the wave propagation in three-dimensional space there exist only five typical types of caustics, i.e. generic catastrophes (see [10], Chapter 12). These caustics are listed in the first column of Table 1. Thus, in the typical case of wave propagation it is possible to perform a local (near the point $(0,0)$) coordinate transformation $(x, y) \rightarrow (q_1, q_2)$ so that the function $\tilde{\tau}(x, y)$ is reduced to the canonical polynomials $\Phi(q_1, q_2)$ given in the second column of Table 1 (see [10], Chapter 7). Then, from the integral in (34.9) we come to

$$I(t) = \int_{-\infty}^{\infty} \int_{-\infty}^{\infty} \tilde{L}(q_1, q_2)\delta(t - \Phi(q_1, q_2))\, dq_1\, dq_2. \tag{34.10}$$

The integral (34.10) is a time-domain equivalent of the oscillatory integral (34.1). It gives the uniformly valid wave field representation near and at the caustics mentioned in Table 1.

Let us make some remarks about double integration of δ-function. The power of the variable q_1 is not higher then two in all polynomials $\Phi(\mathbf{q})$ from the Table 1. Thus, the integration over q_1 in (34.10) can be evaluated analytically (one has to find the roots of the square equation). Again, it is possible to apply the operator $(\mathbf{D}_t^{1/2})^{-1}\mathbf{D}_t^{1/2}$ to the resultant integral,

$$I(t) \sim \mathbf{D}_t^{-1/2} \int_{-\infty}^{\infty} \hat{L}(q_2)\delta(t - \phi(q_2))\, dq_2, \tag{34.11}$$

where $\phi(q_2)$ is a polynomial of the order less then 6.

The right-hand integral in (34.11) can be easily calculated (see [6]).

5.   RESULTS AND DISCUSSION

Procedure (34.7)–(34.11) can be used for the modelling of seismic wave propagation. Its advantage is that the problem of wave dynam-

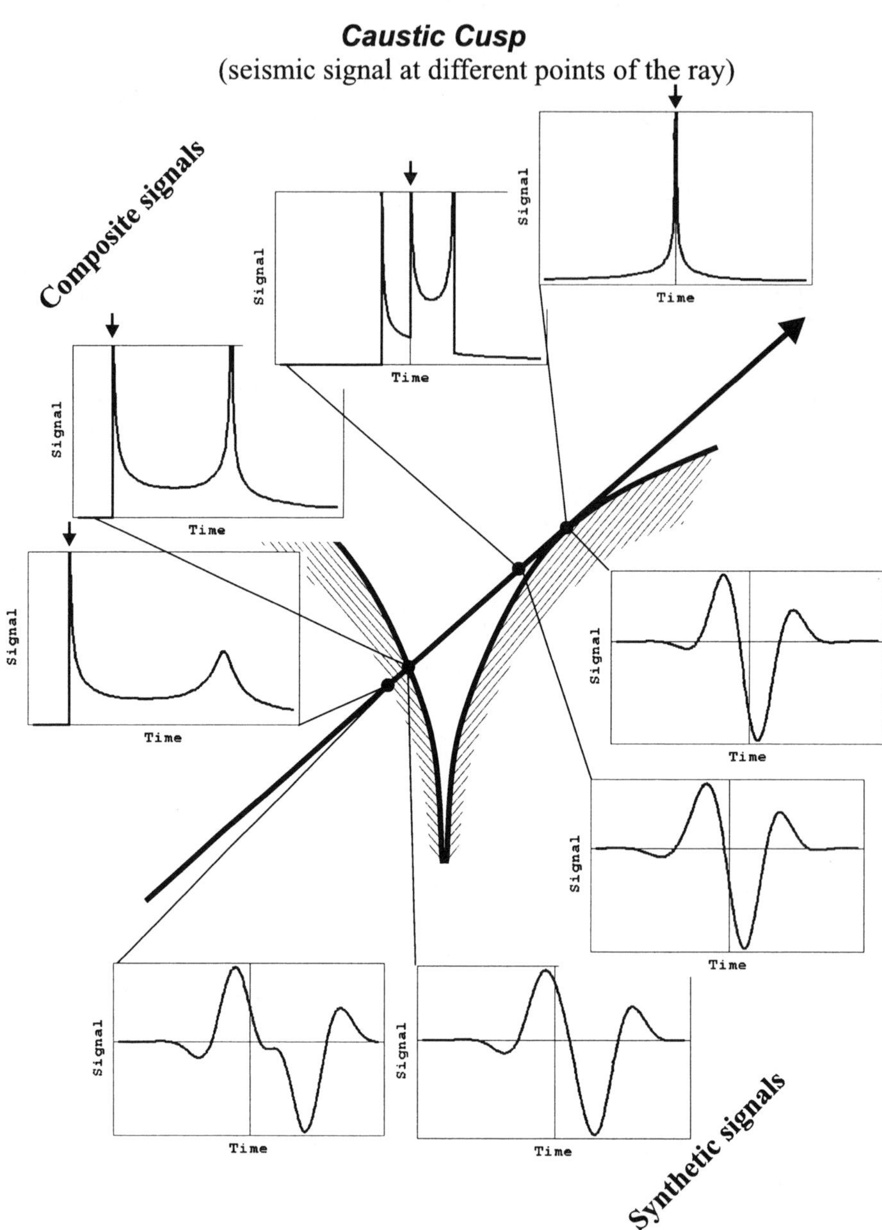

Figure 1 Computation of the seismic signal at different points of the ray. Composite signal is shown on the graphs above the ray and synthetic signal is shown underneath.

ics computation is reduced to the double integration of δ-function. The price we pay for this simplicity is that we are treating waveforms which, far from the caustic, have the form of inverse square root of the time $R^{(+)}_{-1/2,0}(t)$ (before caustic) and $R^{(+)}_{-1/2,1}(t)$ (after caustic). This is a result of integration of (34.10) for regular case. The waveforms near the caustics can be calculated from (34.10) and expressed in terms of rational functions of solutions to polynomial equations and hence are singular algebraic functions of time.

The technique is illustrated in Fig. 1. Initial data (the incident wave (34.6)) were chosen in such a way that our ray (bold arrow in the figure) goes near the caustic cusp. Fig. 1 shows a plane section of three-dimensional space. Computations have been made at several points on the ray.

The results of integration in (34.10) are refered to as a "composite signal" (shown above the ray in Fig. 1. It describes the changes of the singular part of the wave field that happen during the wave propagation. Each singularity of the composite signal corresponds to signals travelling along different rays. The small arrows correspond to kinematic travel-time along the ray of interest. Different singularities (signals moving along different rays) approach each other and coalesce at the caustic. It is interesting that our technique allows us to compute the wave field in the shadow of the caustic cusp (first point of the ray). Composite signal shows the movement of idealized objects (pulses) and may be treated as an impulse response. Further it can be convolved with the appropriately temporally short wavelet yielding the synthetic seismic signal running along the ray (shown below the ray in Fig. 1).

6.   CONCLUSIONS

We have introduced a time-domain technique describing the propagation of the short signal along the ray. This description is uniformly valid near and at caustics. Before talking about advantages of the proposed technique let us recall the main advantages of ray theory: (i) its results are straightforward and intuitive, i.e. concepts of rays, fronts, reflection and transmission are easily understood; and (ii) it is relatively easy and inexpensive to compute theoretical results. The illustrative power of ray theory is often wrongly neglected. But it seems that ideal computation procedure should include the first illustrative step providing the idea of what happens with the signal (calculation of the composite signal) and the second computational step providing the synthetic signal (convolution of the composite signal).

The ray theory formalism is similar in frequency and time domains and the results are related to each other by means of the Fourier transform. But the frequency-domain results directly describe high-frequency harmonic oscillations whereas the time-domain asymptotics describes propagation of the short impulses (discontinuities) with time. Thus, the time-domain description is more illustrative and attractive for geophysicists who deal with the signal propagation in their research practice.

From the computational point of view the proposed technique is attractive because the problem of wave dynamics computation is reduced to a problem of double integration of δ-function. The latter problem is in turn reduced to a search for the roots of polynomials.

References

[1] Ludwig, D (1966) *Uniform asymptotic expansion at a caustic*, Comm. Pure Appl. Math. **19**, 215–250.

[2] Hanyga, A and Seredynska, M (1991) *Diffraction of pulses in the vicinity of a simple caustic and caustic cusp*, Wave Motion **14**, 101–121.

[3] Hanyga, A (1988) *Asymptotic diffraction theory and its application to ray tracing*, Seismo-Series, **26**, Bergen.

[4] Chapman, C J (1992) *Time-domain asymptotics and the method of stationary phase*, Proc. Roy. Soc. Lond. A **437**, 25–40.

[5] Prentice, P R (1994) *Time-domain asymptotics. I General theory for double integrals*, Proc. Roy. Soc. Lond. A **446**, 341–360.

[6] Burridge, R (1995) *Asymptotic evaluation of integrals related to time-dependent fields near caustics*, SIAM J. Appl. Math. **55**, 390–409.

[7] Alekseev, A S, Babich, V M and Gelchinskii, B Ya (1961) *Ray method calculations of intensity of the wave fronts*, Problems of the dynamic theory of seismic wave propagation, **5**, 3–24, Leningrad: University Press.

[8] Goldin, S V (1989) *Method of discontinuities in problems of geophysics and tomography*, Dokl. Ak. Nauk USSR, **308**(4), 824–827.

[9] Goldin, S V and Duchkov, A A (2000) *Integral representations in geometrical seismics*, Rus. Geol. Geoph., **1**, 142–158.

[10] Poston, T and Stewart, I (1978) *Catastrophe Theory and its Applications*, London: Pitman.

PROPAGATION IN CURVED WAVEGUIDES

J. G. Harris

Center QEFP, Northwestern University

2137 N. Sheridan Road, Evanston, IL 60208-3020, USA

j-harris8@northwestern.edu

Abstract   Asymptotic approximations to the waves guided by a curved, elastic waveguide, whose curvature is small and changes slowly over a wavelength are discussed using the JWKB method. The antiplane shear problem is treated. The asymptotic method is carefully explained and a brief comparison with similar and contrasting asymptotic approaches made. The conclusions reached are: (1) Provided the caustic is outside the guide, the presence of the curvature affects the mode shape and modulates the amplitude, but leaves the dispersion relation almost unchanged from that of a guide having no curvature. (2) A caustic can often form inside the guide. Its presence alters the modes, the modulated amplitude and the dispersion rather dramatically. The region of guided propagation becomes that between the caustic and the upper concave surface, though the lower convex surface continues to influence the propagation, and the wavenumber becomes dependent on the curvature.

1.   INTRODUCTION

Guided waves in curved structures can, for our purposes, be divided into two categories; namely, those guided within a structure, such as a curved elastic shell, by the two surfaces and those guided by the concave (convex and concave are defined from the viewpoint of an observer in the structure) surface of the shell and a caustic. The purpose of the present note is to explain how these two possibilities manifest themselves in a thin waveguide, one in which only the first two modes can propagate.

Many aspects of this problem have been studied by others and the results given here are not radically new. However, the several calculations of propagation in curved guides are done in seemly different ways with sometimes conflicting outcomes or outcomes marred by algebraic errors, that, while insignificant in themselves, can cause confusion. One reason, therefore, for undertaking these calculations was to clarify this

321

I.D. Abrahams et al. (eds.),
IUTAM Symposium on Diffraction and Scattering in Fluid Mechanics and Elasticity, 321–328.
© 2002 *Kluwer Academic Publishers. Printed in the Netherlands.*

earlier work. The problem is initially formulated here as system of two coupled differential equations in a manner analogous to that described in Folguera and Harris [2]. This makes the formulation more complicated than it need be, but this approach can be *generalized to the analogous inplane elastic case.* In fact, the second reason for undertaking the present calculations was to clarify aspects of the corresponding inplane ones.

Ahluwalia *et al.* [3] considered propagation in curved waveguides: asymptotic solutions to the reduced wave equation were sought subject to impedance boundary conditions at the walls of the guide. Two approaches were presented. The first used a ray theory anzatz that relied on using as a reference object a caustic whose position must be found as part of the solution. The second, which is appropriate when the caustic lies within the waveguide, used a boundary layer method (see also [1]), which the writer also uses, though his use of it arises more naturally. This work was followed by a second paper by Keller and Ahluwalia [4] in which certain topics of the first paper were expanded upon. Throughout, it was assumed that there is one small parameter, the wavelength. The writer's approach is to find rays along the length of the guide and modes at each cross-section, a method described by Burridge and Weinberg [5]. This approach leads the writer to an asymptotic expansion in two dimensionless small parameters: the curvature scaled by the wavelength and the rate of change of the curvature over a wavelength. These two small parameters are independent of one another.

Smith [6], building upon his earlier work on surface waves [7], also studies propagation in curved waveguides. In a broad sense the approach of Smith [6, 7] and the approach taken here are similar: both use a ray description along the arc of the guide and then use an approximation to the eigenvalue problem through the thickness. Nevertheless the writer believes that his implementation of the asymptotic ansatz unfolds in a more natural way.

2.   THE ASYMPTOTIC APPROACH

Figure 1 shows the geometry and defines the dimensioned coordinates. Note that s is the arclength along the centerline of the guide. The asymptotic expansion is one in two dimensionless parameters: δ the rate at which the curvature changes over a wavelength and χ the curvature expressed in reciprocal wavelengths. The dimensionless coordinates and parameters are defined as

$$x = \delta k s, \; z = kn, \; \chi = 1/(k\rho), \; h = kH,$$

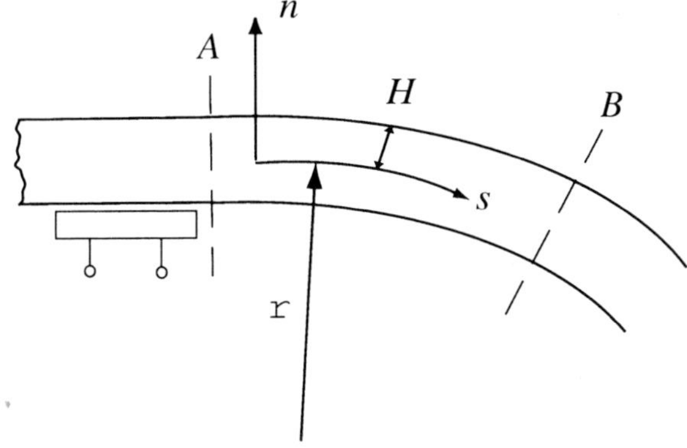

Figure 1   The geometry of the problem indicating the dimensioned coordinates (s, n). s is the arclength along the centerline of the guide, n the normal distance from the centerline, $2H$ the constant thickness and ρ the variable radius of curvature. Not shown is the coordinate b chosen so that (s, n, b) form a right-handed coordinate system. To the left is a schematic sketch of a transducer that resonantly excites a particular mode. The dashed lines A and B indicate two cross-sections referred to in the text.

where k is the wavenumber. The dimensionless curvature χ is a function of x. It is readily seen that

$$[\chi(x + \delta 2\pi) - \chi(x)]/2\pi \approx \delta d\chi/dx.$$

Hence, provided the derivative is O(1), δ measures the rate of change of the curvature over a wavelength. We shall work in that part of the parameter space where $\delta \approx \chi$ and $h \leq \pi$.

The two dependent variables that we work with are τ_{sb} and u_b. The former is the b component of traction acting on the s equal a constant surface and the latter the particle displacement in the b direction. The corresponding dimensionless terms are

$$\tau = \tau_{sb}/\mu, \; u = ku_b,$$

where μ is the shear coefficient. The equations of antiplane motion are written as

$$\begin{bmatrix} 0 & (1 + \chi z) \\ -\partial_z[(1 + \chi z)\partial_z] - (1 + \chi z) & 0 \end{bmatrix} \begin{bmatrix} u \\ \tau \end{bmatrix} = \delta\partial_x \begin{bmatrix} u \\ \tau \end{bmatrix}. \quad (35.1)$$

The boundary conditions to be satisfied are $\partial_z u = 0$ at $z = \pm h$. Though the τ can be eliminated to give the reduced wave equation, (35.1) is a useful way to write the equations when solving the analogous inplane problem [2]. To shorten the notation it is useful to represent the differential operator as $\mathcal{L}$, the vector as $|\mathbf{U}\rangle$, so that the above equation can be symbolically written as

$$\mathcal{L}|\mathbf{U}\rangle = \delta\partial_x|\mathbf{U}\rangle. \tag{35.2}$$

The writer follows [5] almost exactly throughout this stage of the note. To begin, an expansion of the form

$$|\mathbf{U}\rangle = e^{i\theta(x)/\delta}\sum_{\alpha\geq 0}|\mathbf{U}_\alpha(x,z)\rangle\,\delta^\alpha, \tag{35.3}$$

is assumed and a recursive system of equations found. Examining the first of these equations, it is readily seen that by setting $d\theta/dx = \beta_n$ and then $|\mathbf{U}_0\rangle = a_0(x)|\mathbf{u}^n\rangle$, the eigenvalue problem

$$\mathcal{L}|\mathbf{u}^n\rangle = i\beta_n|\mathbf{u}^n\rangle, \tag{35.4}$$

for the nth mode is arrived at. To complete the eigenvalue problem the boundary condition $\partial_z u^n = 0$ at $z = \pm h$ must be satisfied. The $\beta_n(x)$ depend on x through the curvature $\chi(x)$. Thus we are approximating the nth mode $|\mathbf{U}^n\rangle$ of the curved guide.

To proceed further an inner product, based upon the reciprocity relation (this is not quite the usual inner product), is introduced

$$\langle\mathbf{U}^1|\mathbf{U}^2\rangle := i\int_{-h}^{h}(u^1\tau^{2*} - \tau^1 u^{2*})dz.$$

Writing

$$P_{nm} = \langle\mathbf{u}^n|\mathbf{u}^m\rangle, \tag{35.5}$$

it is readily found that by suitably normalizing the eigenfunctions $P_{nm} = \delta_{nm}$.

Using the second differential equation of the recursive system, the term

$$a_0^n(x) = |a_0^n|e^{i\phi_0^n}$$

is calculated to be

$$|a_0^n(x)| = |a_0^n(0)|, \quad \phi_0^n(x) = \phi_0^n(0) + e^{-\int_0^x \Im[W_{nn}(\xi)]d\xi}, \tag{35.6}$$

where $W_{nn} = \langle\partial_x\mathbf{u}^n|\mathbf{u}^n\rangle$. The eigenfunctions can usually be chosen so that $\Im[W_{nn}(x)]$ is 0. The fact that the magnitude of a_0^n remains constant is a manifestation of the *adiabatic invariance* inherent in slowly varying propagation environments. Note, however, that the normalization (35.5) will change as the curvature changes so that (35.6) is *not* a statement that the overall amplitude remains constant.

Figure 1 shows a transducer exciting a disturbance in a flat section of the waveguide. Such transducers can be tuned to excite predominantly one particular mode, say the nth. Modeling this kind of excitation, for the inplane case, is explained in [2]. The outcome of an analogous analysis for this problem gives

$$a_0^n(0) = \frac{-i\,2Cae^{i\beta_n(0)a}}{[2\beta_n(0)h]^{1/2}} \left\{ \begin{array}{l} \cos(n\pi/2), \ n \text{ even} \\ -\sin(n\pi/2), \ n \text{ odd} \end{array} \right. , \qquad (35.7)$$

where C is a constant set by the source and a is one half the width of the source's aperture. Thus $|a_0^n(0)|$ can be calculated at cross-section A (Fig. 1) and be carried by the adiabatic invariance to cross-section B, even though the wavefield between A and B cannot be easily worked out.

The curved waveguide is approximated at each point by an annulus. The radius of its centerline is the radius of curvature ρ at the point x of the guide's centerline (Fig.1). Thus, after eliminating τ, the eigenvalue problem (35.4) becomes that for Bessel's equation. However, to understand this equation's structure, new scaled quantities given by

$$r = [1 + \chi(x)z]/\beta_n(x), \quad \nu = \beta_n(x)/\chi(x) \qquad (35.8)$$

are introduced. The solutions to the resulting equation are linear combinations of the Hankel functions $H_\nu^{(1,2)}(\nu r)$. Uniform asymptotic expansions for $\nu \to \infty$ and arbitrary r are given in [8], items No. 9.3.37 and No. 9.3.45. Satisfying the boundary conditions gives ν_n and hence β_n.

Therefore, the general asymptotic solution for the nth mode propagating in a curved waveguide is

$$|\mathbf{U}^n\rangle = \exp\left[\frac{i}{\delta}\int_0^x \beta_n(\xi)d\xi\right] a_0^n(x)|\mathbf{u}^n(z;x)\rangle. \qquad (35.9)$$

The $;x$ indicates that the mode at each position x is needed. Note that to make the scheme give practical results, $\beta_n(x)$ and the normalization $P_{nn}(x)$ must be found as functions of $\chi(x)$. Rather than continue with this general asymptotic solution, two special cases are considered next.

3.   A THIN CURVED WAVEGUIDE

The caustic is outside the guide: The waveguide is imagined to be such that $h \approx \pi$. Using (35.8), $r_\pm$ are defined as

$$r_\pm = [1 \pm \chi(x)h]/\beta_n(x).$$

The domain $r \in [r_-, r_+]$ is such that Bessel's equation contains no turning point (no caustic in the waveguide). In this case the order ν of the Hankel functions is large while $|r-1|$ is always finite. The eigenfunctions can be approximated by working with the $H_\nu^{(1,2)}(\nu r)$ or by carrying out a JWKB approximation directly on Bessel's equation. Either way the approximate eigenfunctions are

$$u^n = \frac{[-i]_n}{2[P_{nn}(x)]^{1/2}(r^2 - 1)^{1/4}} \left\{ e^{i\nu[S(r)-S_m]} + e^{in\pi} e^{-i\nu[S(r)-S_m]} \right\}, \tag{35.10}$$

where

$$S(r) = (r^2 - 1)^{1/2} - \cos^{-1}(1/r), \quad S_m = (1/2)[S(r_+) + S(r_-)]. \tag{35.11}$$

The special symbol

$$[-i]_n := \left\{ \begin{array}{ll} 1, & n \text{ even} \\ -i, & n \text{ odd} \end{array} \right. .$$

Recall that r is defined in (35.8) and $r_\pm$ indicate that $z = \pm h$. The dispersion relation is given by

$$\nu[S(r_+) - S(r_-)] = n\pi. \tag{35.12}$$

Its solution can be approximated as

$$\beta_n = [1 - (n\pi/2h)^2]^{1/2} + \mathrm{O}(\chi^2), \tag{35.13}$$

which equation is identical to that for a flat waveguide, to the order indicated. A calculation of the boundary conditions to second order, seeking a correction to β_n at $\mathrm{O}(\chi)$, shows that no such term is present. If $h \approx \pi$ then only the two lowest modes $n = 0, 1$ propagate. If $n = 0$, then $r = 1$ at $z = 0$ so that (35.10) cannot describe a mode in a curved guide. However, for $n = 1$, provided $\chi < 1/(8\pi)$ (or equivalently the radius of curvature is greater than four wavelengths) (35.10) is a reasonable approximation.

Completing the calculation for $n = 1$, one finds that $\Im(W_{nn}) = 0$ so that $\dot{\phi}_0^n(x) = \phi_0^n(0)$. The normalization, $P_{nn}(x) = 1$, is calculated from

$$2\nu \int_{r_-}^{r_+} \frac{u^n u^{n*}}{r} dr = 1, \tag{35.14}$$

where the argument r is given by (35.8). This integral must usually be evaluated numerically at each x, for a general $\chi(x)$. Note that $P_n n(x)$ is even in χ so that it does not change very much from its value at $x = 0$.

The caustic is in the guide: The $n = 0$ mode is considered next. The waveguide is now imagined to be sufficiently thin that

$$h \approx 1/\chi^{1/3}.$$

This may not be true in the transition from the flat to the curved section, say between A and B in Fig.1. However, recall that the adiabatic condition ensures that $|a_0^n(x)|$ remains constant so that its value at A equals that at B.

The domain $r \in [r_-, r_+]$ contains the turning point $r = 1$. Again the eigenfunctions can be approximated by working directly with $H_\nu^{(1,2)}(\nu r)$ or by carrying out a boundary layer analysis about the turning point. The case investigated here is such that not only is $r \approx 1$, but also $\nu|r_\pm - 1|$ is large. Described in more physical terms, the whole guide is engulfed by the boundary layer formed about the caustic. To satisfy the boundary conditions, the inner solution, accurate in the neighborhood of $r = 1$, is expanded outward to satisfy the boundary conditions, rather than to match an outer wavefield. The approximate eigenfunctions are

$$
\begin{aligned}
u^n = {} & [P_{nn}(x)]^{-1/2} r^{-1/2} \Big\{ \Big[(r_+ - 1)^{1/4} e^{-i\pi/4} e^{i(2^{3/2}/3)\nu(r_+ - 1)^{3/2}} \\
& - (1 - r_-)^{1/4} e^{(2^{3/2}/3)\nu(1 - r_-)^{3/2}} \Big] A_2[-2^{1/3}\nu^{2/3}(r - 1)] \\
& + \Big[(r_+ - 1)^{1/4} e^{i\pi/4} e^{-i(2^{3/2}/3)\nu(r_+ - 1)^{3/2}} \\
& - (1 - r_-)^{1/4} e^{(2^{3/2}/3)\nu(1 - r_-)^{3/2}} \Big] A_1[-2^{1/3}\nu^{2/3}(r - 1)] \Big\}.
\end{aligned}
$$

The functions $A_{1,2}(z) := \mathrm{Ai}(z) \mp i\mathrm{Bi}(z)$, where the 1 corresponds to the $-$ and the 2 to the $+$. $\mathrm{Ai}(z)$ and $\mathrm{Bi}(z)$ are the Airy functions. Recall that r is defined by (35.8) and $r_\pm$ indicate that $z = \pm h$. The dispersion relation is given by

$$
\cos\left[\frac{2^{3/2}}{3}\nu(r_+ - 1)^{3/2} + \frac{\pi}{4} \right] = 0 \tag{35.15}
$$

Note that r_- does not appear at this level of approximation indicating that the guiding is done primarily between the caustic and the upper concave surface. The wavenumber β_0 can be approximated as

$$
\beta_0(x) = 1 - [\chi(x)]^{2/3}(3\pi/8)^{2/3}2^{-1/3} + \chi(x)h + o(\chi^{2/3}). \tag{35.16}
$$

Recall that $h \approx 1/\chi^{1/3}$.

Completing the calculation one finds that $\Im(W_{nn}) = 0$ so that $\phi_0^n(x) = \phi_0^n(0)$. The normalization is given by

$$
2\nu \int_{\nu^{2/3}(r_- - 1)}^{\nu^{3/2}(r_+ - 1)} \frac{u^n u^{n*}}{\nu^{3/2} + y} dy = 1. \tag{35.17}
$$

The argument $y = \nu^{2/3}(r - 1)$, where r is given by (35.8). Again the integral must be evaluated numerically at each x for a general $\chi(x)$. Recall that both χ and hence ν are functions of x.

4.   CLOSING REMARKS

The writer has described a method for treating propagation in curved waveguides that is more consistent in its outcomes than have his predecessors. However, though he has worked two special cases, the general uniform asymptotic solution is complex enough that simple analytic expressions for the dispersion and normalization are hard to obtain. Lastly, when the caustic lies inside the guide the dispersion and hence the propagation is dramatically affected by the curvature.

Acknowledgments

The support of AFOSR grant number F49620-96-1-0190 is gratefully acknowledged.

References

[1] Babič, V M and Kirpičnikova, N V (1979) *The Boundary-Layer Method in Diffraction Problems*, 37–59, Berlin: Springer.

[2] Folguera, A and Harris, J G (1999) *Coupled Rayleigh surface waves in a slowly varying elastic waveguide*, Proc. Roy. Soc. Lond. A **455**, 917–931.

[3] Ahluwalia, D S, Keller, J B and Matkowsky, B J (1974) *Asymptotic theory of propagation in curved and nonuniform waveguides*, J. Acoust. Soc. Amer. **55**, 7–12.

[4] Keller, J B and Ahuwalia, D S (1975) *Uniform asymptotic solutions of eigenvalue problems for convex plane domains*, SIAM J. Appl. Math. **25**, 583–591.

[5] Burridge, R & Weinberg, H (1977) *Horizontal rays and vertical modes*, in *Wave Propagation and Underwater Acoustics* (eds. J. B. Keller & J. S. Papadakis), 86–152, New York: Springer.

[6] Smith, R (1977) *Propagation in slowly-varying wave-guides*, SIAM J. Appl. Math. **33**, 39–50.

[7] Smith, R (1970) *Asymptotic solutions for high-frequency trapped wave propagation*, Phil. Trans. Roy. Soc. Lond. A **268**, 289–324.

[8] Olver, F W J (1965) *Bessel functions of integer order*, in *Handbook of Mathematical Functions* (ed. M. Abromowitz and I. A. Stegun), 355–433, New York: Dover.

EIGENSOLUTIONS FOR RAYLEIGH WAVE ANALYSIS

A. V. Kaptsov, S. V. Kuznetsov

Institute for Problems in Mechanics, RAS

Vernadskogo pr. 101 block 1, Moscow 117526, Russia

Kaptsov@ipmnet.ru

Abstract   Eigenvalues and eigenvectors of the Christoffel equation for the problem of Rayleigh waves propagating in media with arbitrary anisotropy are analyzed on the basis of three-dimensional complex formalism.

1.   INTRODUCTION

Spectral properties of the Christoffel equation are needed for construction of the exponentially decaying surface (Rayleigh) waves. The following analysis covers theorems of root characterization, structure of eigenvalues and corresponding eigenspaces of the Christoffel equation. The main results are obtained on the basis of three-dimensional complex formalism, which goes back to Rayleigh [1].

For anisotropic media Rayleigh's approach was exploited by Farnell [2] in numerical study of speed of Rayleigh waves, by Stoneley [3], and later by Royer and Dieulesaint [4] in analytical analysis of Rayleigh waves.

Six-dimensional formalism for Rayleigh wave analysis was proposed by Stroh [5]. This is based on similarity of solutions for moving line dislocations in unbounded media and propagation of Rayleigh waves. For Rayleigh wave analysis this approach was developed in [6, 7, 8, 9, 10]. Under assumptions of this formalism theorems of uniqueness and existence for Rayleigh waves were proved.

Yet, some omissions of both three- and six-dimensional approaches did not give solutions to two main problems of Rayleigh wave analysis: (i) existence of 'forbidden directions' along which Rayleigh waves cannot propagate; and (ii) uniqueness of Rayleigh wave propagating in an arbitrary direction. The present analysis does not pretend to solve the problems mentioned above, but it does give complete characterization of spectral properties of the Christoffel equation for Rayleigh waves propagating in media with arbitrary elastic anisotropy.

I.D. Abrahams et al. (eds.),
IUTAM Symposium on Diffraction and Scattering in Fluid Mechanics and Elasticity, 329–336.
© 2002 *Kluwer Academic Publishers. Printed in the Netherlands.*

2.   BASIC NOTATIONS

The equations of motion for anisotropic elastic media are

$$\mathrm{div}\,\mathbf{C} : \nabla\mathbf{u} - \rho\ddot{\mathbf{u}} = \mathbf{0} \qquad (36.1)$$

where $\mathbf{C}$ is the fourth-order elasticity tensor assumed to be positive definite, $\mathbf{u}$ is the displacement field, and ρ is the density of the medium.

Rayleigh wave is composed of partial waves having simple exponential decay in depth (partial waves of different structure are not considered generally):

$$\mathbf{u}(\mathbf{x}) = \mathbf{m}e^{i(\gamma\nu\cdot\mathbf{x}+\mathbf{n}'\cdot\mathbf{x}-ct)} = \mathbf{m}\,e^{-\alpha\nu\cdot\mathbf{x}}e^{i(\beta\nu\cdot\mathbf{x}+\mathbf{n}'\cdot\mathbf{x}-ct)}, \qquad (36.2)$$

where $\beta = \mathrm{Re}\,(\gamma)$, $\alpha = \mathrm{Im}\,(\gamma)$, $\mathbf{m}$ is the amplitude vector, ν is the unit normal to a plane boundary Π_ν, $\gamma = \beta + i\alpha$ is the unknown complex parameter, $\mathbf{n}' \subset \Pi_\nu$ is the unit vector defining direction of propagation of the Rayleigh wave and c is the wave (phase) speed. Substitution of representation (36.2) into equations (36.1) produces the Christoffel equation,

$$\left[(\gamma\nu + \mathbf{n}') \cdot \mathbf{C} \cdot (\mathbf{n}' + \gamma\nu) - \rho c^2\mathbf{I}\right] \cdot \mathbf{m} = 0, \qquad (36.3)$$

or equivalently

$$\det\left[(\gamma\nu + \mathbf{n}') \cdot \mathbf{C} \cdot (\mathbf{n}' + \gamma\nu) - \rho c^2\mathbf{I}\right] = 0 \qquad (36.4)$$

where it is assumed that the phase speed c is already specified. It is clear that the left side of (36.4) is a polynomial of degree 6 in γ.

Simple analysis of (36.4) shows that if c is zero, then there are no real roots of (36.4), since matrix $(\gamma\nu+\mathbf{n}')\cdot\mathbf{C}\cdot(\mathbf{n}'+\gamma\nu)$ is positive definite for real γ and its determinant cannot vanish. Similarly, asymptotic analysis produces

$$\gamma_i^2 \sim \frac{\rho c^2}{\lambda_i(\nu \cdot \mathbf{C} \cdot \nu)}, \quad \text{as} \quad c \to \infty \qquad (36.5)$$

where λ_i, $i = 1, 2, 3$ are eigenvalues (necessarily positive) of the matrix $\nu \cdot \mathbf{C} \cdot \nu$. Expression (36.5) ensures that all roots of (36.4) are real at high values of c.

Suppose that γ is a real root corresponding to a certain value of the phase speed c. Introduction of a new parameter $\varphi = \arctan(\gamma)$ allows the rewriting of (36.4) in terms of parameter φ as

$$\det\left[(\tan(\varphi)\nu + \mathbf{n}') \cdot \mathbf{C} \cdot (\mathbf{n}' + \tan(\varphi)\nu) - \rho c^2\mathbf{I}\right]$$
$$= \det\left[\cos^{-2}(\varphi)\mathbf{w} \cdot \mathbf{C} \cdot \mathbf{w} - \rho c^2\mathbf{I}\right] \quad = \quad 0, \quad (36.6)$$

where $\mathbf{w}$ is a unit vector given by

$$\mathbf{w} = \sin(\varphi)\nu + \cos(\varphi)\mathbf{n}'. \tag{36.7}$$

Definition 1 *Limiting speeds which correspond to the unit vectors ν and $\mathbf{n}'$, are defined by*

$$c_i^{\lim} = \inf_{\varphi \in \left[-\frac{\pi}{2};\frac{\pi}{2}\right]} \left(\cos^{-1}(\varphi)\sqrt{\rho^{-1}\lambda_i(\mathbf{w} \cdot \mathbf{C} \cdot \mathbf{w})}\right), \quad i = 1,2,3 \tag{36.8}$$

where λ_i, $i = 1,2,3$ are eigenvalues arranged in the descending mode.

Expression (36.8) gives:

$$c_i^{\lim} \leq \sqrt{\rho^{-1}\lambda_i(\mathbf{w} \cdot \mathbf{C} \cdot \mathbf{w})}, \quad i = 1,2,3. \tag{36.9}$$

Analysis of (36.8) shows that there are three different cases, when complex roots exist: (i) if $c_2^{\lim} < c < c_1^{\lim}$, then Christoffel's equation has only one pair of complex-conjugate roots; (ii) if $c_3^{\lim} < c < c_2^{\lim}$, there are two pairs of complex-conjugate roots; and (iii) if $c < c_3^{\lim}$, there are three pairs of complex-conjugate roots. Since for existence of Rayleigh wave at least one pair of complex conjugate roots is needed, we arrive at the following result.

Theorem 1 *The speed of the Rayleigh wave cannot exceed $c_1^{\lim}$.*

We remark that exponential decay in the lower half-space ($\nu \cdot \mathbf{x} < 0$) is ensured by complex roots with negative imaginary part: $\alpha < 0$; these roots will be chosen precisely in the subsequent analysis.

It will be assumed that the following condition of supplementary positive definiteness holds:

$$s_{in}C^{ijmn}s_{jm} > 0 \tag{36.10}$$

for any symmetric non-zero second order tensor s.

3.   PROPERTIES OF THE CHRISTOFFEL EQUATION

Appropriate complex root γ being substituted in Christoffel's equation produces following equation for the unknown (complex) eigenvector:

$$(\mathbf{G} - \rho c^2 \mathbf{I}) \cdot \mathbf{m} = 0, \quad \mathbf{G} \equiv \mathbf{A} + i\mathbf{B} \tag{36.11}$$

where Hermitian components $\mathbf{A}$, $\mathbf{B}$ of the matrix $\mathbf{G}$ have the form

$$\mathbf{A} \equiv \left[\mathbf{n}' \cdot \mathbf{C} \cdot \mathbf{n}' + (\beta^2 - \alpha^2)\, \nu \cdot \mathbf{C} \cdot \nu + \beta\, (\mathbf{n}' \cdot \mathbf{C} \cdot \nu + \nu \cdot \mathbf{C} \cdot \mathbf{n}') \right]$$
$$\mathbf{B} \equiv \alpha \left[2\beta\, \nu \cdot \mathbf{C} \cdot \nu + (\nu \cdot \mathbf{C} \cdot \mathbf{n}' + \mathbf{n}' \cdot \mathbf{C} \cdot \nu) \right].$$

Lemma 1 *Matrices $\mathbf{A}$ and $\mathbf{B}$ are both non-zero and do not commute with each other.*

Let $\mathbf{m}' \equiv \mathrm{Re}(\mathbf{m})$ and $\mathbf{m}'' \equiv \mathrm{Im}(\mathbf{m})$. The following proposition flows directly from the analysis of (36.11) and Lemma 1.

Theorem 2 (a) *If $\mathbf{m}'$, $\mathbf{m}'' \notin \ker\left(\mathbf{A} - \rho c^2 \mathbf{I}\right)$, then $\mathbf{m}'$, $\mathbf{m}'' \notin \ker\left(\mathbf{B}\right)$ and*

$$\left[\mathbf{B}^{-1} \circ \left(\mathbf{A} - \rho c^2 \mathbf{I} \right) \right]^2 = -\mathbf{I} \tag{36.12}$$

in the real (necessarily two-dimensional) space generated by vectors $\mathbf{m}'$ and $\mathbf{m}''$; (b) If $\mathbf{m}'$, $\mathbf{m}'' \in \ker\left(\mathbf{A} - \rho c^2 \mathbf{I}\right)$, then $\mathbf{m}'$, $\mathbf{m}'' \in \ker\left(\mathbf{B}\right)$; (c) If $\mathbf{m}' \in \ker(\mathbf{A} - \rho c^2 \mathbf{I})$ and $\mathbf{m}'' \notin \ker(\mathbf{A} - \rho c^2 \mathbf{I})$, then $\mathbf{m}'' \in \ker \mathbf{B}$ and $\mathbf{m}' \notin \ker \mathbf{B}$ (and vice versa).

It is obvious that Theorem 2 covers all possibilities for Hermitian components $\mathbf{m}', \mathbf{m}''$ of the eigenvector $\mathbf{m}$ of matrix $\mathbf{G}$.

Corollary 1 *Condition of Theorem 2(a) ensures:*
   (a) Both matrices $\mathbf{B}$ and $(\mathbf{A} - \rho c^2 \mathbf{I})$ are not of fixed sign in the two-dimensional subspace $Z \subset R^3$ generated by the components $\mathbf{m}'$ and $\mathbf{m}''$;
   (b) $\mathbf{B}(Z) \subset Z$, $(\mathbf{A} - \rho c^2 \mathbf{I}) \subset Z$;
   (c) Unimodal matrix $\mathbf{B}^{-1} \circ (\mathbf{A} - \rho c^2 \mathbf{I})$ in any basis in Z has the form

$$\mathbf{B}^{-1} \circ (\mathbf{A} - \rho c^2 \mathbf{I}) = \begin{pmatrix} \delta & \chi \\ -(1 + \delta^2)/\chi & -\delta \end{pmatrix} \tag{36.13}$$

where δ and χ are real, and $\chi \neq 0$;
   (d) Matrices $\mathbf{A}$ and $\mathbf{B}$ do not commute with each other.

Corollary 2 *Condition of Theorem 2(c) ensures (a) vectors $\mathbf{m}'$ and $\mathbf{m}''$ are not collinear; (b) Tensors $\mathbf{A} - \rho c^2 \mathbf{I}$ and $\mathbf{B}$ have the following structure:*

$$\mathbf{A} - \rho c^2 \mathbf{I} = 0\mathbf{m}' \otimes \mathbf{m}' + c_1 \mathbf{v}_1 \otimes \mathbf{v}_1 - c_1 \mathbf{v}_2 \otimes \mathbf{v}_2,$$
$$\mathbf{B} = 0\mathbf{m}'' \otimes \mathbf{m}'' + c_2 \mathbf{w}_1 \otimes \mathbf{w}_1 - c_2 \mathbf{w}_2 \otimes \mathbf{w}_2$$

where c_1 and c_2 are nonzero real numbers, vectors $\mathbf{m}'$, $\mathbf{v}_1$, $\mathbf{v}_2 \in R^3$ are mutually orthogonal, vectors $\mathbf{m}''$, $\mathbf{w}_1$, $\mathbf{w}_2 \in R^3$ are also mutually orthogonal.

Corollary 3 *Real part β of the complex root satisfies the inequalities:*

$$-\frac{\lambda_{\max}(\nu \cdot \mathbf{C} \cdot \mathbf{n}' + \mathbf{n}' \cdot \mathbf{C} \cdot \nu)}{2\lambda_{\max}(\nu \cdot \mathbf{C} \cdot \nu)} \leq \beta \leq -\frac{\lambda_{\min}(\nu \cdot \mathbf{C} \cdot \mathbf{n}' + \mathbf{n}' \cdot \mathbf{C} \cdot \nu)}{2\lambda_{\min}(\nu \cdot \mathbf{C} \cdot \nu)}.$$

$$(36.14)$$

Proof. The following inequalities flow directly from the definition of matrix $\mathbf{B}$, which gives, for all $\mathbf{w} \in R^3$ with $|\mathbf{w}| = 1$,

$$\alpha(2\beta \, \lambda_{\min}(\nu \cdot \mathbf{C} \cdot \nu) + \lambda_{\min}(\nu \cdot \mathbf{C} \cdot \mathbf{n}' + \mathbf{n}' \cdot \mathbf{C} \cdot \nu))$$
$$\leq \mathbf{w} \cdot \mathbf{B} \cdot \mathbf{w} \leq \alpha(2\beta \, \lambda_{\max}(\nu \cdot \mathbf{C} \cdot \nu) + \lambda_{\max}(\nu \cdot \mathbf{C} \cdot \mathbf{n}' + \mathbf{n}' \cdot \mathbf{C} \cdot \nu)).$$

Corollary 1 ensures that in any case $\lambda_{\min}(\mathbf{B}) \leq 0$, so there exist vectors $\mathbf{w} \in R^3$ with $|\mathbf{w}| = 1$ such that $\mathbf{w} \cdot \mathbf{B} \cdot \mathbf{w} = 0$. Taking into account that α, $\lambda_{\max}(\nu \cdot \mathbf{C} \cdot \nu)$ and $\lambda_{\min}(\nu \cdot \mathbf{C} \cdot \nu)$ are positive, we arrive at (36.14).

Theorem 3 *For any complex root γ, matrix $\mathbf{G}$ is not normal.*

Proof of this proposition follows from consideration of the commutator $[\mathbf{A}, \mathbf{B}]$, which is not zero for complex γ, provided (36.10) holds.

We remark that, for real γ, matrix $\mathbf{G}$ *is* normal. This is due to its symmetric structure, which follows from (36.11) and the subsequent formulas for matrices $\mathbf{A}$ and $\mathbf{B}$.

Theorem 4 *Under the condition of Theorem 2(b), eigenspace in R^3 generated by vectors $\mathbf{m}'$ and $\mathbf{m}''$ is one dimensional.*

Proof. If this eigenspace is two- or three-dimensional, the symmetric matrices $\mathbf{A} - \rho c^2 \mathbf{I}$ and $\mathbf{B}$ could be reduced to diagonal form by the same orthogonal transformation (in such a case $\mathbf{A} - \rho c^2 \mathbf{I}$ and $\mathbf{B}$ have common eigenspace), this leads to their commutation, but these matrices cannot commute since matrix $\mathbf{G}$ is not normal.

4.   SPECTRAL PROPERTIES OF THE CHRISTOFFEL EQUATION

Let $W_{\mathbf{G}}$ be eigenspace of matrix $\mathbf{G}$ corresponding to the eigenvalue ρc^2. Theorems 2–4 produce

Theorem 5 (a) $\dim W_{\mathbf{G}} = 1$, *provided that both matrices $\mathbf{A}' \equiv \mathbf{A} - \rho c^2 \mathbf{I}$ and $\mathbf{B}$ do not have zero eigenvalues, and relation (36.12) holds;* (b) $\dim W_{\mathbf{G}} = 2$, *provided that relation (36.12) holds, and both matrices $\mathbf{A}' \equiv \mathbf{A} - \rho c^2 \mathbf{I}$ and $\mathbf{B}$ have zero eigenvalue;* (c) $\dim W_{\mathbf{G}} = 1$, *provided that relation (36.12) does not hold, and both matrices $\mathbf{A}' \equiv \mathbf{A} - \rho c^2 \mathbf{I}$ and $\mathbf{B}$ have zero eigenvalue.*

Proof.   (a) Corollary 1 shows that real space Z is two-dimensional. Let real vectors $\mathbf{e}_1$ and $\mathbf{e}_2$ form an orthonormal basis in Z, then (complex) eigenvectors $\mathbf{m}'$ and $\mathbf{m}''$ admit representation

$$\mathbf{m}' = \mathbf{e}_1 + i(\mathbf{B}^{-1} \circ \mathbf{A}') \cdot \mathbf{e}_1, \quad \mathbf{m}'' = \mathbf{e}_2 + i(\mathbf{B}^{-1} \circ \mathbf{A}') \cdot \mathbf{e}_2. \quad (36.15)$$

Substitution of relation (36.13) in (36.15) gives

$$\mathbf{m}' = \mathbf{e}_1 + i[\delta\mathbf{e}_1 - \chi^{-1}(1 + \delta^2)\mathbf{e}_2], \quad \mathbf{m}'' = \mathbf{e}_2 + i(\chi\mathbf{e}_1 - \delta\mathbf{e}_2). \quad (36.16)$$

Now, representation (36.16) shows that complex vectors $\mathbf{m}'$ and $\mathbf{m}''$ are linearly dependent: $\mathbf{m}'' = a\mathbf{m}'$ with complex constant

$$a = \chi(\delta + i)/(1 + \delta^2).$$

(b) Linearly dependent complex vectors $\mathbf{m}'$ and $\mathbf{m}''$ satisfying (36.16) form one-dimensional space $K \subset C^3$. Now, eigenvector $\mathbf{m} = \mathbf{e}_3 + i\mathbf{0}$, where $\mathbf{e}_3$ is common eigenvector of matrices $\mathbf{A}'$ and $\mathbf{B}$, is orthogonal to K.

(c) Hermitian components of eigenvectors of the matrix $\mathbf{G}$ corresponding to ρc^2, belong to $\ker(\mathbf{A} - \rho c^2\mathbf{I}) \cap \ker(\mathbf{B})$ due to Theorem 2(b). If $\dim\left(\ker(\mathbf{A} - \rho c^2\mathbf{I}) \cap \ker(\mathbf{B})\right) > 1$, then both matrices $\mathbf{A}$ and $\mathbf{B}$ have common eigenvectors, and therefore they commute with each other, but this contradicts Theorem 3 (if Hermitian components $\mathbf{A}$ and $\mathbf{B}$ of the matrix $\mathbf{G}$ commute, then the latter matrix becomes normal).

Lemma 2 *If λ is a complex eigenvalue of matrix $\mathbf{G}$ and $\mathbf{m}$ is corresponding eigenvector with Hermitian components $\mathbf{m}'$ and $\mathbf{m}'' \in Z \subset R^3$, then $i\lambda$ is eigenvalue of $\mathbf{G}$ with another eigenvector with Hermitian components that also belong to Z.*

Proof is identical to proof of Theorem 2(a) with the only difference that for the eigenvalue $i\lambda$ relation (36.12) changes to the following:

$$\left[\mathbf{A}^{-1} \circ (\mathbf{B} - \lambda\mathbf{I})\right]^2 = -\mathbf{I}.$$

Similarly, values δ and χ from expression (36.13) change.

Theorem 6 *The algebraic multiplicity of the eigenvalue ρc^2 equals its geometric multiplicity.*

Proof of the case corresponding to Theorem 5. Let $Z \subset R^3$ be the two-dimensional space from Corollary 1, then matrix $\mathbf{G}$, regarded as an operator acting in C^3, can be decomposed in the direct sum:

$$\mathbf{G} = (\mathbf{A}'_Z + i\mathbf{B}_Z) \oplus (\mathbf{A}'_{CZ} + i\mathbf{B}_{CZ}) \quad (36.17)$$

where $\mathbf{A}'_Z$, $\mathbf{B}_Z$, $\mathbf{A}'_{CZ}$ and $\mathbf{B}_{CZ}$ are restriction of the operators $\mathbf{A}'$ and $\mathbf{B}$ on Z and $CZ = R^3\backslash Z$. Since CZ is one-dimensional, its algebraic multiplicity coincides with its geometric multiplicity, and direct verification shows that the corresponding eigenvalue does not equal ρc^2 or $i\rho c^2$. Then, Theorem 5(a) shows that CZ is one-dimensional, if regarded as complex space. To complete the proof of this case it is sufficient to note that there is another eigenvector in Z, which corresponds to eigenvalue $i\rho c^2$. It should be noted that eigenvectors from Z are not orthogonal. Proof of the cases corresponding to Theorem 5(b) and (c) are mainly identical to the regarded case.

Proofs of Lemma 2 and Theorem 6 ensure that for all regarded cases matrix $\mathbf{G}$ has three linearly independent complex eigenvectors and these eigenvectors form a basis in C^3. However, these eigenvectors are not orthogonal, otherwise matrix $\mathbf{G}$ would be normal. Existence of three linearly independent eigenvectors gives us:

Matrix $\mathbf{G}$ is similar to a complex diagonal matrix.

Theorem 7 *Let parameter ρc^2 be fixed, then (complex) eigenvectors corresponding to different complex roots with positive imaginary part, are linearly independent.*

Proof. Suppose that common eigenvector $\mathbf{m}$ corresponds to two different complex roots γ_1 and γ_2, then

$$(\mathbf{G}(\gamma_1) - \mathbf{G}(\gamma_2)) \cdot \mathbf{m} = 0. \tag{36.18}$$

By the use of (36.3), (36.18) can be transformed to the following form

$$((\gamma_1 + \gamma_2)\nu \cdot \mathbf{C} \cdot \nu + (\nu \cdot \mathbf{C} \cdot \mathbf{n}' + \mathbf{n}' \cdot \mathbf{C} \cdot \nu)) \cdot \mathbf{m} = 0 \tag{36.19}$$

Since $\mathrm{Im}(\gamma_1 + \gamma_2) \neq 0$, $\ker\left((\gamma_1 + \gamma_2)\nu \cdot \mathbf{C} \cdot \nu + (\nu \cdot \mathbf{C} \cdot \mathbf{n}' + \mathbf{n}' \cdot \mathbf{C} \cdot \nu)\right)$ is nontrivial if and only if matrix $\nu \cdot \mathbf{C} \cdot \nu$ is degenerate or sign indefinite, but this is impossible.

We remark that analysis of eigenspaces of the Christoffel equation shows that when the phase speed belongs to the interval $(0, c_3^{\mathrm{lim}})$ and hence, there are no multiple roots, there exist exactly three linearly independent complex eigenvectors forming partial waves in the Rayleigh wave representation (regular case). This flows from Theorems 6 and 7.

The same theorems guarantee that degenerate case in the interval $(0, c_3^{\mathrm{lim}})$ can only occur when two conditions hold simultaneously: (i) there are multiple roots and, (ii) multiplicity of the eigenvalue ρc^2 is one at any of the fixed roots. It should be noted that in the isotropic case, multiplicity of ρc^2 at multiple roots is two, and no degeneracy occurs.

References

[1] Strutt, J W (Lord Rayleigh) (1885) *On waves propagating along the plane surface of an elastic solid*, Proc. Lond. Math. Soc. **17**, 4–11.

[2] Farnell, G W (1970) *Properties of elastic surface waves*, Phys. Acoust. **6**, 109–166.

[3] Stoneley, R (1955) *The propagation of surface elastic waves in a cubic crystal*, Proc. Roy. Soc. Lond. A **232**, 447–458.

[4] Royer, D and Dieulesaint, E (1985) *Rayleigh wave velocity and displacement in orthorombic, tetragonal, and cubic crystals*, J. Acoust. Soc. Amer. **76**, 1438–1444.

[5] Stroh, A N (1962) *Steady state problems in anisotropic elasticity*, J. Math. Phys. **41**, 77–103.

[6] Barnett, D M and Lothe, J (1973) *Synthesis of the sextic and the integral formalism for dislocations, Green's functions, and surface waves in anisotropic elastic solids*, Phys. Norv. **7**, 13–19.

[7] Barnett, D M and Lothe, J (1974) *Consideration of the existence of surface wave (Rayleigh wave) solutions in anisotropic elastic crystals*, J. Phys. F: Metal Phys. **4**, 671–686.

[8] Lothe, J and Barnett, D M (1976) *On the existence of surface wave solutions for anisotropic elastic half-spaces with a free surface*, J. Appl. Phys. **47**, 428–433.

[9] Chadwick, P and Smith, G D (1977) *Foundations of the theory of surface waves in anisotropic elastic materials*, Adv. Appl. Mech. **17**, 303–376.

[10] Chadwick, P and Jarvis, D A (1979) *Surface waves in a prestressed elastic body*, Proc. Roy. Soc. Lond. A **366**, 517–536.

WAVES IN WOOD

P. A. Martin
Department of Mathematical and Computer Sciences
Colorado School of Mines, Golden, CO 80401-1887, USA
pamartin@mines.edu

J. R. Berger
Division of Engineering
Colorado School of Mines, Golden, CO 80401-1887, USA
jberger@mines.edu

Abstract   Elastic waves in materials with cylindrical orthotropy are considered, this being a plausible model for waves in wood. For time-harmonic motions, the problem is reduced to some coupled ordinary differential equations. Methods for solving these equations are discussed. These include the method of Frobenius (power-series exapansions) and the use of Neumann series (expansions in series of Bessel functions of various orders).

1.   INTRODUCTION

As children, we learn that we can determine the age of a sawn log or tree by counting the annual rings visible at a sawn end. The presence of these rings influences the mechanical properties of the wood, of course. This observation leads to a constitutive model in which the wood is assumed to be an elastic solid with *cylindrical orthotropy*. Thus, we assume that the elastic stiffnesses are constants when referred to cylindrical polar coordinates, r, θ, z. (In general, the material will be inhomogeneous when described in Cartesian coordinates.)

We are interested in the propagation of elastic waves in wooden poles, using the cylindrical-orthotropy model. One application is to understand the use of ultrasonic devices for determining whether wooden telegraph poles have decayed internally [1]. Application of these techniques to live trees has also been made [2]. We are also interested in methods for inspecting wire ropes and overhead transmission lines; wire rope,

337

I.D. Abrahams et al. (eds.),
IUTAM Symposium on Diffraction and Scattering in Fluid Mechanics and Elasticity, 337–344.

consisting of several helically wound strands, may be 'homogenised' [3], giving a similar mathematical model for the material.

Further information on waves in wood can be found in the book by Bucur [4].

In this paper, we begin by formulating the problem of wave propagation in a material with cylindrical anisotropy, using matrix notation where possible. We look for time-harmonic solutions, with a prescribed dependence on θ and z, leading to a 3×3 system of coupled ordinary differential equations in the radial direction. To simplify further, we suppose that the material is cylindrically orthotropic. We review the known exact solutions, corresponding to isotropy, axisymmetry (no dependence on θ) or no dependence on z. We then discuss two methods for solving the remaining situations. These are the method of Frobenius (expansions in powers of r) and a generalization using Neumann series (expansions in Bessel functions of various orders). The two methods are compared and contrasted. The method based on Neumann series seems to be better suited to problems involving wave propagation, and should find further applications.

2.   GOVERNING EQUATIONS

In the absence of body forces, the governing equations of motion are

$$\frac{\partial}{\partial r}\left(r\mathbf{t}_r\right) + \frac{\partial}{\partial \theta}\mathbf{t}_\theta + \mathbf{K}\mathbf{t}_\theta + r\frac{\partial}{\partial z}\mathbf{t}_z = \rho r \frac{\partial^2}{\partial t^2}\tilde{\mathbf{u}}, \tag{37.1}$$

where ρ is the density,

$$\mathbf{t}_r = \begin{pmatrix} \tau_{rr} \\ \tau_{r\theta} \\ \tau_{rz} \end{pmatrix}, \quad \mathbf{t}_\theta = \begin{pmatrix} \tau_{\theta r} \\ \tau_{\theta\theta} \\ \tau_{\theta z} \end{pmatrix}, \quad \mathbf{t}_z = \begin{pmatrix} \tau_{zr} \\ \tau_{z\theta} \\ \tau_{zz} \end{pmatrix},$$

$$\mathbf{K} = \begin{pmatrix} 0 & -1 & 0 \\ 1 & 0 & 0 \\ 0 & 0 & 0 \end{pmatrix}, \quad \tilde{\mathbf{u}} = \begin{pmatrix} u_r \\ u_\theta \\ u_z \end{pmatrix}$$

is the displacement vector and τ_{ij} are the stress components. In what follows, we generalise the matrix formulation of Ting [5] for static problems; he gives expressions for the traction vectors $\mathbf{t}_i$ in terms of $\tilde{\mathbf{u}}$.

We look for time-harmonic solutions of (37.1) in the form

$$\tilde{\mathbf{u}}(r, \theta, z, t) = \mathrm{Re}_\mathrm{i}\left\{\mathbf{u}_m(r)\,\mathrm{e}^{\mathrm{j}m\theta}\,\mathrm{e}^{\mathrm{i}\xi z}\mathrm{e}^{-\mathrm{i}\omega t}\right\},$$

where i and j are two non-interacting complex units, m is an integer, ξ is the axial wavenumber, ω is the radian frequency, and Re_i denotes the real part with respect to i. Use of $e^{jm\theta}$ rather than $\cos m\theta$ and $\sin m\theta$ allows us to retain the nice matrix notation in what follows. Thus, we find that $\mathbf{u}_m(r)$ solves

$$r\mathbf{Q}\left(r\mathbf{u}'_m\right)' + r\mathbf{A}\mathbf{u}'_m + \mathbf{B}\mathbf{u}_m = \mathbf{0}, \tag{37.2}$$

where

$$
\begin{aligned}
\mathbf{A} &= \mathbf{R}\mathbf{K}_m + \mathbf{K}_m\mathbf{R}^T + i\xi r(\mathbf{P} + \mathbf{P}^T),\\
\mathbf{B} &= \rho\omega^2 r^2\mathbf{I} - \xi^2 r^2\mathbf{M} + \mathbf{K}_m\mathbf{T}\mathbf{K}_m + i\xi r(\mathbf{P} + \mathbf{K}_m\mathbf{S} + \mathbf{S}^T\mathbf{K}_m),
\end{aligned}
$$

$$
\mathbf{Q} = \begin{pmatrix} C_{11} & C_{16} & C_{15} \\ C_{16} & C_{66} & C_{56} \\ C_{15} & C_{56} & C_{55} \end{pmatrix}, \qquad
\mathbf{R} = \begin{pmatrix} C_{16} & C_{12} & C_{14} \\ C_{66} & C_{26} & C_{46} \\ C_{56} & C_{25} & C_{45} \end{pmatrix},
$$

$$
\mathbf{T} = \begin{pmatrix} C_{66} & C_{26} & C_{46} \\ C_{26} & C_{22} & C_{24} \\ C_{46} & C_{24} & C_{44} \end{pmatrix}, \qquad
\mathbf{P} = \begin{pmatrix} C_{15} & C_{14} & C_{13} \\ C_{56} & C_{46} & C_{36} \\ C_{55} & C_{45} & C_{35} \end{pmatrix},
$$

$$
\mathbf{M} = \begin{pmatrix} C_{55} & C_{45} & C_{35} \\ C_{45} & C_{44} & C_{34} \\ C_{35} & C_{34} & C_{33} \end{pmatrix}, \qquad
\mathbf{S} = \begin{pmatrix} C_{56} & C_{46} & C_{36} \\ C_{25} & C_{24} & C_{23} \\ C_{45} & C_{44} & C_{34} \end{pmatrix},
$$

$\mathbf{I}$ is the identity, $\mathbf{K}_m = \mathbf{K} + jm\mathbf{I}$, $\mathbf{R}^T$ is the transpose of $\mathbf{R}$, and we have used the contracted notation $C_{\alpha\beta}$ for the elastic stiffnesses. Note that $\mathbf{Q}$, $\mathbf{T}$ and $\mathbf{M}$ are symmetric matrices.

For two-dimensional motions independent of z ($\xi = 0$), we recover the equations studied in [1]. If we also put $m = 0$ (axisymmetry) and $\omega = 0$ (static), we obtain the equations solved by Ting [5].

Setting $\mathbf{u}_m = (u_m, v_m, w_m)$, (37.2) gives three coupled ordinary differential equations for the three components of $\mathbf{u}_m$. In general, these equations do not decouple.

3.   CYLINDRICAL ORTHOTROPY

For materials with cylindrical orthotropy, there are nine non-trivial stiffnesses, namely C_{11}, C_{12}, C_{13}, C_{22}, C_{23}, C_{33}, C_{44}, C_{55} and C_{66}. The matrices $\mathbf{Q}$, $\mathbf{R}$, $\mathbf{T}$, $\mathbf{P}$, $\mathbf{M}$ and $\mathbf{S}$ simplify to

$$
\mathbf{Q} = \begin{pmatrix} C_{11} & 0 & 0 \\ 0 & C_{66} & 0 \\ 0 & 0 & C_{55} \end{pmatrix}, \qquad
\mathbf{R} = \begin{pmatrix} 0 & C_{12} & 0 \\ C_{66} & 0 & 0 \\ 0 & 0 & 0 \end{pmatrix},
$$

$$
\mathbf{T} = \begin{pmatrix} C_{66} & 0 & 0 \\ 0 & C_{22} & 0 \\ 0 & 0 & C_{44} \end{pmatrix}, \qquad
\mathbf{P} = \begin{pmatrix} 0 & 0 & C_{13} \\ 0 & 0 & 0 \\ C_{55} & 0 & 0 \end{pmatrix},
$$

$$\mathbf{M} = \begin{pmatrix} C_{55} & 0 & 0 \\ 0 & C_{44} & 0 \\ 0 & 0 & C_{33} \end{pmatrix}, \qquad \mathbf{S} = \begin{pmatrix} 0 & 0 & 0 \\ 0 & 0 & C_{23} \\ 0 & C_{44} & 0 \end{pmatrix},$$

and the system (37.2) simplifies accordingly.

Note that isotropy is a special case of cylindrical orthotropy. For isotropic materials, $C_{11} = C_{22} = C_{33} = \lambda + 2\mu$, $C_{12} = C_{13} = C_{23} = \lambda$ and $C_{44} = C_{55} = C_{66} = \mu$, where λ and μ are the Lamé moduli. Exact solutions of (37.2) are well known for isotropic solids. They are given in textbooks on elastic waves; see, for example, Graff [6, §8.2]. Some of these solutions will be mentioned below.

4.   AXISYMMETRIC MOTIONS

For axisymmetric motions ($m = 0$) of a cylindrically orthotropic solid, we have $\mathbf{K}_0 = \mathbf{K}$, $\mathbf{RK} = -\mathbf{KR}^T$,

$$\mathbf{A} = \mathrm{i}\xi r \begin{pmatrix} 0 & 0 & C_{13} + C_{55} \\ 0 & 0 & 0 \\ C_{13} + C_{55} & 0 & 0 \end{pmatrix}$$

and

$$\mathbf{B} = \begin{pmatrix} B_5 r^2 - C_{22} & 0 & \mathrm{i}\xi r(C_{13} - C_{23}) \\ 0 & B_4 r^2 - C_{66} & 0 \\ \mathrm{i}\xi r(C_{55} + C_{23}) & 0 & B_3 r^2 \end{pmatrix},$$

where

$$B_i = \rho\omega^2 - \xi^2 C_{ii} \qquad \text{(no sum)}.$$

So, the (axisymmetric) torsional component v_0 decouples from the radial and axial components, u_0 and w_0, respectively. We find that v_0 satisfies

$$rC_{66}\left(rv_0'\right)' + (B_4 r^2 - C_{66})v_0 = 0.$$

This is Bessel's equation; solutions are

$$J_1(\ell r) \quad \text{and} \quad Y_1(\ell r),$$

where $\ell = \sqrt{B_4/C_{66}}$.

The remaining pair of equations for u_0 and w_0 is as follows:

$$rC_{11}\left(ru_0'\right)' + \mathrm{i}\xi r^2(C_{13} + C_{55})w_0'$$
$$+ (B_5 r^2 - C_{22})u_0 + \mathrm{i}\xi r(C_{13} - C_{23})w_0 \;=\; 0, \qquad (37.3)$$
$$rC_{55}\left(rw_0'\right)' + \mathrm{i}\xi r^2(C_{13} + C_{55})u_0'$$
$$+ \mathrm{i}\xi r(C_{55} + C_{23})u_0 + B_3 r^2 w_0 \;=\; 0. \qquad (37.4)$$

Note that axisymmetric motions in a cylindrically orthotropic material do not depend on the stiffness C_{12}.

Equations (37.3) and (37.4) decouple when $\xi = 0$ [1]. Typical solution pairs are found to be

$$u_0 = 0 \quad \text{and} \quad w_0 = J_0(\kappa_a r),$$

and

$$u_0 = J_\gamma(\kappa_1 r) \quad \text{and} \quad w_0 = 0,$$

where $\gamma = \sqrt{C_{22}/C_{11}}$, $\kappa_1 = \omega\sqrt{\rho/C_{11}}$ and $\kappa_a = \omega\sqrt{\rho/C_{55}}$.

For isotropic materials, one solution pair for (37.3) and (37.4) is

$$u_0 = \xi J_1(Kr) \quad \text{and} \quad w_0 = iK J_0(Kr),$$

where $K^2 = \rho\omega^2/\mu - \xi^2$. Another is

$$u_0 = k J_1(kr) \quad \text{and} \quad w_0 = -i\xi J_0(kr),$$

where $k^2 = \rho\omega^2/(\lambda + 2\mu) - \xi^2$. See, for example, [6, p. 471].

5. TWO-DIMENSIONAL MOTIONS

For motions of a cylindrically orthotropic material that are independent of z ($\xi = 0$), we again find that the 3×3 system (37.2) partially decouples. However, now we obtain a pair of equations for u_m and v_m, and a single equation for w_m. The latter can be solved exactly: two independent solutions are

$$J_\beta(\kappa_a r) \quad \text{and} \quad Y_\beta(\kappa_a r),$$

where $\beta = m\sqrt{C_{44}/C_{55}}$ and $\kappa_a = \omega\sqrt{\rho/C_{55}}$.

The pair of equations for $u_m(r)$ and $v_m(r)$ is as follows [1]:

$$rC_{11}\left(ru'_m\right)' + jmr(C_{66} + C_{12})v'_m$$
$$+ (\rho\omega^2 r^2 - m^2 C_{66} - C_{22})u_m - jm(C_{66} + C_{22})v_m \;=\; 0, \quad (37.5)$$
$$rC_{66}\left(rv'_m\right)' + jmr(C_{66} + C_{12})u'_m$$
$$+ (\rho\omega^2 r^2 - C_{66} - m^2 C_{22})v_m + jm(C_{66} + C_{22})u_m \;=\; 0. \quad (37.6)$$

Note that motions independent of z in a cylindrically orthotropic medium depend on only six of the nine stiffnesses, namely C_{11}, C_{12}, C_{22}, C_{44}, C_{55} and C_{66}.

For a homogeneous isotropic elastic solid, solutions of (37.5) and (37.6) are known. Let (u_m, v_m) be a solution pair. There are four independent solution pairs. The first two pairs are

$$u_m = J'_m(k_P r) \quad \text{and} \quad v_m = jm(k_P r)^{-1} J_m(k_P r) \qquad (37.7)$$

and

$$u_m = m(k_S r)^{-1} J_m(k_S r) \quad \text{and} \quad v_m = \mathrm{j}\, J_m'(k_S r), \tag{37.8}$$

where $k_P = \omega \sqrt{\rho/(\lambda + 2\mu)}$ is the compressional wavenumber and $k_S = \omega\sqrt{\rho/\mu}$ is the shear wavenumber. The second two pairs are obtained by replacing J_n in these expressions by Y_n.

6.   COUPLED SYSTEMS

In special cases, we have seen that we can find explicit solutions of (37.2). However, in general, we are left with a coupled system of ordinary differential equations to solve. This will be a 2×2 system if $m = 0$ ((37.3) and (37.4)) or if $\xi = 0$ ((37.5) and (37.6)), but it will be a 3×3 system in the general case.

How can we solve such systems? A standard technique for solving ordinary differential equations is the method of Frobenius, in which one looks for solutions in the form of power series. The method proceeds by writing

$$
\begin{aligned}
u_m(r) &= \sum_{n=0}^{\infty} \hat{a}_n \, (\kappa r)^{2n+\alpha}, \\
v_m(r) &= \mathrm{j}\sum_{n=0}^{\infty} \hat{b}_n \, (\kappa r)^{2n+\alpha}, \\
w_m(r) &= \sum_{n=0}^{\infty} \hat{c}_n \, (\kappa r)^{2n+\alpha},
\end{aligned}
$$

where the coefficients $\hat{a}_n$, $\hat{b}_n$ and $\hat{c}_n$, and the exponent α are to be determined. It turns out that there is no loss of generality in using $(\kappa r)^{2n}$ rather than $(\kappa r)^n$.

Substituting these expansions in (37.2) and collecting terms leads to an indicial equation for α and coupled recursion relations for the coefficients. This yields an efficient method for computing the coefficients. In the present context, it has been used by many authors, including Ohnabe and Nowinski [7], Chou and Achenbach [8], Markuš and Mead [9] and Yuan and Hsieh [10].

The main drawback of the method of Frobenius is that it is essentially a *static* method: power series in κr are only expected to be good for small values of κr.

If we examine the known exact solutions for isotropic solids, such as (37.7) and (37.8), we see that they can be written as linear combinations of two Bessel functions, with orders differing by two. For example, (37.7) can be written as

$$u_m(r) = \tfrac{1}{2}\left\{ J_{m-1}(k_P r) - J_{m+1}(k_P r) \right\},$$
$$v_m(r) = \tfrac{1}{2}\mathrm{j}\left\{ J_{m-1}(k_P r) + J_{m+1}(k_P r) \right\}.$$

This suggests that we should use a generalization of the method of Frobenius, in which u_m, v_m and w_m are expanded as *Neumann series*,

$$u_m(r) = \sum_{n=0}^{\infty} a_n\, J_{2n+\alpha}(kr),$$
$$v_m(r) = \mathrm{j}\sum_{n=0}^{\infty} b_n\, J_{2n+\alpha}(kr),$$
$$w_m(r) = \sum_{n=0}^{\infty} c_n\, J_{2n+\alpha}(kr),$$

where the coefficients a_n, b_n, c_n and α are to be determined. Note that the parameter k is to some extent at our disposal; its choice is discussed in [1], making use of the asymptotic behaviour of solutions of the governing differential equations for *large r*.

In [1], we have used Neumann series to solve (37.5) and (37.6), corresponding to $\xi = 0$. We investigated two methods for finding a_n and b_n, which we call *direct* and *indirect*. In the direct method, we substitute the Neumann-series expansions directly into (37.5) and (37.6) and then group terms. This requires manipulating series of Bessel functions, and so is more complicated than at the analogous stage of the method of Frobenius. It turns out that α solves the same indicial equation as before. Eventually, we obtain some recurrence relations for a_n and b_n; they are fairly complicated but they are well behaved numerically [1].

For the indirect method, we begin with the standard method of Frobenius, leading to the computation of the coefficients $\hat{a}_n$ and $\hat{b}_n$. From these, we then compute the coefficients a_n and b_n, using the known expansion of an arbitrary power in terms of Bessel functions:

$$(\tfrac{1}{2}kr)^\nu = \sum_{n=0}^{\infty} \frac{(2n+\nu)\,\Gamma(n+\nu)}{n!}\, J_{2n+\nu}(kr).$$

(Compare this with the definition of a Bessel function,

$$J_\nu(kr) = \sum_{n=0}^{\infty} \frac{(-1)^n}{n!\,\Gamma(n+\nu+1)}(\tfrac{1}{2}kr)^{2n+\nu},$$

which can itself be obtained by the method of Frobenius.)

We have found that the use of Neumann series is more efficient for the problems of wave propagation in wood that we have described above. We think that the method will find application to other problems involving wave propagation in materials with cylindrical anisotropy.

References

[1] Martin, P A and Berger, J R (2001) *Waves in wood: free vibrations of a wooden pole*, J. Mech. Phys. Solids **49**, 1155–1178.

[2] Dolwin, J A, Lonsdale, D and Barnett, J (1999) *Detection of decay in trees*, Arboricultural J. **23**, 139–149.

[3] Cardou, A and Jolicoeur, C (1997) *Mechanical models of helical strands*, Appl. Mech. Rev. **50**, 1–14.

[4] Bucur, V (1995) *Acoustics of Wood*, Boca Raton, Florida: CRC Press.

[5] Ting, T C T (1996) *Pressuring, shearing, torsion and extension of a circular tube or bar of cylindrically anisotropic material*, Proc. Roy. Soc. Lond. A **452**, 2397–2421.

[6] Graff, K F (1975) *Wave Motion in Elastic Solids*, Oxford: University Press.

[7] Ohnabe, H and Nowinski, J L (1971) *On the propagation of flexural waves in anisotropic bars*, Ingenieur-Archiv **40**, 327–338.

[8] Chou, F -H and Achenbach, J D (1972) *Three-dimensional vibrations of orthotropic cylinders*, Proc. ASCE, J. Engng. Mech. Div. **98**, 813–822.

[9] Markuš, Š and Mead, D J (1995) *Axisymmetric and asymmetric wave motion in orthotropic cylinders*, J. Sound Vib. **181**, 127–147.

[10] Yuan, F G and Hsieh, C C (1998) *Three-dimensional wave propagation in composite cylindrical shells*, Composite Structures **42**, 153–167.

BOUNDARY INTEGRAL METHODS FOR ELASTIC LAYERED MEDIA

M. Stojek

Department of Robotics and Machine Dynamics, University of Mining and Metallurgy
PL-30-059 Kraków, Poland

ghstojek@cyf-kr.edu.pl

A. T. Peplow

Department of Mathematical Sciences
University of the West of England, Bristol, BS16 1QY, UK

Andrew.Peplow@uwe.ac.uk

I. Karasalo

National Defence Research Institute
Dept 64, S-172 90, Sundbyberg, Sweden

ilkka@sto.foa.se

Abstract   Green's tensors are determined for a point force in layered horizontally homogeneous solid media at fixed horizontal wavenumber k due to a Fourier-Bessel series expansion. The media may consist of an arbitrary mixture of solid layers, bounded by homogeneous half-spaces at both ends. A finite element technique is proposed for solving a resulting two-point ordinary differential equation (ODE) boundary value problem for the wavefield using either a variational form or a propagator matrix approach. A brief discussion of the boundary element method for elastic-wave scattering from a bounded object using new Green's tensors is outlined.

1.   INTRODUCTION

For solid media, conventional general purpose numerical methods have been applied successfully in seismological modelling, including the finite difference method [1], the polynomial collocation method [2] and the finite element method [3]. The method described here is designed to combine the favourable features of layer dependent elastic media [4, 5] and general purpose numerical techniques to generate efficiently the Green's

I.D. Abrahams et al. (eds.),
IUTAM Symposium on Diffraction and Scattering in Fluid Mechanics and Elasticity, 345–352.
© 2002 *Kluwer Academic Publishers. Printed in the Netherlands.*

tensors for boundary element calculations. Two methods are described, the first is based on finite elements and is derived from a variational formulation of the system of second order ODE's. The second derives exact stiffness matrices and equivalent nodal forces by means of propagator matrices for first order ODE's.

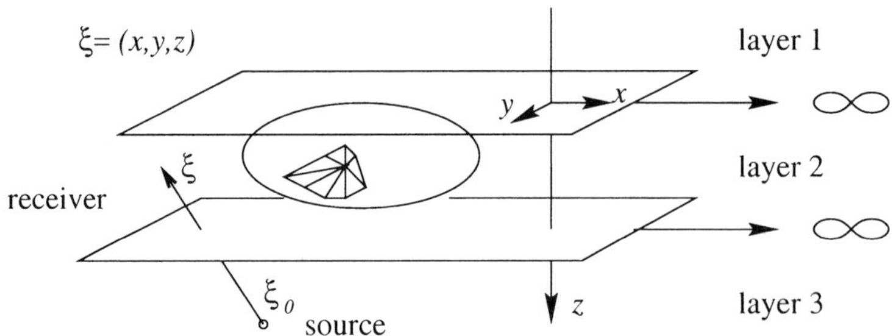

Figure 1   Vertically layered medium

2.   PROBLEM FORMULATION

Assuming harmonic time dependence, $e^{-i\omega t}$, throughout, the vector equation of motion for infinitesimal displacements is given by [6]:

$$\nabla\lambda(\nabla \cdot \mathbf{u}) + \nabla\mu \times (\nabla \times \mathbf{u}) + 2(\nabla\mu \cdot \nabla)\mathbf{u}$$
$$+ (\lambda + \mu)\nabla(\nabla \cdot \mathbf{u}) + \mu\nabla^2\mathbf{u} = -\rho\omega^2\mathbf{u} - \mathbf{b}. \quad (38.1)$$

Note that elastic constants, λ, μ and ρ may vary with position. Separability of (38.1) in a homogeneous layer is often used to obtain the elastic response of vertically layered structures [3, 5, 6].

Consider surface waves in a vertically heterogeneous, isotropic, elastic medium in which elastic constants $\lambda(z)$, $\mu(z)$ and density $\rho(z)$ are arbitrary functions of z. It can be shown that the following vectors

$$\mathbf{T}_{km}(r, \phi) = \frac{1}{k}\nabla \times (0, 0, Y_k^m),$$

$$\mathbf{S}_{km}(r, \phi) = \frac{1}{k}\nabla Y_k^m, \quad (38.2)$$

$$\mathbf{R}_{km}(r, \phi) = (0, 0, -Y_k^m),$$

where

$$Y_k^m(r, \phi) = J_m(kr)e^{im\phi} \quad m = 0, \pm1, \pm2, \ldots$$

form a complete set of orthogonal vector harmonics with a certain weight function. Hence, we may expand any vector $\mathbf{v}(r, \phi, z)$ with respect to the surface harmonics (38.2):

$$\mathbf{v}(r,\phi,z) \;=\; \frac{1}{2\pi}\sum_m \int_0^\infty [V_T(k,m,z)\mathbf{T}_{km}(r,\phi) + \tag{38.3}$$

$$V_S(k,m,z)\mathbf{S}_{km}(r,\phi) + V_R(k,m,z)\mathbf{R}_{km}(r,\phi)]k\,dk.$$

Separable solutions to the equations of motion (38.1) are possible using a cylindrical coordinate system in which the material parameters are permitted to vary only in the vertical z direction. The horizontal dependence of the solution, r and ϕ, is known and involves radial wavenumber k and azimuthal frequency, m. The vertical dependence, z remains to be found by solving a system of ordinary differential equations.

Expanding $\mathbf{u}(r,\phi,z)$ with respect to the surface harmonics (38.2), $\mathbf{u}(k,m,z) = [U_T(k,m,z),\; U_S(k,m,z),\; U_R(k,m,z)]^T$, we obtain from the left-hand side of (38.1), the stress-divergence,

$$\triangle(k,m,z) \;=\; \triangle_T(k,m,z)\mathbf{T}_{km}(r,\phi) +$$

$$\triangle_S(k,m,z)\mathbf{S}_{km}(r,\phi) + \triangle_R(k,m,z)\mathbf{R}_{km}(r,\phi),$$

where omitting the k, m indices:

$$\begin{aligned}
\triangle_T(z) &= \left[\mu U_T'\right]' - k^2\mu U_T, \\
\triangle_S(z) &= \left[\mu(U_S' - kU_R)\right]' - k\lambda U_R' - k^2(\lambda + 2\mu)U_S, \quad (38.4) \\
\triangle_R(z) &= \left[(\lambda + 2\mu)U_R' + k\lambda U_S\right]' + k\mu U_S' - k^2\mu U_R.
\end{aligned}$$

Note, that we have decoupling of $\mathbf{T}_{km}$ and $(\mathbf{S}_{km}, \mathbf{R}_{km})$ harmonics. The square bracketed terms in equations (38.4) are equal to the respective traction components,

$$\mathbf{t} = \mathbf{t}(\mathbf{u}) = \mathbf{M}_0\,\mathbf{u}' + k\mathbf{M}_1\mathbf{u}. \tag{38.5}$$

Finally, the equations of motion, (38.1), in the transform domain [7], are written conveniently in matrix form as:

$$\mathbf{Lu} = \left[\mathbf{M}_0\,\mathbf{u}' + k\mathbf{M}_1\mathbf{u}\right]' - k\mathbf{M}_1^T\mathbf{u}' - k^2\mathbf{M}_2\mathbf{u} + \rho\omega^2\mathbf{u} = \mathbf{f}, \tag{38.6}$$

where $\mathbf{f}(k,m,z) = [f_T(k,m,z),\, f_S(k,m,z),\, f_R(k,m,z)]^T$ denotes the expanded force term $-\mathbf{b}$.

The second order ODE's (38.6) may be rewritten as a first-order matrix system [8]

$$\mathbf{y}' = \mathbf{Ay} + \mathbf{f}_1, \tag{38.7}$$

where $\mathbf{y}(z) = [\mathbf{u}(z), \mathbf{t}(z)]^T$ and $\mathbf{f}_1 = [\mathbf{0}, \mathbf{f}]^T$. For the decoupled system $(SH$ and $(SV\text{--}P))$ we may write 2×2 and 4×4 first-order systems as (38.7) with respective matrices,

$$\begin{pmatrix} 0 & \mu^{-1} \\ k^2\mu^2 - \rho\omega^2 & 0 \end{pmatrix}, \quad \begin{pmatrix} 0 & k & \mu^{-1} & 0 \\ -k\frac{\lambda}{\lambda+2\mu} & 0 & 0 & \frac{1}{\lambda+2\mu} \\ k^2\frac{4\mu(\lambda+\mu)}{\lambda+2\mu} - \rho\omega^2 & 0 & 0 & k\frac{\lambda}{\lambda+2\mu} \\ 0 & -\rho\omega^2 & -k & 0 \end{pmatrix}.$$

3.   EXACT FINITE ELEMENT APPROACH

Consider a medium composed of horizontal layers of solid materials, with parameters that depend on the vertical z coordinate only. The computational domain is subdivided into n elements with nodes $z_0 < z_1 < \ldots < z_n$, layer interfaces, where element i is located in (z_{i-1}, z_i).

The system of ODE's (38.6) may be written in variational form appearing as a weak formulation of the equation [9]. The global finite element equation system may then be written as

$$\mathbf{K}(k)\,\mathbf{x} = \mathbf{c}(k), \tag{38.8}$$

where $\mathbf{x}$ is the vector of unknowns at the nodes z_i, $\mathbf{x} = [\mathbf{u}_0, \mathbf{u}_1, \ldots, \mathbf{u}_n]^T$. The system of equations (38.8) may be solved as a generalized eigenvalue problem or directly for each wavenumber k. For this formulation, assuming homogeneity of the layer, a particularly useful set of basis functions for the solid media is provided by the exact local solutions to the governing source-free ODE's.

In the following section we revisit exact expressions for the local (each layer) stiffness matrix given in [9], and formulate equivalent nodal forces and radiation condition at infinity by means of solutions to the first-order system (38.7).

4.   PROPAGATOR MATRIX APPROACH

Solution to the system (38.7) at arbitrary z may be found as a function of known solution at z_0 and propagator matrix $\mathbf{P}$ as follows:

$$\mathbf{y}(z) = \mathbf{P}(z, z_0)\mathbf{y}(z_0) + \int_{z_0}^{z} d\xi\, \mathbf{P}(\xi, z_0)\mathbf{f}_1(\xi), \tag{38.9}$$

where $\mathbf{P}$ is a fundamental matrix of the system (38.7) with following additional properties:

$$\det \mathbf{P}(z, z_0) = 1,$$

$$\mathbf{P}(z, z_0) = \mathbf{P}(z, \xi)\,\mathbf{P}(\xi, z_0). \tag{38.10}$$

Observe that, if $\xi = z_i$, $i = 1, \ldots, n$, then (38.10) is equivalent to imposing continuity of traction and displacement between the layers.

For the sake of simplicity consider only the SV–P formulation (38.7). Assuming matrix $\mathbf{A}$ has constant entries (material parameters are piecewise constant), the eigenvalues are chosen such that real parts are non-negative. Thus, with eigenvalues

$$\gamma = \sqrt{k^2 - \rho\omega^2/\mu}, \quad \nu = \sqrt{k^2 - \rho\omega^2/(\lambda + 2\mu)},$$

and respective eigenvector matrix, $\mathbf{D}$, one may construct the 4×4 propagator matrix, for the layer of thickness $d = z_i - z_{i-1}$, as follows [6]:

$$\mathbf{P}(z_i, z_{i-1}) = \mathbf{D} \, \mathrm{diag} \left[e^{-\gamma d}, e^{-\nu d}, e^{\gamma d}, e^{\nu d} \right] \mathbf{D}^{-1}. \tag{38.11}$$

The diagonal matrix in (38.11) suggests a natural decoupling of the *downgoing* ($d > 0$) and *upgoing* ($d < 0$) waves. Hence the propagator matrix may be decomposed into P_D and P_U matrices:

$$\mathbf{P}(z_i, z_{i-1}) = \mathbf{P}_D(z_i, z_{i-1}) + \mathbf{P}_U(z_i, z_{i-1}).$$

In addition the propagator matrix possesses a convenient partitioned form

$$\mathbf{P}(z_i, z_{i-1}) = \begin{pmatrix} \mathbf{P}_{uu}(z_i, z_{i-1}) & \mathbf{P}_{ut}(z_i, z_{i-1}) \\ \mathbf{P}_{tu}(z_i, z_{i-1}) & \mathbf{P}_{tt}(z_i, z_{i-1}) \end{pmatrix} \tag{38.12}$$

and a corresponding inverse, equal to the propagator in the opposite direction, $\mathbf{P}^{-1}(z_i, z_{i-1}) = \mathbf{P}(z_{i-1}, z_i)$, is easily obtained [10]. Assuming the body force, $\mathbf{f}$, in (38.9) to be a point force, we have

$$\mathbf{y}(z) = \mathbf{P}(z, z_0)\mathbf{y}(z_0) + \mathbf{P}(z, z_f) \left[\mathbf{0}, \mathbf{f} \right]^T H(z - z_f), \tag{38.13}$$

where z_f is the location of the force and H denotes the Heaviside function. In one layer, $z_{i-1} \le z \le z_i$, (38.13) may be represented as a sum of homogeneous and particular solutions, respectively, as follows:

$$\begin{bmatrix} \mathbf{u}_i \\ \mathbf{t}_i \end{bmatrix} = \mathbf{P}(z_i, z_{i-1}) \begin{bmatrix} \mathbf{u}_{i-1} \\ \mathbf{t}_{i-1} \end{bmatrix},$$

$$\begin{bmatrix} \mathbf{0} \\ \mathbf{t}_i^f \end{bmatrix} = \mathbf{P}(z_i, z_{i-1}) \begin{bmatrix} \mathbf{0} \\ \mathbf{t}_{i-1}^f \end{bmatrix} + \mathbf{P}(z_i, z_f) \begin{bmatrix} \mathbf{0} \\ \mathbf{f} \end{bmatrix} H(z - z_f).$$

Thanks to partitioning of propagator matrices (38.12), the above system of equations may be rearranged to obtain expressions for the local stiffness matrix $\mathbf{K}_{i-1,i}$

$$\mathbf{K}_{i-1,i} \begin{bmatrix} \mathbf{u}_{i-1} \\ \mathbf{u}_i \end{bmatrix} = \begin{bmatrix} -\mathbf{t}_{i-1} \\ \mathbf{t}_i \end{bmatrix}, \tag{38.14}$$

and equivalent nodal forces $\mathbf{t_f}$, respectively,

$$\mathbf{K}_{i-1,i} = \begin{pmatrix} \mathbf{P}_{ut}^{-1}(z_i, z_{i-1})\mathbf{P}_{uu}(z_i, z_{i-1}) & -\mathbf{P}_{ut}^{-1}(z_i, z_{i-1}) \\ -\left(\mathbf{P}_{ut}^{-1}\right)^T(z_i, z_{i-1}) & \left(\mathbf{P}_{ut}^{-1}\right)^T(z_i, z_{i-1})\mathbf{P}_{tt}^T(z_i, z_{i-1}) \end{pmatrix},$$

$$-\mathbf{t}_{i-1}^f = \mathbf{P}_{ut}^{-1}(z_i, z_{i-1})\mathbf{P}_{ut}(z_i, z_f)\,\mathbf{f},$$

$$\mathbf{t}_i^f = \left[\mathbf{P}_{tt}(z_i, z_f) - \mathbf{P}_{tt}(z_i, z_{i-1})\,\mathbf{P}_{ut}^{-1}(z_i, z_{i-1})\,\mathbf{P}_{ut}(z_i, z_f)\right]\,\mathbf{f}.$$

Hence, by means of the direct stiffness method, we obtain a linear system of equations (38.8). An important property of the propagator matrix approach is, assuming that for $z \geq z_n$ the homogeneous infinite layer is source free, the essential radiation condition is equivalent to the suppression in (38.11) of upgoing waves

$$\mathbf{0} = \mathbf{D}\,\text{diag}\left[0, 0, e^{\gamma(z-z_n)}, e^{\nu(z-z_n)}\right]\mathbf{D}^{-1}, \qquad \forall\ z \geq z_n.$$

Thus, at $z = z_n$ we have $\mathbf{0} = \mathbf{P}_U(z_n, z_n)\left[\mathbf{u}(z_n), \mathbf{t}(z_n)\right]^T$ which leads to the exact Dirichlet to Neumann boundary condition:

$$-\mathbf{t}(z_n) = \mathbf{K}_{nn}\mathbf{u}(z_n). \tag{38.15}$$

While the local stiffness matrix for the finite layer, (38.14), is single valued with isolated simple poles at $\det \mathbf{P}_{ut} = 0$, matrix $\mathbf{K}_{nn}$ in (38.15) has one or two branch points for SH or SV–P waves, respectively.

5.   GREEN'S TENSORS

Taking the point of application, in cylindrical coordinates as $r = 0$, $z = h$ and $\phi = 0$, for the body force (per unit volume),

$$\mathbf{F} = (F_x, F_y, F_z)\frac{\delta(r)}{2\pi r}\delta(\phi)\delta(z - h),$$

only three terms in (38.3) should be calculated [11], namely, $m = 0, \pm 1$. To obtain Green's tensors we proceed as follows: first for given (k, m) we solve (38.8), next we calculate nodal tractions by means of local stiffness matrix, then calculate $\mathbf{u}, \mathbf{t}$ for given z, from original propagator matrix for respective layer and finally re-transform the solution. For our application we are able to make use of the identity,

$$\int_0^\infty J_m(kr)f(k)k\,dk = \frac{1}{2}\int_{-\infty}^\infty H_m^{(1)}(kr)f(k)k\,dk,$$

where m is integer and f satisfies the relation $(-1)^m f(-k) = f(k)$. Hence we may obtain the displacement vector $\mathbf{u}(r, \phi, z)$ either from residue calculus,

$$\mathbf{u}(r, \phi, z) = \frac{1}{2\pi} \sum_{m=0,\pm 1} e^{im\phi} \left[\mathbf{u}_b(r, z) + \mathbf{u}_{res}(r, z)\right]_m,$$

where $\mathbf{u}_b(r, z)$ denotes integration along branch cuts and $\mathbf{u}_{res}(r, z)$ a finite sum of residues at the modal wavenumbers, k, or by direct numerical inverse transform (38.3) [12].

6.   BOUNDARY INTEGRAL EQUATIONS

Boundary integral methods for modelling elastic wave scattering from an irregular shaped 3D obstacle inside a layer dependent medium (Fig. 1) are particularly useful where Green's tensors, described in previous section, may be computed efficiently. Assuming that the bounded obstacle, Ω, may be modelled as an elastic object and the exterior is influenced by an incident wave, we derive interior and exterior boundary integral formulations.

Applying reciprocity theorems [8] to the interior of the object, using free-field Green's tensors, we obtain an integral formulation over the boundary, $\Gamma = \partial\Omega$. Once again applying reciprocity theorems to the exterior layered medium, using the calculated respective Green's tensors, an integral formulation over the obstacle surface is obtained. On combining both formulations, on Γ, one can take advantage of the same discretised mesh for the coupled exterior-interior problem.

The numerical method of solution of the BIE implemented assumes that the scattering object may be described in elliptic coordinates. With this choice of coordinate system collocation with B-spline base functions have been shown to be an efficient scheme for computation purposes for acoustic scattering within a fluid layer [13]. This scheme is currently being extended to model elastic solid layer problems outlined in this paper.

7.   CONCLUSIONS

It has been shown that the Green's tensors, necessary for a boundary integral formulation, for stratified layered media may be found via a Fourier-Bessel series expansion. Only three terms, in the azimuthal direction, are necessary in the expansion. In the vertical direction a system of ordinary differential equations has been formulated. Using exact local solutions an efficient finite element formulation has been proposed for each horizontal wavenumber using properties of a propagator matrix approach.

Acknowledgments

This work has been carried out under EPSRC grant GR/M65267.

References

[1] Alekseev, A S and Mikhailenko, B G (1980) *The solution of dynamic problems of elastic wave propagation in inhomogeneous media by a combination of partial separation of variables and finite-difference methods*, J. Geophys. **48**, 161–172.

[2] Spudich, P and Ascher, U (1983) *Calculation of complete theoretical seismograms in vertically varying media using collocation methods*, Geophys. J. Roy. Astr. Soc. **75**, 101–124.

[3] Olson, A H, Orcutt, J A and Frazier, G A (1984) *The discrete wavenumber/finite element method for synthetic seismograms*, Geophys. J. Roy. Astr. Soc. **77**, 421–460.

[4] Thomson, W T (1950) *Transmission of elastic waves through a stratified solid*, J. Appl. Phys. **21**, 89–93.

[5] Haskell, N A (1953) *The dispersion of surface waves in multilayered media*, Bull. Seismol. Soc. Amer. **43**, 17–34.

[6] Kennett, B L N (1983) *Seismic wave propagation in stratified media*, Cambridge: University Press.

[7] Keilis-Borok, V I, Neigauz, M G and Shkandinskaya, G V (1965) *Application of the theory of eigenfunctions to the calculation of surface wave velocities*, Rev. Geophys. **3**, 105–109.

[8] Aki, K and Richards, P G (1980) *Quantitative Seismology*, San Francisco: W. H. Freeman & Co.

[9] Karasalo, I (1994) *Exact finite elements for wave propagation in range-independent fluid-solid media*, J. Sound Vib. **173**, 671–688.

[10] Chapman, C H and Woodhouse, J H (1981) *Symmetry of the wave equation and excitation of body waves*, Geophys. J. Roy. Astr. Soc. **65**, 777–782.

[11] Peplow, A T, Stojek, M and Karasalo, I (1999) *Efficient calculation of Green's tensors for elastic layered media*, Proc. 2nd UK Conf on Boundary Integral Methods, 73–80.

[12] Ivansson, S and Karasalo, I (1992) *A high-order adaptive integration method for wave propagation in range-independent fluid-solid media*, J. Acoust. Soc. Amer. **92**, 1569–1577.

[13] Karasalo, I and Mattsson, J (1997) *A high-order boundary-integral method for acoustic scattering by 3-D bodies in a layered fluid-solid medium*, J. Acoust. Soc. Amer. **102**, 3211.

Author Index